Industrial Painting and Powdercoating

Principles and Practices

Third Edition

By

Norman R. Roobol

HANSER GARDNER
PUBLICATIONS

Hanser Gardner Publications
Cincinnati, OH

Library of Congress Cataloging-in-Publication Data

Roobol, Norman R.
Industrial painting and powdercoating : principles and practices / by Norman R. Roobol.—3rd ed.
p. cm.
ISBN 1-56990-338-7
1. Painting, Industrial. I. Title
TT305 R563 2003
667'.9—dc21

2002014172

Previous editions of this book appeared under the title *Industrial Painting: Principles and Practices.*

Hanser Gardner Publications
6915 Valley Avenue
Cincinnati, OH 45244-3029
www.hansergardner.com

Preface to the Third Edition

I have conceived this book on painting and powder coating to be directly useful and practical rather than deeply theoretical. It provides information that will be helpful on the plant floor as well as in the office. You will not need knowledge of chemical facts or theories to understand any section; it is all explained in a straightforward manner. If you find it otherwise in any spot, contact me and I'll take it upon myself to clarify the topic for you. In places where I felt it instructive, modest amounts of basic science were included. My intention was to avoid a "do this" and "do that" and "don't do this or that" presentation while still teaching the underlying principles that readers deserve to learn. This book contains a wealth of information for anyone associated with liquid and powder paint—both veterans with years of coatings experience, and those who may be quite new in the field of coatings.

The chapter order in this third edition follows almost exactly the topic sequence in the three-day Industrial Painting Processes course that I've presented over 250 times, either as a publicly offered course or as in-plant sessions for a single company. The course content has been continually updated to reflect the improvements and changes that occur in painting processes, equipment, and materials, but the general topic order of the course and this book is relatively unaltered from the first time it was given over 25 years ago. In a few instances the sequence was shifted to maximize the learning experience.

Before I wrote the first edition in the mid-1980s, I had looked long and hard without success for a good detailed and accurate industrial painting book. Even though I realized I would probably need to write it myself, the hope remained for some time that another person would go through the long and arduous task of putting a book like this together. When I finally forced myself to begin writing, I was surprised to find it was often enjoyable despite the inevitable delays, frustrations, and the total ineptitude and procrastination of my first editor, who was assigned to me by the publisher. When he was eventually fired, the book quickly proceeded to publication and has been in very capable hands ever since.

It is my hope that this book will assist you to become more knowledgeable in all aspects of painting. Any comments you have on this book will be eagerly received at any time, especially all suggestions for improvement that can be incorporated into future editions. You are welcome to contact me at the addresses below.

Dr. Norman R. Roobol
507 Haddington Lane
Peachtree City, GA 30269
phone: 770-487-9133
fax: 770-487-2792
email: norm@roobol.net
drroobol@flash.net
website: www.roobol.net

Dedication of the Third Edition

This is especially for Joan, still my wonderful wife after 46 years of marriage, and more than ever My Life. It's for my kids and grandkids, too, who show their love and support in so many ways. Thanks, all you guys! It's also for Louise Ezinga, my most loving and loved mother-in-law.

Acknowledgements

Many people helped me by providing all the things that make a book better. I hereby express my deep gratitude to each of them, especially to the kind folks at Hanser Gardner—Jerry Poll, Tom J. Burke, Carl P. Izzo, and Glen Muir. Most of all, I am so happy and thankful for the unceasing support of Woody Chapman. Unless you've worked with different editors, you cannot fully understand how much difference complete competence and total concern can make in a book such as this.

"Cor Meum Tibi Offero Domine, Prompte Et Sincere" is the motto of my undergraduate alma mater, but it becomes the personal vow of most who attend, for ". . . the Calvin 'Spark' you once have caught, shall never die again." In English it reads, "I offer my heart to God, promptly and sincerely." Surely it is true that living any other way is no life at all.

Credits

I would like to extend my sincere appreciation to the many chemical, coating, and equipment suppliers who provided assistance in the preparation of this book, and to the companies and publications listed below for providing illustrations used in this edition.
Binks Manufacturing Co.: Figures 6-7; 6-11; 11-4; 12-1; 14-6; 16-1; 16-3; 16-4; 16-5; 16-6; 16-7; 18-1; 21-2; 21-5; and 22-8.
Dürr Industries, Inc.: Figures 8-6; 10-12; and 22-4.
E-Z Go Textron: Figure 17-4.
Industrial Paint and Powder magazine: Figures 2-6; 3-10; 5-7; 6-20; 7-1; 15-3; 15-4; 17-5; 21-3; and 22-1.
ITW-Ransburg Electrostatic Systems: Figures 13-6; 14-10.
Kolene Corporation: Figure 18-6.
Nordson Corporation: Figures 6-8; 6-9; 6-13; 6-14; 12-3; 12-4; 14-4; 22-5; and 22-9.
Osmonics, Inc.: Figure 7-11.
Perstorp Compounds, Inc.: Figure 18-2.
Powder Coating magazine: Figures 6-3; 6-12; 6-17; 6-18; 18-3; 18-4; and 18-5.
Products Finishing magazine: Figures 3-6; 6-10; 7-2; 7-4; 9-5; and 18-7.
Toyota Motor Manufacturing, USA: Figure 15-7.

Table of Contents

Chapter 1

Paint Components and Their Functions

Introduction to Paint

In any field of study, the meaning of a list of related terms is required preliminary learning for anyone who wishes to understand the material. Industrial painting is no different in this regard; therefore, as the chapters progress, the needed terminology will be given and defined. But three chemical terms are so basic and essential to the explanation of paint adhesion and material compatibility that they need to be explained here. The terms *atom*, *molecule*, and *polarity* are crucial to explaining many painting topics; you will see them used repeatedly throughout the entire text.

At school, you were taught that all materials are composed of atoms, those unbelievably tiny bits of matter that make up absolutely everything in nature. Undoubtedly the names of some of the 98 natural atoms are familiar: carbon, sodium, iron, oxygen, aluminum, chlorine, lead, neon, and others. For our initial purposes, you only have to be aware that the elements exist as atoms; their individual names when given will be recognized as they occur in the text.

It is important for you to remember that nearly all atoms can chemically bond to other atoms, and can thereby form limitless kinds of atomic groupings. The chemically bonded atomic groups are termed "molecules." When you read "H_2O" or hear the sounds "aich-two-ooh," these immediately cause the mental recognition of "water." The formula H_2O is the chemist's shorthand for the molecules of the compound we call water, which is formed when two hydrogen atoms plus one oxygen atom physically merge closely, and then these three atoms chemically bond together into a unit, a molecule of water. Each water molecule is identical to every other water molecule in the world; all have exactly two hydrogen atoms and one oxygen atom. Of course, it takes billions of water molecules to make even a barely visible drop of water. Water is but a single example of how atoms of different elements can unite chemically, forming into molecules of totally new substances. The molecules that are formed by bonded atoms may consist of as few as two atoms per molecule, or hundreds of thousands of atoms (or more) per molecule. Molecules of two substances used in painting, acetone and isopropyl alcohol, are small. These solvents have molecules with only 10 and 12 atoms each, respectively. Medium- and large-sized molecules are also used in coating processes. Examples are dioctyl phthalate (76 atoms) and anti-mar wax agents (500–3500 atoms). Liquid paint resins frequently have molecules that are comprised of 6,000–15,000 or more atoms; after curing, the size may grow to include a million or more atoms. When they get to be this size, the designation "molecule" tends to become unimportant.

Molecules of different substances may or may not mix together. Oil will dissolve in xylene, but not in water. When oil and water are mixed, even if accompanied by vigorous stirring or shaking, the mix is temporary. On standing, the mixture almost immediately separates as the water sinks and the oil floats to the top. On the other hand, table salt, which dissolves quickly when stirred into water, absolutely will not dissolve in xylene. This solubility behavior is explained by "polarity," a magnet-like property possessed in varying degrees by all molecules. Polarity is always constant for each kind of molecule, but among various molecules it ranges from almost no polarity, to very slight polarity, on up to high polarity. Oil and xylene have low polarities; water and salt both have rather high polarities.

The polarity of two materials must be somewhat closely matched for them to be compatible, that is, to dissolve or adhere together. When rain falls on a newly waxed car, the water beads up because the wax has only slight polarity. The highly polar water molecules act like clumps of tiny magnets, drawing themselves into little droplets that attempt to touch the wax as little as possible. By recognizing the significance of polarity matches, you can understand why a highly polar paint will have little or no chance of sticking to a very low-polarity surface such as polypropylene. Yet that same paint will adhere firmly to the surface of a rather polar acrylic plastic such as polymethyl methacrylate.

"Paint" Defined

The term "paint" includes all organic resin-containing materials used as decorative, protective, and functional coatings on any kind of surface. In that sense it includes both wet paints and dry powder coatings. When employed in other contexts, "paint" is often used to distinguish liquid coating formulations from powder coatings. It does not matter how a coating is cured, how it is applied, what other ingredients it contains, whether it is liquid or powder, or whether it is solventborne, waterborne, or solvent free, it can indeed be called "paint." Some people insist that powder coatings are not paint, but powder coatings are actually true paints in every sense of the word. Paint is paint in any of its myriad forms. In this book we may at times refer to both liquid and powder coatings collectively as "paint" without distinction. However, at other times it may be convenient or make better sense to distinguish between them by calling liquid finishes "paint," while referring to powder finishes as "powder" or "powder coats."

I'm frequently asked if the term "coating" means something other than paint. There are no legal or scientific reasons for such a distinction. However, to a degree at least, ordinary usage tends to apply the label "paint" to films that are up to about 25–30 mils (635–760 μm) thick.[1] The label "coatings" is applied to films thicker than about 10–12 mils (250–300 μm) on up to 100–200 mils (2,540–5,080 μm). This is especially true for finishes that are primarily protective or functional in nature rather than decorative. You can see there is an ambiguous thickness overlap. The important point to be seen here is that paint terminology is not always precise. Everyone should strive, then, to make certain the paint terms he or she uses are fully understood by others.

---

[1] Thousandths of an inch are commonly referred to as "mils," and 25–30 mils can be expressed with decimals as 0.025"–0.030." Millionths of an inch are expressed as microinches (μin.). The international measurement that is normally used for paint film measurement is the micrometer, or micron (μm), which is one millionth of a meter. These measurements are used throughout the book.

Paint Composition

Four major classes of materials are commonly present in organic paints, as shown in **Figure 1-1**.

Resins, also called binders, vehicles, polymers, or plastics, form the paint film. Without a resin there is no coating.

Pigments provide, among other functions, opacity and color to the applied film. Pigments may be omitted from coatings such as varnishes and clearcoats.

Solvents (fluidizers) are used in most, but not all, liquid paints; they are not used in powder paints or in some UV-curing liquid paints.

Additives are substances that may be added to provide special properties to an uncured paint, or to the cured paint film. Hundreds of substances have been used as paint additives, and they might be liquid or solid materials.

The resins, pigments, and additives make up the solids portion of a paint; the fluidizers or solvents that evaporate during curing are called the volatiles in the formulation. For EPA compliance, however, you should be aware that volatiles may also be present in the resin component of the paint. Paint is prepared by mixing together a particular resin or resins combination, a solvent or a solvent blend, and most often additives and pigments. This mixing is done according to a specific formulation so that when the applied paint is properly cured, it will possess certain performance characteristics and display specified physical properties such as hardness, color, tensile strength, and gloss.

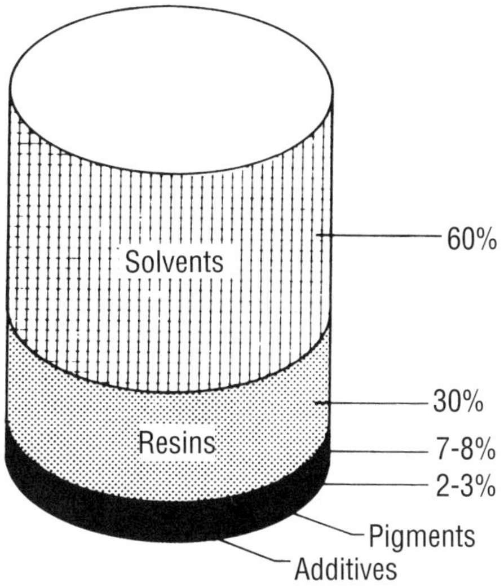

Figure 1-1. Paint composition by volume.

Resins

The resins, also called binders or polymers, are the organic film-former portion of paints. After application onto a surface, the resin enables paint to cure into a continuous uniform layer that encapsulates all the paint's other pigment and additive components. A cured paint film is actually a layer of plastic, although few people think of a paint layer in that way or even recognize that fact. The cured film normally contains all of the ingredients present in the liquid paint except the fluidizers (solvents) portion, which evaporate during the paint application and curing processes. The resin (or resins) therefore must be regarded as the most important component in any paint because resin(s) must be present, although paints need not contain any pigments, additives, or fluidizers.

The resins are usually made up only of polymers, but in some radiation cured paints they are present exclusively as monomers. A polymer in organic chemistry is a network or chain molecule comprised of many individual molecules that have chemically bonded together. The individual molecules from which the large polymer molecules are made are referred to as monomers. Many of the polymers used as resins tend to be highly viscous and therefore are generally thinned to a lower viscosity with solvent. A paint manufacturer will usually buy resins with some solvents already mixed in for easier material handling.

Polymers for paints can be categorized into two discrete types—lacquers and enamels—depending upon how the resins form their relatively hard and dry paint film. Every paint—liquid, powder, waterborne, solventborne, air-dry, heat-curing, radiation-curing, high-solids, or low-solids—can be classified as either a lacquer or an enamel. The type of resin makes no difference; it could be urethane, acrylic, epoxy, alkyd, vinyl, polyester, or whatever. A paint is either a lacquer or an enamel. The essential difference is in how the resin in the applied paint forms a solid paint film.

Lacquers

Lacquers in all types of liquid paint form dry paint films simply by evaporation of the solvent. Loss of solvent from the film permits the polymer molecules to approach each other closely and unite to form a continuous film. The resin molecules in lacquers are always highly polar. When the solvent evaporates, the resin molecules are held together strongly by magnet-like polar dipole attractions. No chemical cross-linking between or among resin molecules occurs. The polar attraction among lacquer molecules is so strong it is able to produce a firm dry paint film. The lack of cross-linking leads some people to claim that lacquers "dry" but do not "cure." That seems to be an unnecessary restriction. Lacquers obviously go from a wet film to a hard dry film. To my mind it is easier to call this "curing" and to explain that lacquers cure by solvent evaporation only. Solvent evaporation, which is needed to permit resin molecules to move close together, may be hastened by heat in some paints. In others, called air-dry paints, the solvents may be allowed to evaporate at ambient conditions.

Liquid lacquers are thermoplastic but are normally called "lacquers" rather than "thermoplastics." Powder coatings that are lacquers, however, are usually termed "thermoplastics" instead of "lacquers." I'm not sure why this should be, as both liquid and powder of this type contain polar resin molecules. Applied powder coat layers melt when heated in the bake oven and then flow out to form a liquid film. The molecules are held together by the magnet attraction of their dipolar forces. When the molten powder cools, the intermolecular attractive forces are strong enough to make the layer into a solid paint film. Again, no cross-linking bonds are formed between any of the resin molecules.

Enamels

In contrast, all enamel paints are characterized by having resin molecules that undergo additional bonding together by chemical polymerization reactions (cross-linking) after the paint has been applied to a substrate. Resin molecules in enamels have only low or moderate polarity. The solvent evaporates as it does with lacquers, but if reactive cross-linking did not also take place, the enamel resin film would remain a soft gummy layer. The chemical cross-linking of enamel resin molecules can occur by a variety of methods, depending on the chemical nature of the resins used. The principal factors involved in these methods will be discussed directly.

Air-drying enamels. Nearly all, if not all, air-dry enamel film-forming resins are composed of molecules that include numbers of carbon-to-carbon double bonds. The double bonds are known as unsaturated bonds. When exposed to air, the double bonds in adjacent molecules react together chemically along with atmospheric oxygen molecules to produce ether linkages. This is a form of resin polymerization that results in a cross-linked paint film. These are identical to the chemical bonds in unsaturated fats and oils recommended by dietary experts in place of the saturated type. In fact, many of the resins that have unsaturated bonds are derived from vegetable oil products such as linseed, tung, soya, and rapeseed (canola) oils. Unsaturated vegetable oils are used to produce alkyd paints primarily, but are also used in some epoxy-polyester paints.

Air-dry enamel paint resins of other types may react with moisture (water molecules) in the air instead of oxygen to cause cross-linking. One-component (1K) polyurethane paints are examples of this type. Unreacted isocyanate groups (also termed free isocyanate groups) in these coating resin molecules react quickly with atmospheric moisture to form carboxylic acid groups. These acid groups then undergo cross-linking reactions that cure the enamel paint film. Such highly reactive paints must be carefully stored so that they are not exposed to moisture before they are applied. This is necessary to prevent premature cross-linking of the paint while it is still in the container.

Baking enamels. Heat is commonly used to activate molecules and cause them to undergo faster cross-linking. Thermal energy is almost always supplied by hot air or by a form of infrared radiation. Isolated instances of microwave and induction heating can be found, but these remain of only minor interest. Baking enamels often contain two distinct types of polymer resin: a backbone polymer resin such as an alkyd, and a cross-linking resin such as melamine or urea-formaldehyde. In the presence of heat, the backbone polymer and cross-linking agent react to form new chemical bonds that link the molecules into a cured paint film. Some baking acrylics have an acrylic polymer backbone plus an acrylic acid or acrylate ester monomer or oligomer that cross-links the polymer molecules during the bake cycle.

Catalytic cure enamels. In many cases, catalysts (also known as accelerators) can be used to hasten the heat activated cross-linking reactions. But catalysts do not automatically require heat. A limited number of catalyst-cured enamels are cross-linked without any energy input at ambient temperatures. In another unheated method using curing catalysts, an enclosed contained chamber filled with catalyst vapor is used. Newly painted parts are placed into the vapor chamber for paint curing. The catalyst greatly increases the rate of cross-link curing without any heat or radiant energy being required. Alternatively, catalyst vapor can be injected into the paint during the coating application process rather than using a catalyst vapor chamber. This technology, as might be expected, is known as vapor injection curing. It is not widely utilized, but catalyst vapor could be introduced easily during spray or rotary bell painting.

Radiation cure (radcure) enamels. When free-radical-curable resins are exposed to electron beams (EB) or to ultraviolet light (UV), they cross-link into a cured paint film. UV light can often accomplish full cure of a freshly applied paint film within just a few seconds; electron beam does the same thing in only a fraction of a second. Since UV and EB cause curing without heating the film, the advantage of such coatings for use on temperature-sensitive substrates is obvious. While some of the radcure paints contain volatile solvents, others are solventless. These have a big environmental advantage in that they produce no volatile emissions. These zero VOC (volatile organic compounds) paints, while actually liquid in form until they are cured, are said to be 100% solids since they have no volatile components. In 100% solids coatings for radiation curing, small molecule polymers with low viscosities are used to avoid any need for solvents.

Paints with zero volatiles are not exclusive to radiation curing. Some heat-cured liquid paints also are 100% nonvolatile, but this net result is achieved in another way. With higher viscosity polymers, various reactive diluents can be added to act in a double role. Reactive diluents act in the container and in the freshly applied paint as solvents to reduce viscosity; they also cross-link with the resin polymer during the curing operations. Rather than evaporating and producing volatile emissions, the reactive diluents become an integral part of the cured paint film itself. Oligomeric (low molecular weight polymer) hydroxyl compounds based on ε-caprolactone, for example, are effective diluents for use in acrylic-melamine resin paints.

Two-part reactive enamels. If two resins are especially highly reactive, they can begin to cross-link spontaneously as soon as they are mixed together. Once mixed, they will begin to react; thus, the mixture must be used within a time limit or else the cross-linking will have proceeded too far to be able to produce an acceptable coating film. Paint that has exceeded its pot life should be disposed. Some two-component (2K) enamels have as long as 12–16 hours pot life, while very reactive types may have only a 5-minute or shorter pot life. These extremely reactive paints are kept in separate containers and mixed only as they are being fed to the paint application device. Proportional mixing equipment to use all types of reactive 2K paints is marketed and used rather widely. Most proportional mix equipment is used for applying 2-part urethane paints.

Resin Qualities and Attributes

The types of polymers used in binders include acrylics, alkyds, aminoplasts, cellulosics, epoxies, chlorofluorocarbons, fluorocarbons, natural plant oils, phenolics, polyesters, coal tars, polyurethanes, silicones, vinyls, and more. The chemistry of these polymers and various cross-linking resins can become exceedingly intricate and esoteric for nonchemists. Some idea of the complexity is seen in the statement that "a polyfunctional homopolymeric backbone polymer needs hydroxy (-OH), carboxy (--COOH), and monosubstituted or unsubstituted amino (--CH_2NH--) functionalities in order to cross-link into polyamide resins." Although this is fascinating to us organic chemists, that much scientific detail is unnecessary here, and is beyond the scope of this book. For those who are interested, cross-linking mechanisms are discussed in wondrous detail in any good organic chemistry textbook. Paint makers need a full understanding of the organic chemistry of resin polymers and their cross-linking systems, but most paint users do not. Therefore, our text will continue with discussions of more practical resin properties.

In addition to serving as the "heart" or primary building block of any paint, resins need to perform a number of other important functions; these include:

- Color and clarity

- Bonding or binding
- Encapsulation of pigment particles
- Flowout uniformity
- Required physical properties.

Color and Clarity

When used in a paint where color is of primary importance, paint resins need to be completely transparent and relatively untinted. This prevents possible interference with the color shade of the pigment. The resins must not be susceptible to yellowing or discoloration with age. Exposure to long periods of sunlight should not slowly shift the color of the film. This is a strict demand on resins because naturally occurring impurities in some resins can contribute to color variation. Impurities that might affect color may need to be removed before some resins can be used in paints where color is particularly critical. Extra purity comes, of course, at an extra cost.

Bonding or Binding

Resins are also called binders because they need to bind or grab hold of the substrate in a bonding action. This is achieved largely by mechanical keying into any slight micro- or macro-roughness of the substrate surface. There is little true chemical bonding that occurs between the substrate and the paint: only a tiny percentage of paint or powder adhesion is due to van der Waals' forces. These minor forces are the weak attractions that develop between all forms of matter. This explains why it can be difficult for paint to adhere well to extremely smooth surfaces such as glass and plated metals.

Encapsulation of Pigment Particles

Each pigment particle must be completely wetted out and surrounded by resin. The resins should encapsulate and hold the pigment particles separate, at the same time not allowing them to cluster (agglomerate). Since pigment particles reinforce the plastic film, pigment clumping or clustering reduces film strength. Agglomeration of the pigment particles also lowers the film's hiding power and dilutes the color.

Flowout Uniformity

Resins in the applied film must be able to flow out and form a smooth, continuous film. This not only provides optimum film properties, but also produces good gloss by maximizing the reflection of incident light rays. Poor flow out weakens the film's physical properties and also distorts and dulls the film surface with an orange peel effect.

Providing Required Physical Properties

Resins must be chosen for the most desirable physical and chemical properties for their specific end-use application. Important properties may include:

Hardness	Heat resistance	Chemical resistance
Hot/cold resistance	Abrasion resistance	Recoatability
Weather resistance	Stain resistance	Corrosion resistance

Initial gloss and gloss retention	Impact resistance	Detergent resistance
Water resistance	Flexibility	Sunlight resistance.

Manufactured products generally require paint films that have properties meeting the typical demands on the item. For example, automobiles and trucks require finishes that resist road salt, gasoline, oil, grease, car wash strips and cleaners, bird droppings, tree sap, and acidic components in rain and air. Home appliances such as washing machines must have finishes that stand up to hot water, bleach, cloth dyes, and alkaline detergents. Industrial and commercial machinery finishes must resist cleaning agents, heat-transfer fluids, cutting fluids, stamping oils, and a variety of organic and inorganic lubricants.

Pigments

Pigments are tiny solid particles that are used to enhance appearance by providing color and/or to improve the physical (functional) properties of the paint film. Pigments used for color generally range from 7.9–15.7 µin. (0.2–0.4 µm) in diameter; other functional pigments are typically larger at about 78.7–157.5 µin. (2–4 µm) in diameter, but they may range as high as 2 mil (50 µm). The pigments need to be permanently insoluble in the binder and solvents of the coating. If they were to dissolve, they would change the color and strength properties of the binder and lose their capability to provide a particular appearance enhancement and/or functional improvement. They would also let light shine through the film. If they are to provide hiding power, pigment particles must be large enough to block light rays. Otherwise, the paint film would have only weak hiding power and appear as a tinted clearcoat instead of a solid opaque color film completely hiding the underlying substrate. Sometimes this clear effect is wanted. Tinted paints such as tinted clearcoats are used, but here the color in the transparent film comes from soluble colorants called toners, not from insoluble pigments.

Paints must be well stirred before use (and in most instances it is best to stir them during use as well) because the pigments are particulate in nature, and thus by gravity they will always tend to sink to the bottom of their containers. The stirring should not be too vigorous, this will only produce excess foam that can interfere with application, but agitation is important. Too often the need for paint agitation is overlooked. Where color is critical, the paint should also be kept in constant circulation through paint lines and hoses to prevent pigments from settling in those parts of the system.

Appearance Pigments

Among the types of pigments that enhance appearance are white and black pigments, the entire range of colored pigments, pearlescent, metallic, and metal flake pigments. Examples of white pigments are antimony oxide, leaded zinc oxide, titanium dioxide (both anatase and rutile crystal forms), white lead, and zinc oxide. Included within the category of white pigments are extender pigments, which can include barytes, bentonite, calcium carbonate, China clay, mica, silica, and talc. The relatively low-cost extender pigments can be added in small percentages to reduce the amount of white pigment required.

Colored pigments are available in both inorganic and organic material types; this contrasts with most other kinds of pigments that are almost exclusively inorganic types. Inorganic pigments have a number of important and highly desirable properties, including low water solubility, negligible organic solvent solubility, strong ultraviolet light stability, and very high thermal stability. Most organic pigments are much weaker than inorganic pigments in all these attributes; nevertheless, organic pigments are often used because of their superior visual purity, crispness, and brilliance. Organic pigments have great liveliness and excitement when compared with inorganic pigments of the same color, and thus are often called "clean" colors. They display none of the muddy overtones commonly observed in the inorganic type pigments.

Examples of common colored pigments are phthalocyanine blue, cadmium yellow and orange, molybdate orange, chrome orange, toluidine red, cadmium red, phthalocyanine green, chrome green, black iron oxide, yellow iron oxide, red iron oxide, and brown iron oxide. Colored pigments are often used in blends of two or more types to yield specific shades of color. Safety concerns have long been associated with lead-containing paints, but safety in recent years has also affected the choice of pigments containing other elements. Pigment materials containing cadmium, chromium, and most other heavy metals have increasingly been phased out because of their toxicity. The problem is that without them some shades of color are almost impossible to achieve in stable pigment blends.

Examples of metallic pigments are aluminum and bronze flake (shown greatly magnified in **Figure 1-2**), which produce a sparkly effect by randomly reflecting light. The same type of effect can also be achieved with reflective-coated mica-type nonmetallic pigments. The metallic pigments, due to their conductivity, tend to align themselves differently in a paint film when applied by electrostatic versus nonelectrostatic techniques. The mica-type pigments do not have this drawback because micas are nonconductive. Mica flakes can also be coated with various tinted and colored materials to generate interesting optical effects such as color inversions, opalescence, and pearlescence.

Figure 1-2. Photomicrograph of metallic pigments.

Nonleafing grade aluminum flake is coated with oleic acid or a similar unsaturated fatty acid. This enables the flake to be uniformly distributed predominately parallel with the film surface throughout the paint film by total wet-out with the resin. The flat orientation allows maximum light reflection that optimizes the metallic appearance of the paint. This form is the one utilized for decorative metallic paints of any color shade. The even flake distribution throughout the paint allows the film to achieve excellent intercoat adhesion with clearcoats applied over it.

Functional Pigments

Aluminum flake pigments are also available in a leafing form. The leafing type is primarily used in formulating long-lasting weather-resistant maintenance and roof coatings because they reflect all light rays, including UV. Water resistance is also excellent due to the metal flake barrier layer. All such paints will have a distinct metallic aluminum appearance; their high light reflection properties make them useful when this attribute is needed. The metal flake is coated with a thin layer of a saturated fatty acid such as stearic acid to produce a high surface tension. This causes the flakes to resist wet-out and thus float up towards the film surface. The aluminum flakes lie flat and a high concentration of them is located within the topmost portion of the paint layer. The silver-looking paint is not readily topcoated because the aluminum on the surface affords poor adhesion. However, after weathering for several years, a portion of the surface aluminum flake will have oxidized and been eroded away by wind and rain. When less of the aluminum flake is present at the surface, the film becomes more able to be successfully repainted.

Water will slowly react with aluminum so flakes used in waterborne formulations must be stabilized by encapsulation or by forming a coating of an inert substance around each flake.

Zinc dust is another example of a functional pigment that is sometimes added to an epoxy binder formulation, often in extremely large amounts, to improve corrosion resistance. Some zinc-rich coatings may contain as much as 90% zinc by weight. Coatings of this type may be referred to as "zinc-rich" paints. The heavy concentration of zinc dust allows the metal particles to touch each other, providing continuous electrical conductivity. This conductivity permits the zinc to provide sacrificial protection in the corrosion process. When the zinc-rich painted part gets wet, a galvanic couple between steel and zinc is formed in which the zinc is anodic and the steel cathodic. Scientists have known for several hundreds of years that zinc has a higher oxidation potential than steel and will give cathodic protection to steel. As a result, even if the zinc-loaded paint film is scratched completely through to the steel substrate below, the zinc will still corrode preferentially in place of the steel. In a sense this acts as a painted-on galvanizing which similarly provides corrosion protection to steel. When zinc reacts with water, strong alkaline products are generated. Because no comparably priced resin possesses the alkali resistance of epoxies, the zinc-rich liquid and powder paints are almost exclusively epoxy products.

These are but a few of the functional pigment types. Other functional properties of a paint film that can be affected by pigments include adhesion, corrosion resistance, film strength, water resistance, and gloss reduction. The type of functional pigment(s) to be selected for inclusion in a particular paint formulation will obviously depend upon the desired performance properties for the paint. Remember that a large group of low cost substances such as silicates and carbonates are used as extender pigments in coatings. Since they are less expensive than color pigments, it is easy to think they are used only for cost reduction reasons. This assumption is only partially true; paint formulas (formulae for language purists) may include extenders for legitimate functional purposes.

Pigment Milling and Dispersion

Paint is prepared by mixing together a selected resin or combination of resins with a selected solvent or solvent blend, then the pigments and additives called for in the formulation are mixed in. The goal is to produce a specific paint composition that, when properly applied and cured, will possess certain performance characteristics and display specified physical properties. Most dry pigments purchased by paint manufacturers tend to contain numerous tightly adhering clusters of particles. If not broken up, these agglomerates will appear in the paint film as unsightly specks. The clusters are broken down into their original separate particles in one of several types of milling and dispersing processes. One mill for this process consists of a container with a motor-driven fan-type blade shaped for optimum shearing and circulation of the material being milled. The material as fed into the mill includes a precisely measured amount of binder and pigment. Solvents may also be present to ease the blending procedure. Ball mills, pebble mills, sand mills, bead mills, and roller mills are all types used to create various pigment dispersions and coatings.

With dry pigments, in addition to breaking up pigment clusters, the milling/dispersing process has another function: to completely "wet-out" each pigment particle with binder. If pigment particles are not properly wetted and dispersed, clustering will result. Clustering (clumping) reduces the effectiveness of the pigment particles, as shown in **Figure 1-3**. The maximum amount of pigment wettable by the binder is known as the critical pigment volume concentration (CPVC). Paints need not contain this much pigment, but the CPVC

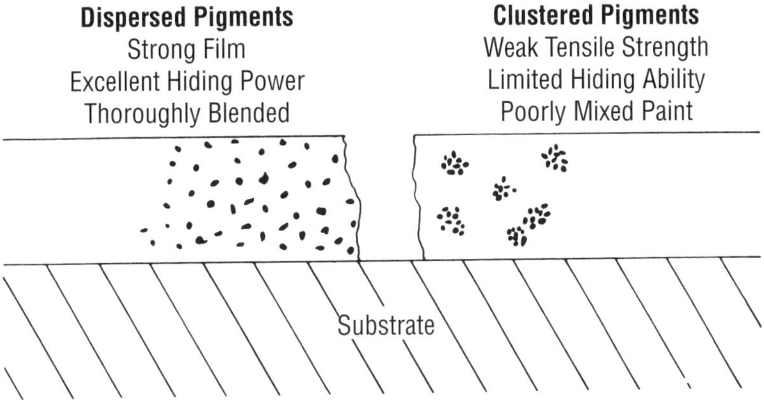

Figure 1-3. Clustering of pigment particles adversely affects paint film characteristics.

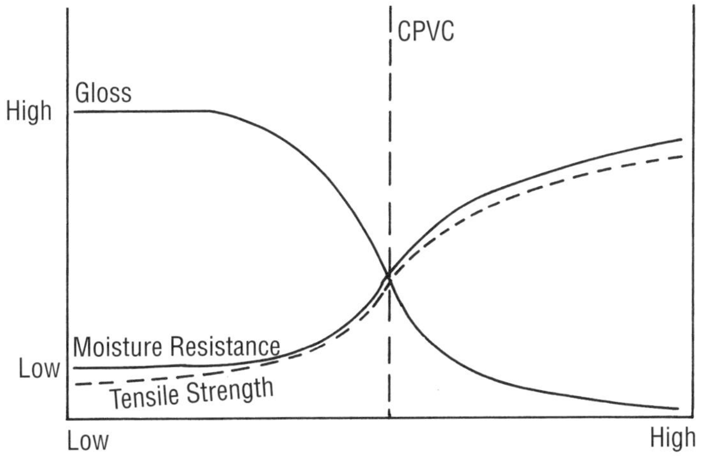

Figure 1-4. Paint film characteristics versus pigment volume concentration (PVC).

is the upper limit the blend should never exceed. Pigments may also be available to paint makers prewetted in slurry form for easier and faster dispersion and mixing into the resins. Many of the more common pigments are purchased in predispersed form.

Decreasing or increasing the amount of pigment in a binder-pigment dispersion will bring striking changes in the paint film properties. These changes can be graphed and will produce characteristic curves for each property. Each graph of pigment-to-volume concentration versus a particular paint film property will show a dramatic change in direction at the CPVC, as shown in **Figure 1-4**. Obviously the object is to make sure the CPVC is not exceeded.

When the pigment loading in paint is high, pigment particles tend to protrude from the applied paint film, creating a microscopically rough surface. A coarsely ground pigment accentuates the gloss reduction. This roughness at the film surface scatters incident light, giving a low gloss appearance to the film. In contrast, a low pigment concentration, especially in combination with a finer pigment grind, will tend to produce a smoother top paint film surface. This results in a coating with higher gloss, as shown in **Figure 1-5a**. The effect of pigment particle size can be seen in **Figure 1-5b**.

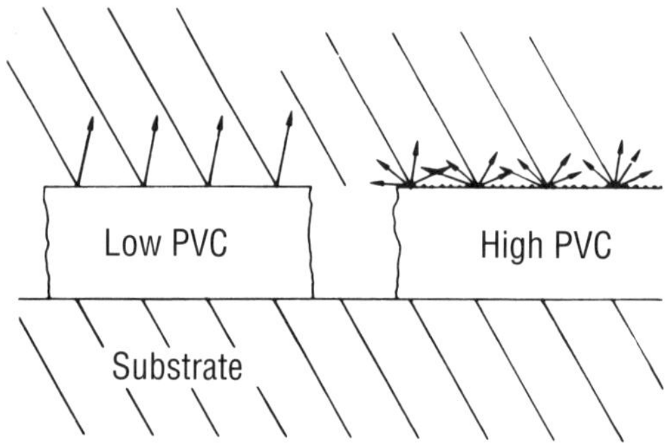

Figure 1-5a. Light diffusion from a high PVC paint reduces surface gloss.

Gloss Level	Average Diameter of Particle in Microns
63	5
48	10
42	15
34	20
31	25
24	30
19	35
16	40
12	45
7	50

Figure 1-5b. Silicate pigment particle size effect on gloss in a clear varnish.*

**All varnishes at equal pigment concentration by weight.*

Solvents (Fluidizers)

The fluidizer component of paint is usually a blend of several liquids, often simply termed "solvent," although this is more for convenience than technical accuracy. This terminology is simple and frequently used, so we will often follow it here also. In some paints the fluidizer truly dissolves the resins; in others it merely acts as a carrier or diluent without actually dissolving the resin molecules. In all cases, the fluidizer, or solvent, is a liquid that is able to lower the viscosity of a paint formulation sufficiently to allow application onto a product. It then slowly evaporates to permit the formation of a cured paint film. The fluidizer's function in paint is, therefore, transitory. It is a vital part of the paint to keep it liquid until the paint is applied, then it leaves so the paint film can solidify and the resin can properly cure.

Solvents can be categorized into five types:

- Petroleum hydrocarbon solvents
- Chlorinated hydrocarbon solvents
- Oxygenated solvents
- Terpene solvents
- Aqueous solvents.

Petroleum Hydrocarbon Solvents

This category consists of organic aliphatic (straight-chained) solvents and aromatic (containing benzene rings) solvents. Aliphatic solvents, because of their low polarity, tend to be limited in their ability to dissolve many resins. Examples of aliphatic solvents are mineral spirits and Varnish Makers' & Painters' (VM&P) naphtha that are isolated from a narrow boiling range fraction of petroleum. The boiling point range for each is about 93–149° F (34–65° C). Aromatic solvents, however, generally have higher resin-dissolving power than aliphatic. Only three aromatic solvents are commonly used in coatings: medium-flash aromatic naphtha, high-flash aromatic naphtha, and xylene. Three isomeric (related) xylene compounds are found in petroleum, and blends of these are simply labeled "xylene" since their solvent characteristics are almost the same.

Chlorinated Hydrocarbon (CHC) and Chlorofluorocarbon (CFC) Solvents

High solvent power, very fast evaporation rates, and virtually no flash point (relatively nonflammable) characterize chlorinated hydrocarbon and CFC solvents used in paints. Examples of CHC and CFC solvents that have been used in coatings and parts cleaning are trifluoromethyl para-chlorobenzene (Oxsol 100), 1,1,1-trichloroethane (methyl chloroform), trichloroethylene or "trichlor" (1,1,2-trichloroethane), and methylene chloride (dichloromethane). Initially, in the mid- to late-1980s, CHCs and CFCs were of great interest as paint solvents because they did not contribute to air pollution and thus were not considered as VOC under many (but not all) state and provincial rules. However, in many countries most chlorinated solvents are, or will be, outlawed for use or manufacture because of their role in the depletion of the earth's ozone shield in the upper atmosphere. Some are also listed as Hazardous Air Pollutants (HAPs). As a result, their importance as paint solvents is now becoming negligible except in certain countries. (See also the section on solvent cleaning in Chapter 7 for more details on which countries may continue to use halogenated solvents for cleaning and in paints.)

Oxygenated Solvents

Oxygenated solvents are available in such a range of solvent powers that a blend can be adjusted to almost any polarity value by judiciously combining various individual solvents. The main oxygenated types used as solvents in coatings include alcohols, ketones, esters, glyceryl- and glycol-ethers (and also their acetates). In June 1995, the U.S. Environmental Protection Agency (EPA) declared acetone (dimethyl ketone) to be nearly photochemically inactive and thus no longer to be classified as a VOC or HAP. However, the individual states still have the legal power to restrict its use.

Terpene Solvents

Terpene solvents are derived from pine tree sap and are characterized by high solvent power. Common terpene solvent types are turpentine, dipentene, and pine oil. Most people find their piney odor somewhat pleasant or at least unobjectionable.

Aqueous Solvents

Since water is a highly polar substance, its use is limited to dissolving resins that are also quite polar. To better dissolve mildly polar paint resins, the first heavily used industrial waterborne paints were usually mixed with organic solvents. The organic liquids are termed cosolvents because they assist water in solubilizing the resin. The mixture of water and organic solvent also has another advantage—it evaporates much more readily than water alone. Even today most so-called waterborne paints contain at least a modest amount of organic cosolvent as well. A number of waterborne paints actually contain more organic solvent than water, some to the degree that they are actually flammable. For improved film flowout and greater resin solubility, the majority of paints with water as their principal solvent also include from 0.5% to as much as 20% oxygenated organic solvent.

Polar groups that are used on resins to enhance water solubility are generated by acid-base reactions. Some resins are designed to react with aqueous hydrogen ion donors (acids), while other resins are prepared to react with hydroxide ion donors (bases). For example, in mildly acidic aqueous solutions, tertiary amine groups change into highly polar quaternary ammonium cation groups; in mild aqueous-base solutions, carboxylic groups along a resin molecule backbone will form polar carboxylate anions. Cathodic and anodic electrodeposition coatings (explained in Chapter 10) are examples of paints that use (respectively) tertiary amine groups and carboxylic acid groups to achieve water solubility. Many early versions of waterborne solvent paints used ammonia and amines as bases (alkaline substances) to solubilize resin molecules that had carboxylic acid groups on them.

Selecting a Solvent

Chemists have devised many parameters for the properties of solvents, but the two most important for coatings are solvent power (more precisely, "solubility parameter," but that term is best left to chemists) and solvent evaporation rate. Solvent power is the ability of a solvent or solvent blend to dissolve a particular binder (resin), and is related to the relative polarity of the material. The different solvent blends will each have a range of solvent powers; and therefore each of the numerous types of resins can only be dissolved by those solvents (or solvent blends) that fall within a distinct range of solvent power values. Not surprisingly, then, a good solvent for an epoxy resin may be ineffective as a solvent for an acrylate resin.

The evaporation rate of a solvent from a paint film is of crucial importance—the paint must cure at an acceptably rapid rate. The solvent needs to leave the paint at a controlled rate that is neither too slow nor too fast. This will allow the paint film to flow and crosslink without forming solvent vapor bubbles (solvent popping) in the film. A blend of solvents is able to achieve this balance. Solvents with fast evaporation rates include acetone, methyl alcohol, methyl ethyl ketone (MEK), methyl isobutyl ketone (MIBK), and ethyl

Solvent Type	**Common Solvents**	**lb/gal**	**g/l**	**Evaporation Rate BuAc = 1**
Aliphatics	VM&P Naptha	6.24	747	01.60
	Mineral Spirits	6.63	793	00.10
Aromatics	Toluol	7.25	868	02.00
	Xylol	7.25	868	00.60
	Aromatic 100	7.28	872	00.20
	Aromatic 150	7.49	897	00.04
Alcohols	Ethanol	6.78	812	02.80
	Isopropanol (99%)	6.58	788	02.10
	n-Butanol	6.78	812	00.45
Esters	Ethyl Acetate (99%)	7.55	904	05.50
	Isopropyl Acetate	7.30	874	04.60
	Butyl Acetate (BuAc)	7.38	884	01.00
	2-Ethoxyethyl Acetate	8.15	976	00.21
Ketones	Acetone	6.63	794	11.60
	MEK	6.75	808	05.80
	MIBK	6.70	802	01.60
	Diacetone Alcohol	7.84	939	00.14
Glycol-Ethers	2-Methoxyethanol	8.07	966	00.47
	2-Ethoxyethanol	7.77	930	00.32
	2-Butoxyethanol	7.54	903	00.06
	2-Methoxyethoxethanol	8.56	1,025	00.01
	2-Ethoxyethoxethanol	8.55	1,023	00.01
	2-Butoxyethoxethanol	7.94	951	00.01
Miscellaneous	Methylene Chloride	11.11	1,330	14.50
	1,1,1,-trichloroethane	10.99	1,316	03.50
	N-Methyl-2-pyrrolidone	8.55	1,023	00.01
Water		8.33	997	00.36

Figure 1-6. Weight and evaporation rate of common solvents. Weights are given in pounds per gallon (lb/gal) and grams per liter (g/l), and butyl-acetate (BuAc) is the reference solvent for evaporation rate.

acetate. Solvents with evaporation rates between fast and slow include mineral spirits and xylene. Some paint solvents with slow evaporation rates are ethylene glycol monobutyl ether acetate, di-isobutyl ketone, and amyl alcohol. Interestingly, as shown in **Figure 1-6**, some slow solvents have an even slower evaporation rate than water.

Solvents may be further classified as true (or active) solvents, diluents, reducers, and thinners. But the terms are specific to distinct solvent-binder combinations. Thus, MEK may be a true solvent for an acetate resin, yet not be a true solvent for an epoxy-amine resin. A true solvent is able to completely dissolve a resin and thus reduce the viscosity of a solution of that resin. A diluent, however, is unable by itself to dissolve that particular resin. Within limits, a diluent solvent is able to reduce viscosity if added to a concentrated mixture of resin and solvents. But if this diluent is added in too large an amount, it will force the resin out of solution. Various solvents can function either as true solvents or in other instances as diluents, depending on the particular type of resin that is being considered.

The terms "thinner" and "reducer" are less precise. They refer to any volatile organic liquid used in lowering the viscosity of paint to the level appropriate for application without concern about whether the liquid acts as a true solvent or as a diluent. All true solvents are thinners and reducers, but not all thinners and reducers are true solvents.

Coating Solvents and the Environment

In the early 1970s, following California's lead, the EPA began to draft regulations to limit the amount of VOC emissions from coatings. Its main method of doing so was to force paint users to lower the solvent content of the coatings they applied. This trend has continued since then and will not slacken, although instances of vigorous resistance by industrial groups to what are deemed overly restrictive EPA implementation or proposed new EPA regulations are increasing. Both the provincial and federal agencies in Canada have increasingly similar restrictions on VOC emissions. The U.S. Clean Air Act Amendments of 1990 heavily restricted HAP use after 1994. The Title V definition of a plant's "potential to emit" was based upon a 365 days per year, 24 hours per day capability to cause HAP and VOC emissions. This obviously tightened the compliance picture. For some plants that had a history of a 6 days per week, 2 shift per day painting operation, this was perhaps a reasonable calculation method. It was unduly harsh, however, for the plants that historically had never run their paint shops more than 1 shift per day. Relief from this unfair method of VOC and HAP calculation was correctly instituted several years ago.

The Title V provisions of the 1990 amendments to the Environmental Protection Act are by far the most significant clean air regulations that face painting operations. Federal rules designate a "major source" as one that has the potential to emit annually more than 100 tons VOC, or 10 tons of a single HAP, or more than 25 tons of all HAPs. This category is one most plants strongly wish to avoid. The category of "synthetic minor source" is for plants that would otherwise be major sources, but they either accept limitations on VOC in paint in addition to time restrictions on painting operations, or they add abatement control equipment to capture and/or destroy VOC-HAP emissions. Any plant that falls below the emissions tonnage for a major source is classified as a "minor source."

Environmental laws do make sense. A typical low-solids paint loses more than half of its volume as it changes from a wet film to a dry film due to evaporation of volatile solvents, as depicted in **Figure 1-7**. In addition to causing air pollution, this obviously is wasteful of petroleum products. The method of paint application determines to some extent how much solvent is needed in the coating to make it suitable for use. For example, far more solvent is required to air spray a given paint than to apply it by rotary atomizer. Yet most federal and state EPA regulations make little if any attempt to encourage reductions in net usage of organic solvents and resins, but only on the total net emissions of organics. Plants are generally allowed to use as much solvent as they choose, just as long as they do not emit more than the legal limit of solvent into the environment. That is not unexpected. Conservation of industrial resources is nearly always initiated by cost reduction factors, rather than by legislative mandate. As petroleum becomes increasingly costly, we will see more impetus to conserve in coating operations.

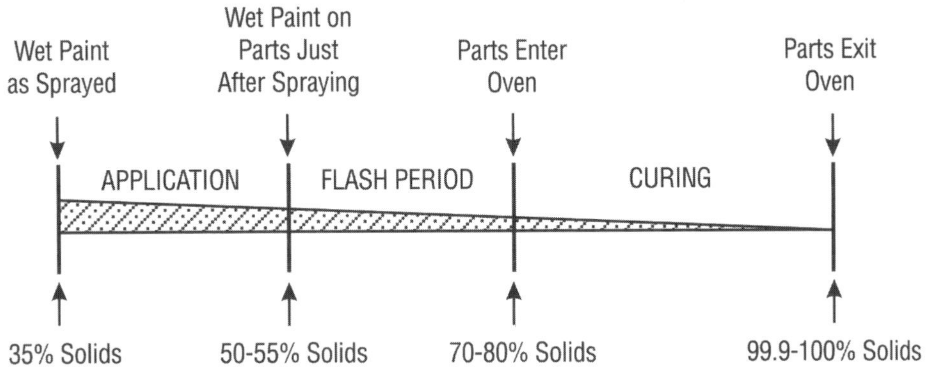

Figure 1-7. Effects of solvent evaporation.

Additives

Literally hundreds of chemicals can be added to both liquid and powder paint formulations to improve some aspect of coating performance. These additives are usually considered to be a distinct category of paint ingredients. Most additives have only a single function. Sometimes, however, an additive can also act as a resin or a pigment. A unique type of binder can in some instances be considered an additive, such as when some low molecular weight epoxy resin is included to increase the smoothness of the coating film. This is not common, however.

Some additive types, which are discussed below, include:

- Antiblock agents
- Antifreeze
- Blending aids
- Curing agents
- Antimicrobial agents
- Defoamers (antifoamers)
- Flow control agents
- Gloss modifiers
- Plasticizers (film softening agents)
- Storage stabilizers
- Thixotropes
- UV stabilizers
- Antimar and antistick agents
- Antifloat agents

Antiblock Agents

Antiblock agents reduce the tendency for paint films to stick tightly together (blocking) or show marks (marring) if they are stacked. Most plants want to pack and ship parts as quickly as possible after painting. If parts are handled while still a bit warm or when the paint for some reason is not quite fully cured, chances of blocking and marring are increased. A longer cool down section may be needed. If this is not possible, antiblock materials are commonly added so that the newly coated parts can be stacked together or packaged soon after they have exited the cure oven.

Antifreeze

Waterborne emulsions can be ruined if they freeze. The water separates as ice, and when the paint thaws the resins float like cream to the top. No amount of stirring can reemulsify the mixture and the paint must be discarded if this happens. Antifreeze materials are added to nearly every waterborne emulsion paint to reduce the likelihood of this damage occurring.

Blending Aids

These are designed to increase the efficiency and speed of paint manufacturing. Mixing aids such as dispersants, emulsifiers, surfactants, and related compounds help simplify production of paint by reducing mixing and blending times.

Curing Agents

This group of additives can improve paint curing properties. These catalysts (also termed driers, or activators) are used to increase the rate or extent of resin cross-linking during enamel curing. The goal is to shorten the time interval needed between painting and the handling, packaging, and shipping of painted parts. Plants would prefer to go right from the paint line into the box if possible. If the paint is still too soft at that point, the parts cannot be wrapped or packaged. They first have to be taken off the paint line and stored somewhere in the plant for a time to let the paint fully harden. Then they can be packaged. Extra expenses in handling are incurred, extra plant storage space is needed, and more opportunities for damage are present. By catalyzing the cure it may be possible to harden the paint sufficiently to permit immediate packaging after cool-down.

Antimicrobial Agents

Bacteria are able to use the organic resin portions of paint films as nutrients for growth and replication. Sea-going vessels are subject to snails, barnacles, and related marine organisms that foul ship hulls and increase fuel consumption. Painting with coatings resistant to unwanted organisms helps to reduce this problem. House and building paints, particularly in hot humid regions, should include fungicide additives in their formulation to prevent them turning an unsightly bilious green from mildew growth. Alkaline zinc oxide has been used in house paints to inhibit fungal growth because it is not toxic. Barium metaborate and organo-tin are more effective biocides but are less desirable since they are quite toxic.

Defoamers (Antifoamers)

Foam eliminators are frequently necessary with waterborne emulsion paints. Foaming is especially likely to be a problem in dipping and flow-coating operations. The emulsifiers used by paint makers in the manufacture of waterborne paints tend to create foam when these paints are vigorously stirred and pumped. This action is also seen with emulsifying detergents, which emulsify kitchen grease and oils. The emulsifier (detergent) will generate foam in dish cleaning water if the liquid is agitated.

Flow Control Agents

These can ease paint application or improve paint flowout characteristics. Sag balancers, bodiers, flow-control additives, leveling agents, and thixotropes will all affect the rheology (flow properties) and viscosity of wet paint films. Polyethylene oxide is used in waterborne paints to enhance atomization permitting good sprayability at lower air pressures than would otherwise be required. Spraying with lower air pressures can sharply reduce paint overspray waste. Acrylic lacquers often contain cellulose acetate butyrate (CAB), which eases breakup into spray droplets and improves wet paint flowout to minimize orange peel. CAB also holds the paint film "open" to reduce the tendency for solvent popping. Flow agents can aid as well in preventing paint wrinkling and lessening paint drain-off accumulations.

Gloss Modifiers

Various additives can reduce gloss levels without affecting film strength if low gloss or completely nonglossy surfaces are desired. Clays and finely ground silicates are frequently used for this purpose. They tend to float to the paint film surface and kill gloss by dispersing incident light rays so that few if any rays are reflected back.

Plasticizers (Film Softening Agents)

These film softening agents generate limited flexibility in hard brittle polar resins such as certain acrylic and vinyl lacquers. The plasticizer molecules act as internal lubricants between the much larger resin molecules, thereby increasing the freedom of resin molecular movement when the paint film is stressed by flexure of the substrate.

Storage Stabilizers

Paint manufacturers may use additives to increase paint storage stability. These additives will reduce the tendency of some paints to skin over; they also can reduce pigment settling during storage. Stabilizers may also allow a uniform viscosity to be maintained for longer periods.

Thixotropes

A so-called Newtonian fluid such as water has the same viscosity at rest as it does immediately after it has been stirred vigorously (sheared). This is not always desirable in paints. Thixotropes are employed to change a given paint's shear/viscosity relationship. A thixotrope would be added to a paint in order to make it viscous at rest, but significantly less viscous when pumped, sprayed, or stirred, as shown in **Figure 1-8**. When the paint is applied and no longer being mechanically worked, its viscosity rises again rapidly. Thixo-

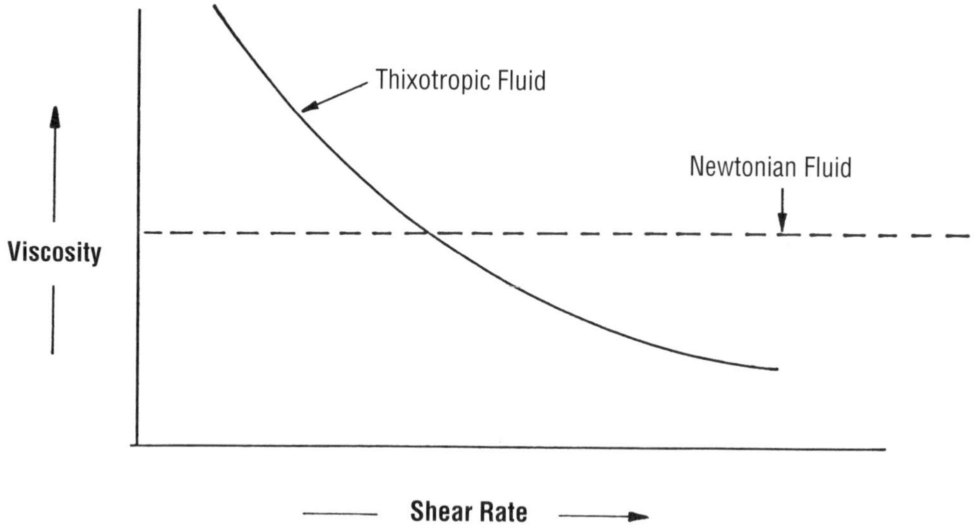

Figure 1-8. Paint's shear/viscosity relationship. Any paint's shear/viscosity curve will vary depending on its composition and formulation.

tropes used in waterborne finishes work extremely well to stop the problems of paint runs and sagging on vertical surfaces.

Ultraviolet Light Stabilizers

This category of additives can increase a paint film's resistance to ultraviolet (UV) light by several mechanisms, all of which prevent free radical chemical breaking by UV rays. Nearly all organic polymers are somewhat susceptible to attack and damage by exposure to UV light. Even natural protein polymers such as human skin can be harmed by sunlight. Doctors warn us that frequent exposure to sunlight can damage skin due to the breakdown of organic cellular materials, and cause cancerous tissue growth. Skin pigments and sun screening lotions lessen this danger. Nonpigmented paint films lack sun-blocking protection. In unpigmented coatings, such as spar varnish for wood in marine applications or automotive clearcoats, screeners, quenchers, or selective UV light-ray absorbers provide UV light protection. UV light is very powerful and tends to degrade exterior finishes on cars, buildings, boats, and homes noticeably more rapidly in sunny regions such as San Antonio and Miami than in the temperate clime of Saskatoon and Minneapolis.

Antimar and Antistick Agents

Finely divided teflon powder and various waxes may be utilized in paints to reduce the surface friction on paint surfaces. The potential for scratching and marring is thereby lessened. Some folding furniture items are painted with such coatings because mating surfaces will rub together often in use, and these paints minimize paint rub-off. Nonstick cookware is coated for easy cleaning. Antigraffiti paints allow any defacing materials to be readily washed away.

Antifloat Agents

Some pigments in certain formulations tend to float to the surface of freshly applied films, leaving visible streaks or blotches. This unsightly defect is more prevalent with dark blue and black pigments. One frequent cause is the overaddition of water to waterborne emulsion paints or excess solvent addition to solventborne coatings. When it arises from causes other than this, it is the fault of the paint maker's dispersion process. In powder coatings, a similar-appearing defect may be noticed if the powder has been stored too warm or for too long past its shelf live.

The list of "anti" agents for paint is too extensive to be included here, but the basic principles are clear by now. Other additive types include antioxidants, antiskin agents, antistatic agents, catalysts, coupling agents, dispersants, driers, flame retardants, friction reducers, conductive agents, reflective agents, and many more.

Solvent Properties and Safety Chart

Name(s)	HAP?	Evaporation vs n-BuAc = 1.0	OSHA ppm 8 hr limit	Boiling Point (°F)	Flash Point (°F)
Acetone	NO	14	1000	134	-2
Methyl Ethyl Ketone (MEK)	YES	6.3	100	170	26
Methyl Acetate (MeAc)	NO	5.3	200	135	14
Toluene (Toluol)	YES	2.0	200	225	45
Oxsol-100 (p-chloro trifluoromethyl benzene)	NO	0.9	100	139	109
Methyl Isobutyl Ketone (MIBK)	YES	1.6	100	240	75
n-Butyl Acetate (n-BuAc)	NO	1.0	150	257	81
Xylene (Xylol)	YES	0.8	100	283	76
Water	NO	0.36	No Limit	212	NONE
Cellosolve	YES	0.32	No Limit	275	118
Cellosolve Acetate	YES	0.2	No Limit	320	124
Mineral Spirits (Stoddard)	NO	0.1	100	172	111
Butyl Cellosolve	YES	0.07	No Limit	335	141

Final thoughts: WHAT ARE EXEMPT SOLVENTS?

Although most solvents are classified as VOCs, several solvents have by study been shown not to harm the environment when used in normal coating operations. Their usage quantities need not be tabulated against statutory VOC and HAP limits. Remember, all solvent classified as HAP are also VOCs, but not all VOCs are HAPs. Understandably, paint users are very reluctant to use coatings that contain HAPs. Oxsol-100, acetone, and methyl acetate are not HAPs and thus find usage as paint solvents. Acetone, aka dimethyl ketone, propanone, and 2-propanone, are highly volatile and extremely flammable, so they are used only in limited amounts as paint thinners. Methyl acetate is considerably less volatile and makes a good solvent component in a blend. Oxsol-100, a trade name for para-chlorobenzotrifluoride, is even less volatile and it is used both as a cleaning solvent and a coating solvent. Mineral spirits also fit this category, but evaporates rather slowly, so only a limited amount can be put into a liquid coating formulation.

Carbon dioxide is neither a VOC nor an HAP. Fluidized carbon dioxide under pressure has been used in expensive special application equipment as paint thinner, but only a few plants have tried it due to cost and usage restrictions.

Chapter 2

Classifying Industrial Paints

Three Major Groups of Paint

Although paints could be categorized in several ways, it is probably most useful to segregate them into three major groups: trade sale or consumer paints, maintenance paints, and industrial paints. Under each of these major groupings, numerous subcategories are commonly defined.

Trade Sale or Consumer Paints

Trade sale paints are those purchased primarily in relatively small quantities by dwelling occupants for residential painting, or by home and office decorators for commercial painting. Because of their easy application and cleanup, the vast majority is waterborne emulsion paints for painting dwellings and commercial buildings. Older oil-based paints for this purpose are also available, however, and some people still prefer them. Both oil-based and waterborne are formulated for use either as interior and exterior coatings. The paint quality is usually reflected by the price. Nearly all trade sale paints are either bought from a large retail outlet such as Home Depot or Wal-Mart, or from a local paint store or hardware dealer. The shelves of such stores are likely to carry paints that could fall into many classifications including:

• High-gloss	• Floor	• Lacquer	• Semigloss	• Oil-based
• House	• Flat	• Latex	• Trim	• Deck
• Varnish	• Enamel	• Marine	• Shellac	• Wall.

Brush and roller are used to apply trade sale paints in large measure, although low cost airless sprayers are on sale for home use. Airless spray can apply paint rapidly and is frequently used by professional decorators on larger jobs. Trade sale paints will not be covered in any more detail in this text.

Maintenance Paints

Maintenance paints make up an extensive and varied group that is used in exceptionally large volumes each year. Generally without exception all maintenance coatings are air-dry paints only. Considerable amounts of paint are applied onto objects that are immobile or too large to allow force curing in an oven. Examples of their use are traffic-lane markings on roads and parking lots, the interior and exterior of industrial buildings, commercial ships, oil rigs, large construction equipment, and highway bridges and overpasses. Large organizations often purchase their maintenance paints directly from the paint manufacturer through commercial accounts. State and local governments, many of whom purchase thousands of gallons of road-marking paints each year, can receive volume discounts and save money by utilizing direct purchase agreements. There may be one or two exceptions, but otherwise these commercially important products are liquid paints exclusively.

Industrial Paints

Industrial paints, like the maintenance paints, are also normally purchased directly from the paint manufacturer, often in large lot quantities. Package sizes routinely range from single gallons of special colors, to 300-gallon (1135 l) tote tanks of heavily used paints. Fifty-five-gallon (208 l) barrels and 5-gallon (19 l) or 10-gallon (38 l) pails are also prevalent. Industrial paint, the type that is the subject of this book, can be grouped according to three major distinctions:

1. End use characteristics
2. Types of resin
3. Physical makeup.

Because these three classes encompass so many different kinds of paints, each of the categories has been further divided into more descriptive subcategories. While it may at first seem strange, I have repeatedly found that it is more helpful to students to explain the subcategories before going back to the main divisions. For this reason, various end-use characteristics will be discussed next in this chapter, while resin types and physical makeup will be covered in Chapter 3.

Industrial Paint Types According to End-Use Characteristics

Let's now look, then, at the variety of paints classified by their usage. This category identifies industrial paint either according to the substrate onto which it is applied, or by the way the paint is intended to function. The following are some examples:

- Primers
- Sealers
- Surfacers
- Basecoats
- Clearcoats
- Topcoats
- Concrete paints
- Wood finishes
- Marine finishes
- Peelcoats
- Chemical agent resistant coatings (CARC).

Primers

A primer is paint formulated to be applied to a substrate before another paint is applied, often (but not always) directly to an unpainted surface. This, naturally, is suggested by the name, since "prime" connotes "first." Sometimes only primer is ever applied to an object, but more often it is applied and then another layer of paint is put on over it. A major function of primers is to promote lasting adhesion of subsequent paint layers to the substrate. Often, topcoat paint, if applied directly to a surface, will not adhere sufficiently. When a primer is applied first and then a topcoat, film adhesion and durability are substantially enhanced. Another important function of a primer can be to isolate the substrate from the effects of weather more completely than is possible by the topcoat alone. This is probably of greatest importance for any corrosion-prone metal substrates. This seems especially true of ferrous metals, for the red color of rust (iron oxide) seems to be more unsightly and alarms people more than the white oxide products from zinc and aluminum corro-

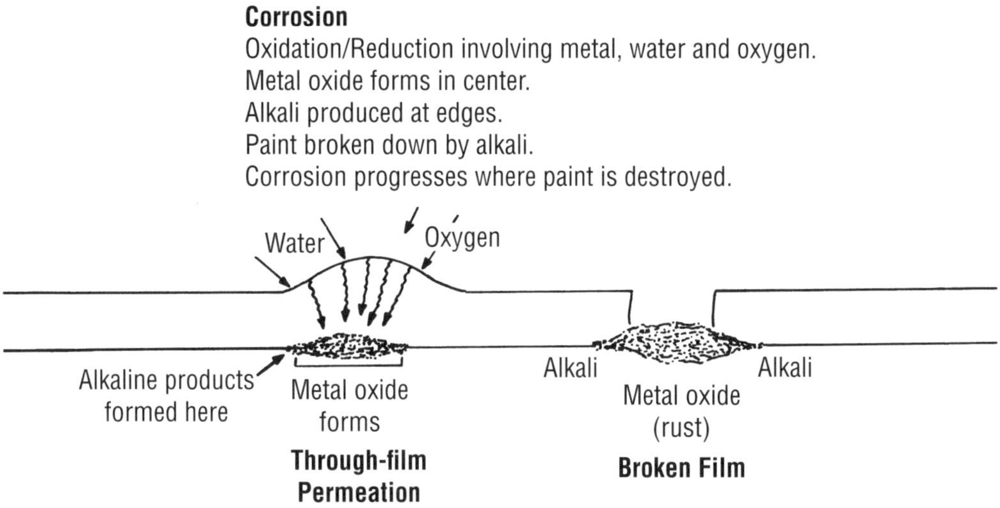

Figure 2-1. How water and oxygen can cause corrosion of painted metal.

sion. **Figure 2-1** shows the basics of metal corrosion processes occurring under a coat of paint.

The most significant types of primers include:

- Wash (etch) primers
- Shopcoat primers
- Shop-plate primers
- Flash primers
- Baking and forced-dry primers
- Spray primers
- Electrocoat (ecoat) primers.

Wash (etch) primers. Despite the name, wash (etch, or acid etch) primers do not perform a washing or cleaning function. The name indicates that they are low in viscosity and provide only a very thin or "wash" coating of about 0.25–0.50 mils (6.0–12.5 μm) thickness. "Etch" means the primers are formulated to chemically attack and roughen metal surfaces. Wash primers are used as the first coating on clean metal surfaces that have not received a phosphate or chromate conversion coating. Then the next coat or coats of paint are applied. If the other paint(s) had been applied without using a wash primer first, their adhesion might be unsatisfactory.

Wash primers are formulated with an etchant such as phosphoric acid to attack and slightly roughen the metal surface, giving the wash-primer binder a surface with "teeth" for mechanical interlocking. Wash primers require an acid-resistant resin component, most commonly vinyl butyrate. In the past, wash primers frequently contained chromate pigments, which are powerful corrosion inhibitors. The characteristic greenish-yellow color (or dark green) of many wash primers was due to the zinc chromate pigment. Due to chromium-related toxicity, the chromates have been largely replaced by zinc phosphate and less environmentally dangerous pigments. Most wash primers are manufactured as two-part systems that are mixed just prior to use—the resin in one part and the acid in the other part. This prevents the resin from being deteriorated by the acid during storage. Once the two parts are mixed together, the wash primer has a pot life of about eight to ten hours; after this period any unused primer begins to lose film strength and should be discarded.

Wash primers are intended for application on clean, bare metal. They are not normally applied over metal that has been phosphated or chromated. The acid must be able to react with bare metal, and a conversion coating will prevent this. If applied over conversion coatings, the acid in the wash primer may not react and will later cause blisters when the painted part is exposed to moisture and humidity. The unreacted phosphoric acid is hygroscopic, that is, it will attract water to itself through the coating layer, resulting in blisters. However, a conversion coating plus wash primer is nevertheless still required in some military specifications. Since 1995 more and more military specifications have been updated to omit the wash primer when the substrate is to receive a conversion coating.

Shopcoat primers. Shopcoat or "shop" primers are usually considered to be only temporary coatings applied at about 1 mil thickness or less. Most times they are used to protect metal products that are too large or too numerous to be stored inside, especially on large steel weldments to prevent rusting. Bulky long-bed truck frames, for example, are put into outdoor storage until they are used to make the final product. The shopcoat primer keeps them from rusting severely during the weeks or months they are stacked outdoors. The shopcoats are frequently totally stripped off before assembly and final painting is done.

Shop-plate primers. In shipbuilding, a shop-plate primer is applied to steel plate at the mill under carefully controlled conditions. This mill-applied coating is superior to the primer coats applied to the steel hull under possible adverse outdoor weather conditions at the shipyard. Shop-plate primers have excellent durability because the controlled mill environment enhances paint film quality. A shop-plate primer is at times also called a shopcoat primer. These primers are often rich in zinc powder for anticorrosion properties. The zinc dust is electrically conductive and enables welding of painted parts, although toxic zinc fumes are generated. The name "weld-through primers" is given to such paints. This can be an important attribute, since welding of metal coated with most organic coatings is not done; it results in inferior weld quality.

Flash primers. An air-dry primer is called a flash primer because part or all of the solvents evaporate or flash off without the use of added heat before the next coat is applied. The term "flash" distinguishes this class from the baked primers. Sometimes flash primers can be topcoated without first being fully cured; this method of applying the topcoat on the primer (or any two paints one on another) is referred to as "wet on wet" application.

Baking and forced-dry (forced-curing) primers. Baking or forced-drying of an applied primer or other paint film can be brought about if a resin would otherwise cure too slowly at ambient conditions. Many paints will not fully cure unless they are baked because the cross-linking reactions do not begin until bake oven temperatures are reached. This is especially true for powder coats and high-solids paints. Paints that must be baked to cure are classified as baking paints.

Other paints will cure slowly at ambient conditions without heat. However, they cure far more quickly at slightly raised temperatures. Forcing can be done with most air-dry paints and two-component (2K) paints. "Forcing" the process with mildly elevated temperatures considerably shortens the cure time for such coatings. Forced drying, also called forced curing, is usually done so that painted items can be taken off a conveyor line and packaged within a shorter period of time. That is important because time is money in manufacturing.

The U.S. EPA has classified bake temperatures below 195° F (90° C) as low- or forced-baking conditions; temperatures at or above 195° F are decreed to be high-baking conditions. This legal dividing line is somewhat arbitrary but nonetheless reasonable; the government needed it to legislatively create two distinct classes of coatings with separate regulations for each class.

Spray primers. This paint type, as the name suggests, includes various kinds of primers that are intended to be spray-applied, as opposed to other means of application. More primers are included in this category than in any other. Despite the name, rotary atomizers also can and do apply primers of this type. The name "spray" serves only to show that the application is by paint droplet formation and not by spreading or an immersion process.

Electrocoat (Ecoat) or electrodeposition primers. Electrocoating is an interesting and widely utilized paint application technique whereby direct current electrochemically causes paint to deposit on electrically conductive parts. Parts are connected either to the anode or cathode while they are immersed in special ecoat paints; the direct current causes paint resin insolubilization and deposition. Depending on the formulation and use, a deposited ecoat may be a singlecoat, a primer, or a topcoat. The electrocoat paint application process is fully detailed in Chapter 10.

Sealers

The term sealer as applied to paints has various meanings in different manufacturing industries. In wood finishing, which makes abundant use of sealers, the sealer is applied over a stain. Wood sealers are used to close the pores of the wood, lock in the color of the stain, and raise tiny wood fibers along the edges of the wood grain. Raising these fibers allows them to be sanded off to produce a smooth wood surface. This step is crucial for fine-quality furniture finishing.

In sheet metal finishing, sealer function is quite different since no pores or fibers are present. Metal sealer can be used for several different purposes. The sealer may be used between a primer and topcoat layers to prevent topcoat solvent migration into the primer. Solvent migration from the topcoat could result in pigment leach-out from the primer, and this would stain or discolor the topcoat. This phenomenon is called pigment bleeding or staining. The leached-out pigment is especially noticeable with dark primers and light color topcoats. Solvent migration can also cause swelling of the underlying paint layer, and solvent absorption occurring along fine sanding scratch lines of a paint layer can produce ridges that are visible through the topcoat. The ridges become magnified when they are painted over due to swelling from absorbed solvent. This defect, known as sand-scratch swelling, is stopped by applying a sealer coat between the coats of paint.

Sealers are also used to bridge large differences in polarity between consecutive coats of paint such as primer and topcoat, shown in **Figure 2-2**. Such large polarity differences reduce intercoat adhesion. A sealer of intermediate polarity can be applied onto low-polarity epoxy enamel ecoat paint so that a highly polar acrylic lacquer topcoat applied next will adhere. If the highly polar acrylic were applied directly over the low-polarity epoxy, loss of adhesion would readily occur between these layers. The situation is similar to the

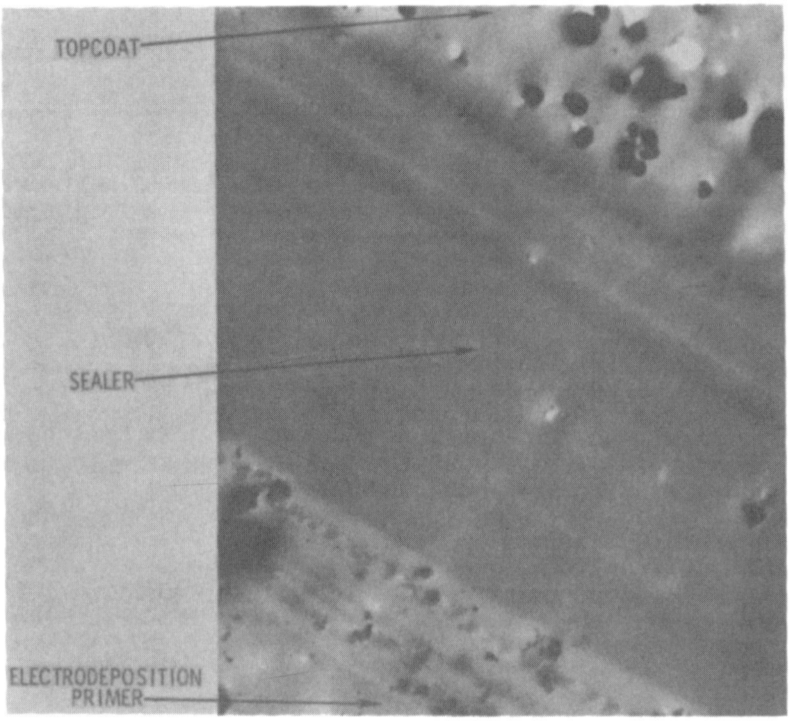

Figure 2-2. Photomicrograph showing use of sealer as bridge between primer and topcoat.

tendency for water to form beads on a waxed surface. Because of their mutual attraction, the highly polar molecules of water tend to pull themselves together into individual drops rather than adhere uniformly to the nonpolar waxed surface. Since lacquers are no longer in heavy use due to their high solvent content, sealer for this purpose is only rarely required anymore.

Surfacers

Surfacers are applied to provide a smooth base for a topcoat that must exhibit a high gloss. Smoothness is one of the characteristics of a coating with high gloss. Unless the surface is already smooth, surface sanding is often required to level out minor roughness prior to applying high-gloss finishes. Few paints have enough flow to level out and fill surface irregularities completely. Paints are usually easier to sand than metals, but not all paints can be designed as surfacers. Pigments and resins that permit surfacers to be readily sanded are used in their formulation. By the proper choice of resin and pigments, paints designed as surfacers are made easier to sand. Because sanding is slow and costly, it tends to be done only where necessary and then only on parts where the added cost can be justified. A related class of paints is termed primer/surfacer because in addition to easy sanding they act to improve corrosion resistance as well. Primer/surfacers should be considered primarily as surfacers, but the added rust protection they afford is significant enough to justify prefixing "primer" to their name.

Basecoats

Any paint layer applied under the topcoat has long been considered in a general sense to function as a basecoat. The term "basecoat" has changed over the last 20 years or so. It now more specifically refers to a pigmented color topcoat that will receive a clearcoat over it. An automotive standard topcoat finish now is the color basecoat plus clearcoat. A large number of industries whose products are enhanced by high gloss also employ basecoat/clearcoat (bc/cc) technology to enhance the brilliance of their products' finish. The basecoat provides the requisite color; the overlying clearcoat amplifies the film gloss, or shine. The basecoat/clearcoat pair does not provide a lot of rust protection for steel so it is nearly always applied over anticorrosion primer coat(s) rather than directly to the substrate.

European auto manufacturers developed basecoat/clearcoat technology in an attempt to match the fine gloss of metallic finishes on cars produced in North America. In the 1970s, North American cars were topcoated with acrylic enamels and lacquers, while European cars were finished mostly with alkyd melamine topcoats. The acrylic metallic enamels and lacquers satisfactorily buried most of the metal flake added into the paint. Alkyd melamines, however, tended to leave microscopic portions of the flakes protruding out of the paint, which lowered the gloss. When compared to North American car metallic finishes, the Asian and European metallic color cars looked drab; they badly needed more gloss.

The European car manufacturers, in cooperation with their paint suppliers, developed an unpigmented paint to solve this problem. When applied as an extra coat over the pigmented alkyd melamine metallic topcoats, the clearcoat effectively buried the metallic flake, creating a smooth glossy paint surface. The smooth clearcoat surface exhibited exceptionally high gloss. It was, in fact, noticeably higher in gloss than the North American topcoat acrylic finishes. Japanese car producers immediately adopted this clearcoat technology for their metallic finishes. Not to be outdone, North American auto and truck manufacturers were forced to follow suit and began adding clearcoats to their products in the late 1970s.

Most automakers worldwide have used clearcoats for the past 15–20 years on both nonmetallic and metallic colors because of the excellent gloss they provide. Acrylic, polyester, and urethane topcoat materials have been used for the base colorcoat and the clearcoat finishes. The film thickness on North American vehicles typically is 0.75 mil (19 µm) of metallic color basecoat plus approximately 1.2–1.6 mils (30–40 µm) of clearcoat applied over it. (These values are for liquid paints; film thickness is about 30–40% higher if powdercoats are used.) Nonmetallic colors normally use more basecoat thickness with slightly less clearcoat.

A development in low-VOC basecoat/clearcoat technology from the 1990s and still employed today uses a thixotropic waterborne color basecoat. The paint is dehydrated in a quick bake, usually by infrared, to remove nearly 95% of the water. The waterborne color basecoat is then topcoated with a liquid waterborne or high-solids clearcoat, although some powder clearcoats are being applied also. Powder clearcoats and even powder basecoats have been utilized, but powder for this use has thus far been quite limited. In the future much more use of powder for automotive basecoats and clearcoats can undoubtedly be expected. Waterborne clearcoats have not been (and likely never will be) particularly successful.

While many metallic powders are available, they are not yet fully able to capture the automotive color basecoat market. A characteristic strong point of powdercoats, namely low film shrinkage, becomes a disadvantage when used for a metallic basecoat. The best metallic appearance is achieved with color basecoats that have a large shrink factor, that is, paints that contain high percentages of volatile materials. The film shrinkage helps to force the metal flakes into the preferred orientation. Metal flakes aligned parallel to the surface are best because this affords maximum light reflection. Since VOC restrictions make the possibility of using organic solvents in large quantities unsuitable or too costly, the method of choice must be to use significant amounts of a safe and nonpolluting solvent to achieve the necessary high film shrinkage. Since water is the only solvent that meets these criteria, waterborne color basecoat technology has been perfected. Clearcoats using waterborne formulations can even further reduce VOC emissions, but, naturally, the ultimates are the new powder clearcoats.

Clearcoats

Clearcoats are defined as paints without pigments. **Figure 2-3** shows a brass bed coated with a clearcoat to prevent tarnish. **Figure 2-4** shows drawings of metal flake projecting through the surface, reducing film gloss. A clearcoat applied over the projections would increase gloss. A clearcoat applied to wood is usually termed a varnish, but some BMW car advertisements seen on British television call their exterior-finish material "a varnish." To reduce VOC emissions from the currently used and relatively high solvent content clearcoats, powder clears are coming into use. Considerable work has been and is still being done to perfect lower cost application and powder paints for automotive clearcoats. Toners, which are soluble paint resin tinting dyes, can be added to clearcoat formulations without changing the "clear" aspect of the coating. Special vehicle paint effects are achieved using a tinted clearcoat plus a metallic or pearlescent basecoat. One of the first companies to do this was Harley-Davidson to provide distinction on featured motorcycle models. This technique is beginning to be used for some monochrome colors also. The effects can be most striking and are highly glamorous.

Topcoats

The topcoat's major function is to provide a pleasing appearance to the surface it covers. The gloss of a topcoat may vary from almost none for photographic darkroom paints to the very high gloss of automotive finishes. **Figure 2-5** shows a photomicrograph cross-section of a topcoat and a primer. Finely ground pigments in the topcoat enhance topcoat gloss; coarse pigments are employed to reduce gloss.

Topcoats must also exhibit the proper shade and intensity of color. The topcoat, being the most exposed surface, also needs to protect the surface during exposure to the normal service environment. This might require a particularly strong ability to fight abrasion, withstand sunlight, resist attack by chemicals, or be extra durable under whatever exposure conditions are anticipated. Sometimes a textured or wrinkled topcoat surface is desired. These effects can be achieved using various resins that undergo surface distor-

Figure 2-3. Brass bed coated with clearcoat.

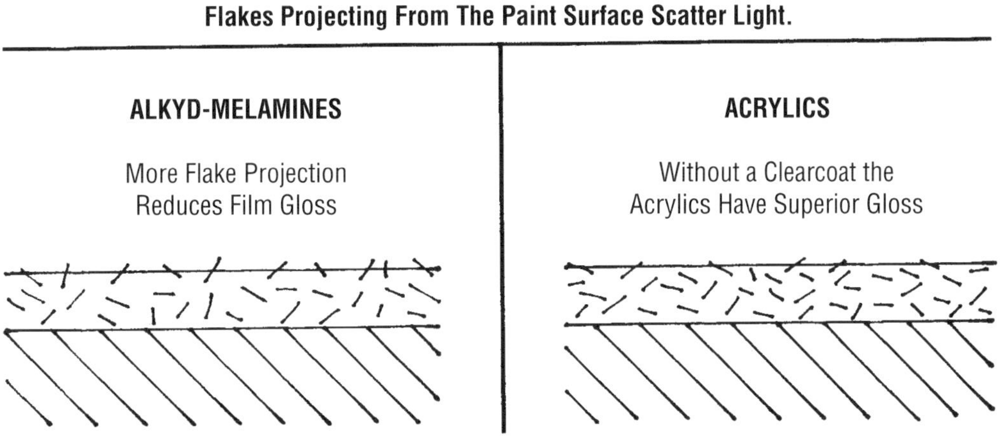

Figure 2-4. Typical metal flake orientation patterns.

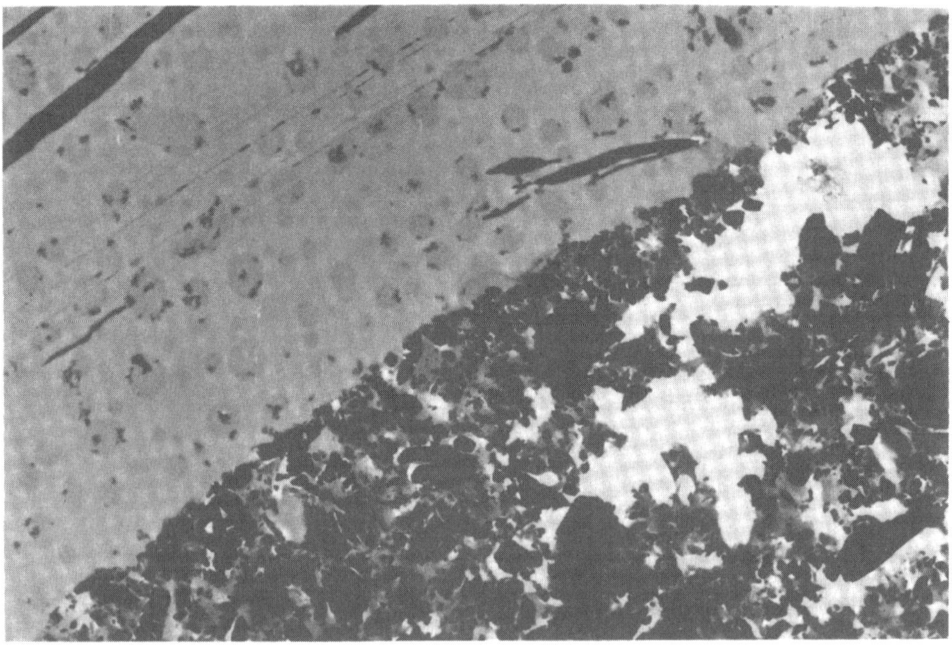

Figure 2-5. Photomicrograph showing difference between finely ground topcoat pigment (at top) and coarsely ground, heavily concentrated primer pigment.

tion during cross-link curing. Another texture, known as "spot," is produced by special spray application techniques. The object is painted completely using normal airspray gun settings. When the paint has dried to a degree, the object is again sprayed, this time with sharply reduced air pressure. The paint doesn't atomize but rather forms modest-sized droplets that create a pattern of spots. (I have heard this texture called "spot-on-plain" or "spot-on-plane" at several plants, but am unsure if they meant "plain" as in ordinary paint finish, or "plane" in the flat geometric sense.)

Concrete Paints

Some paints are formulated especially for strengthening concrete floors and roadways. Slowing the rate of concrete curing with these coatings can significantly extend road life. This creates a harder material. Coatings also help roadways last longer by reducing moisture absorption from rain or melted snow and ice into the concrete. Absorbed water can cause severe road damage in cold climates from progressive cracking during freeze-thaw cycles. These types of coatings must be extremely durable to withstand constant weather exposure and traffic abrasion.

Other coatings, also labeled concrete paints, are intended for much less severe indoor use to improve appearance or to seal the surface so that concrete floors are easier to clean. Sealing the floor inside plants reduces the tendency for fine dust formation when concrete floors undergo abrasive wear from pallets, forklifts, and related items. The caustic nature (pH below 7) of concrete requires the resins to be resistant to attack by alkaline materials. Urethane and epoxy resins are the ones most used for brick, mortar, and concrete walls, floors, or blocks.

Wood Finishes

As the name indicates, these are formulated for wood including chipboard and fiberboard composites. Wood finishes may be pigmented, clear, or tinted clear coatings. Various types of coatings for these products include fillers, stains, toners, varnishes, sealers, and lacquers. To avoid water loss from the wood item being coated, these finishing materials are normally cured at temperatures at or below 180° F (82° C), and only rarely at higher temperatures. UV curing powders have successfully been utilized to finish medium-density fiberboard. The low heat requirement with UV curing is clearly a big advantage in wood finishing, but the fast curing speed is also important.

Marine Finishes

Coatings used at sea and in ocean shore locations must have extraordinary resistance to sun, saltwater, and humidity. Marine paints are applied to pilings and offshore platforms, boats, buoys and beacons, docks and ancillary dock equipment, harbor warehouses and wharves, etc. Epoxy polyamides, urethanes, and alkyd-silicone, which are all strongly weather-resistant resins, predominate in this class of paints.

Peelcoats

Unlike most paints that are designed to stick well to the substrate, peelcoats are designed for easy removal. Peelcoats are typically applied to surfaces that require frequent quick and easy cleaning. The interiors of paint booths, for example, are often protected with a white pigmented coating of this type to provide better visibility for sprayers inside the booth. Paint, especially dark colors of paint that builds up on booth walls, makes it harder to see well. Paint buildup on booth lights really darkens the inside, so clear peelcoats are often used over the glass covers of light housings. After a period of use, the peelcoat can be readily stripped off, removing the built-up overspray and accumulated dirt along with it. The surface is given a new peelcoat immediately after the old one is removed. Fast and easy periodic stripping and recoating of the surface with fresh peelcoat keeps the booth clean and helps visibility.

Surplus military items that are too large to be stored indoors, such as ships, tanks, helicopters, and howitzers, are protected during long outdoor storage with peelable coatings. "Mothballing" in this fashion combines a sheltered storage condition with easy removal of the protective coating shield for quick reactivation of the equipment when needed.

Chemical Agent Resistant Coatings (CARC)

Special paints were developed for military vehicles and a wide range of combat equipment to permit rapid decontamination with aqueous cleaners in case of chemical, biological, or nuclear incidents. Equipment used in the nuclear power industry is also coated with these finishes. They are formulated to resist penetration by the dangerous materials mentioned above, and also to be minimally affected by strong decontamination solutions. Many tend to be formulated with polyurethane resins.

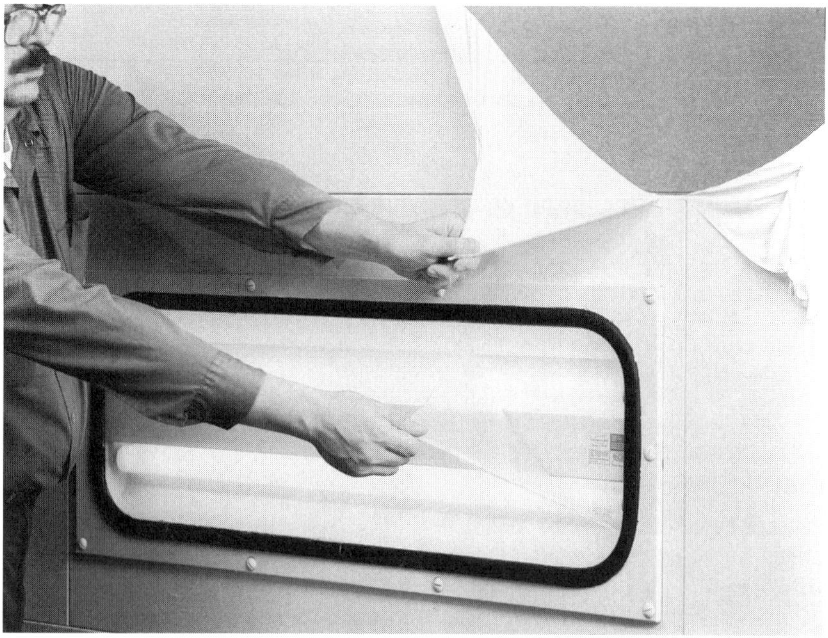

Figure 2-6. Two versions of peelcoat being removed. Clear peelcoat has been used on the glass covering the light, and opaque white peelcoat was used on the spraybooth walls.

Other End-Use Classifications

Singlecoat

The term singlecoat plainly indicates that only one coat of paint is used on the item because that is a sufficient finish. For some items, one coat of paint is able to perform all of the functions expected of the coating on that particular product. The paint might be expected, for example, to have an attractive color plus be able to resist marring and scratching. It is always less costly to use just a single application whenever possible. Some products may be quite suitable for singlecoats, especially if the items are low in cost and/or have a short service life expectancy. Singlecoats are appropriate, for example, on lawnmower blades or broom handles. In use, the blades lose their paint after just 2 or 3 uses; the broom handle doesn't receive enough use to need more durable paint. Just one ecoated layer of acrylic paint is used on room air conditioner covers; only one powder coat is applied on microwave oven cavities.

Multicoat

The term multicoat indicates that two or more coatings are applied to a surface. Often the protective and decorative functions required of paint are beyond the capabilities of any single coating formulation. Then two or more different paints might be used for specific aspects of the job. Each paint type has its own special function(s) to perform. By using several different formulations, the overall paint system is stronger and has better performance than any one coating material alone. A wood cabinet may be stained to the proper color, then sealed for sanding smoothness, and then doubly varnished to highlight the finish and protect the wood. Similarly, a bulldozer or farm tractor is epoxy primed for corrosion protection, then topcoated with acrylic enamel for durability in sun and rain and snow. In each case, customer expectations, the cost of the paints and the item being painted, and the service life expected for the product help dictate whether or not a multi-coat system should be used, and if that system will be cost effective.

In another example, a Canadian locomotive manufacturer uses an alkyd primer for adhesion and corrosion resistance. Next, this manufacturer applies a modified acrylic primer-surfacer for smoothness and some additional corrosion protection. The third and final coat used is polyurethane for good gloss and long-term weatherability. As a result of applying three different paints, the overall triple-coat system has superior performance compared to what could be achieved by any single paint standing alone.

United States Metric Tons Coating Usage Per Year Estimates

MARKET SEGMENT	1990	1995	2000	2005	2010
Maintenance	2320	2490	2915	3300	3770
Automotive – New	393	465	527	568	616
Automotive – Refinish	162	179	217	242	267
Other Manufacturing	3910	4200	5070	5820	6805

Final thoughts: WHAT ENABLES SOLIDS TO DISSOLVE?

How can salt and sugar dissolve in water but oil and sand cannot? Why are paint makers so fussy about what solvents they want us to use when we thin our paints? Well, it all comes down to physical attraction or the lack thereof. That is, how attractive do submicroscopic atoms and molecules find each other. Electrically positive and negative areas of atoms, ions, and molecules will attract opposite charged areas and repel similarly charged areas. Resin molecules will dissolve in a solvent if the solvent molecules attract the resin molecules at least as strongly as they do their own molecules. If that doesn't happen, it will not dissolve. It works for salt in water, but not for grease in water. It doesn't work for salt in alcohol, but works fine for grease in alcohol. Chemists say that "like dissolves like" when they refer to the relative polarities of materials. The solid to be dissolved needs a polarity match to the solvent if solution is to occur.

The challenge for coating manufacturers is to find a solvent blend that can dissolve the resin but is not too expensive or flammable, has no HAPs, and will evaporate out of the applied film at an appropriate rate. Luckily, with judicious selection and formulation, the chances of identifying a suitable blend are good. That is why so many different coating formulations are available for painters to coat just about any item at a diverse range of prices.

Chapter 3

Industrial Paint Categorized by Resin Category, Physical Makeup, and Cure Mechanism

Resin Categories

Naming types of paint after the chemical category of the major resin component was started because the binder, as the film former is also called, constitutes the most essential part of a paint film. The intrinsic chemistry and molecular configuration of any resin, including the extent to which it is cross-linked, will give it (and the formulated coating material) its specific performance properties. For example, the hardness and flexibility of a dry paint film can vary according to the type of resin in the paint, and also with the average size and size-distribution range of the resin molecules in a given paint batch. **Figure 3-1** shows the molecular weight distribution curves of two polyester resins. Notice that the narrower curve exhibits better overall paint properties. **Figure 3-2** depicts how paint film properties will vary according to the film's softness or hardness even though they are made with resins of the same chemical name. Softer lacquer resins have smaller molecules than hard lacquers; and softer enamel resins are less cross-linked than harder ones.

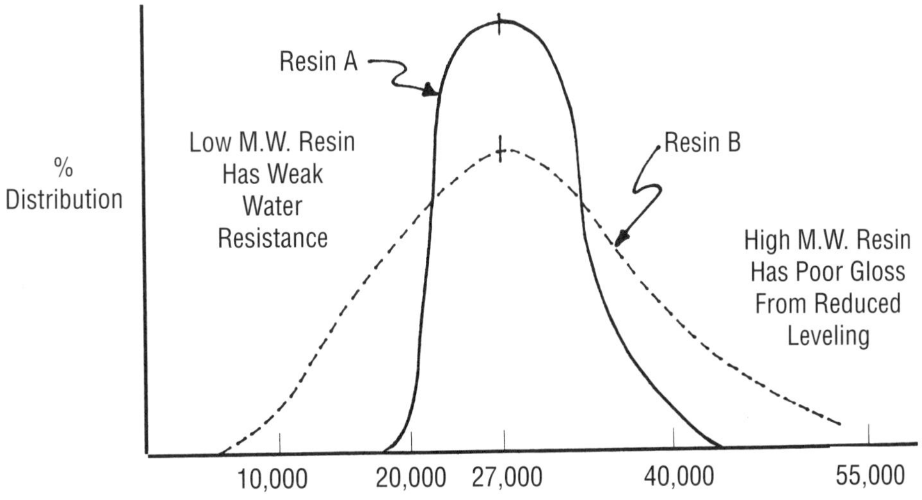

Figure 3-1. Molecular weight distribution curves.

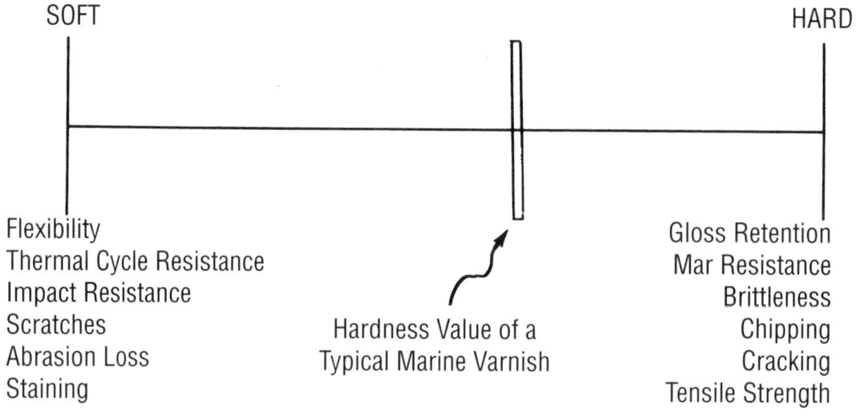

Figure 3-2. Resin soft-to-hard ranges and related properties.

Stating that one certain paint is, for example, an acrylic, reveals a great deal about the coating. An acrylic paint will normally be expected to have the generic characteristics of acrylic resins, possibly modified somewhat by the pigmentation and additives. Within each resin category, several different but chemically related monomeric units (uncross-linked molecules) may bear the same general name; thus, there are almost limitless variations among the many possible acrylic resins. The same variability within each class holds true for all the other resin types such as vinyl types, or alkyd types, or whatever type. Any comprehensive description of resin properties will be typical for the resin as a category only, and actual properties may indeed vary to a degree from one resin of a given molecule type to the next. As an illustration of this, we can consider the alcohols: ethyl alcohol, propyl alcohol, and n-butyl alcohol. Their chemical behavior is similar, but while ethyl and propyl alcohols are completely miscible with water, n-butyl alcohol is only partially soluble in water. The human body tolerates ethyl and n-butyl alcohol if swallowed in small amounts; but ingestion of any amount of propyl alcohol is highly dangerous. Differences also exist among members of the same resin families.

Many chemical varieties of film-forming resins are used as binders, both alone and mixed with other chemical resin types. The following is a list of eleven of the major resin types followed by each of their typical properties, based on the chemical nature and physical structures of the resin molecules. Note the use of plurals to emphasize that not just one but many related resins belong to each class.

- Acrylics
- Alkyds
- Cellulosics
- Epoxies
- Halogenateds
- Oleoresins
- Phenolics
- Polyesters
- Silicones
- Urethanes
- Vinyls.

Each class of resin has some typical properties associated with it. An appreciation of these characteristics is helpful in understanding industrial coatings.

Acrylics

Acrylic resins are able to achieve a high gloss and to form hard, highly weather resistant surfaces. The hardness of acrylics contributes to good abrasion resistance and increases their durability. Acrylic resins are above average in cost. They have very good heat stability and excellent chemical and ultraviolet light resistance. Nearly every motor vehicle produced in North America from about 1950 through 1990 received either an acrylic lacquer topcoat or an acrylic enamel topcoat because of these strong performance properties. Even today the acrylics still make up roughly 90% of automotive colorcoats. Paints with hard surfaces such as the acrylics have maximum abrasion resistance, but if paint films are too hard, they also tend to be brittle. This can result in a predisposition toward chip formation from scrapes, stone bruises, bumps, and related impacts. Acrylics for automotive use are therefore formulated to minimize their tendency to exhibit this defect while still retaining their excellent outdoor stability.

Alkyds

This resin name, often spoken incorrectly, is pronounced, "Al Kid" with either the first or second word slightly accented. Both ways will be heard and both are accepted as correct usage. The name "alkyd" is derived from the "alcohols" (alk) and "acids" (yd) that react to form them. The alkyd family of resins is widely used because it is so varied; and each individual alkyd resin can be coformulated with many different binder resins. Alkyd-melamine and alkyd-epoxy are typical mixed resin coatings. By mixing resins in this way, it is possible to produce paints with an enormous variety of specific performance properties.

Alkyds come close to being "the best general-purpose paint" because of their desirable overall properties plus their moderate cost. Alkyds are chemically modified vegetable (plant) oils, especially soya and linseed oils. Because they incorporate what are called "drying oils" (meaning they will cure when exposed to air), alkyds are often used to produce ambient-cure paints. This is an important factor in why alkyds are one of the most heavily used paint resins. The carbon-to-carbon unsaturated intramolecular bonds (double bonds) in alkyds (and in all drying oils) react with oxygen in the air to produce fully cured finishes. Alkyds are the major resins used in combination with others to introduce air-dry capability in a paint formulation. They can, of course, also be formulated into baking finishes and cured using heat. Heat curing almost always results in a more durable finish and is much faster than air-drying. Most manufacturers use heat since it allows quicker packaging, more expeditious delivery times, and, more importantly, lower inventories.

Alkyds are normally modified with various chemicals to increase specific properties, even though this may also make the coating somewhat higher in cost. Styrenated alkyds, for example, show increased chemical resistance compared with pure alkyds. Alkyds cross-linked with melamine resins produce durable finishes with markedly greater weatherability than pure alkyds. European automobile manufacturers have used alkyd-type paints for both primers and topcoats for many decades. The alkyd-melamine finishes are normally only suitable as oven cured finishes.

Cellulosics

These inexpensive materials are produced by the reaction of nitric acid with cellulose, which forms nitrocellulose lacquer resins. They are highly polar and hence limited in solubility, tend to be brittle, and have quite poor heat resistance. They tend to be used because they are relatively low in cost and fast drying. You may recall that the dangerously flammable film used by the fledgling movie industry was made of nitrocellulose. Although they have poor to fair chemical resistance, nitrocellulose resins are attractive for use in low-priced, quick drying lacquers. They are sometimes used in strippable peelcoats for these reasons.

Because of their limited durability and high solvent content they are now seldom used for industrial painting, but for some wood products they are still used and work well. Nitrocellulose lacquers were first used for automotive topcoats in 1928 because they sprayed well and air-dried in hours. This property made them far more attractive than the conventional oil-based paints that had been brush applied and which required a week or longer to cure completely. Most plants can no longer use them due to the high solvent content; exceptions are plants that can afford to use VOC incineration.

Epoxies

Industrial demand for epoxy just grows and grows, it seems. Production of epoxy resins in 2005 is estimated at well over 700 million pounds (317 million kg), half of which will be used for coatings and paints. This amount is nearly three times greater than it was in 1990 when annual usage reached just a bit less than 250 million pounds (113 million kg).

The majority of epoxy resins are produced by the chemical reaction of epichlorohydrin with any of several different bisphenol compounds. An example of this reaction and the polymer created is seen in **Figure 3-3**. The epoxies have excellent water resistance, plus they rank among the highest in resistance to attack by alkali of any commercial resin. This is a valuable trait since alkali tends to break down paint resins (strong alkali is used in paint strippers). Epoxy paints form hard, tough, salt resistant films and display superior resistance to most chemicals. Their excellent heat resistance and abrasion resistance make epoxies ideal primers. Because they have such strong resistance to the alkaline byproducts formed when metals rust, they are one of the very best primers for metals. The epoxies are in most cases low or moderate in cost. All the one-component enamel epoxies require baking; many formulations require substantial oven times and high bake temperatures. This is not always possible so other curing methods have been developed that require no heating.

Two-part epoxies produce equally durable finishes and have an added advantage over one-part epoxy resins. Two-part epoxy paints can be used for ambient-temperature curing

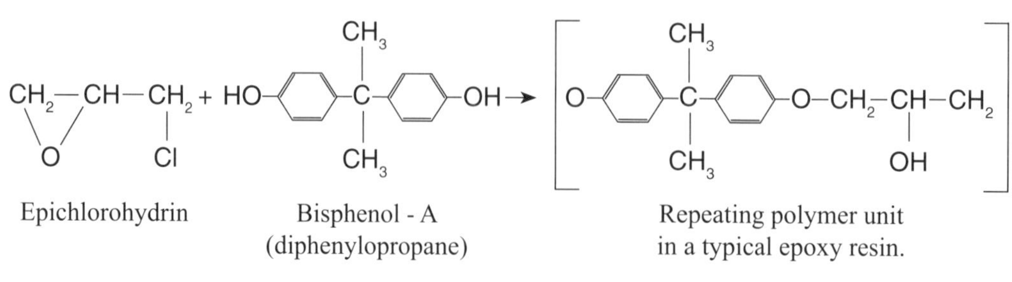

Figure 3-3. Epoxy resin made by reacting epichlorohydrin with bisphenol A.

when it is impossible to bake the part. This capability is especially valuable in fields, such as the marine industry, where extreme corrosion resistance is required and yet painting must be done under any weather conditions during all four seasons of the year. If items are too large to fit into an oven (airplanes, bridges) or too heat sensitive (electronic equipment, plastics) to allow baking, two-component epoxy can be used. Ambient curing can sometimes be hastened by low temperature baking if the substrate can tolerate slightly elevated heat conditions.

The major weakness of epoxies is their tendency to chalk when exposed to ultraviolet light. This chalking contributes to loss of gloss but is rarely extreme enough to interfere with the performance or structural integrity of the coating. Epoxies should not be used outdoors on items whose appearance is critical, or on surfaces that could release chalkiness onto a person's clothing. Combining epoxy and polyester in powder coating resins to increase the chalk resistance of pure epoxy powders creates what are termed "hybrid" powdercoats, which have become quite popular.

Epoxies are ideal resins wherever excellent corrosion resistance and strong chemical resistance are important. Large-scale industrial use of epoxies is directly attributable to their outstanding protective qualities. Epoxies are considered the workhorse among metal primers, especially for use on steel.

Halogenateds

Fluorocarbon, chlorocarbon, and fluorochlorocarbon polymers as well as chlorinated natural rubber resins are employed in the formulation of superior long-term weather-resistant maintenance coatings. They are always expensive, often very difficult to apply, and surprisingly may have relatively low heat resistance. Their chief virtue is their ability to be used in harsh climate environments and be able to provide up to 30 or 40 years of reliable service. Kynar® is one of the more familiar examples in this halogenated resin class. Chlorinated natural rubbers, the lowest cost resins of this type, may be used alone or as a blend with alkyd resins. Halogenated resins are used to decorate and protect exterior surfaces of metal panels and similar decorative structural members in architectural construction. Large, modern office buildings use components coated with these resins for striking visual effects, and these pieces are often attached in hard-to-reach locations or high up on the building. Architects recognize that although the initial cost of halogenated coatings is high, it is still much less expensive than hiring steeplejacks to repaint such difficult-to-reach areas.

Oleoresins (Agricultural Oil-Based)

Oleoresins are some of the oldest resins used to make coatings, but these low-cost binders can still be used to make good paints. Although they dry slowly, drying agents can be added to speed curing. Dryers (or catalysts) are frequently cobalt and manganese salts of naphthenic or tall-oil fatty acids such as octoic, neodecanoic, and other structurally similar fatty acids containing around 8–12 carbon atoms per molecule. Technically these salts are metallic soaps. Drying oils such as linseed oil, soya oil, tung oil, castor oil, cottonseed oil, and similar nonpetroleum oils form the basis for these binders. They are often modified with synthetic resins to improve drying time or to increase ultimate film hardness. Drying oils are sometimes mixed with alkyds to speed ambient curing and to reduce paint costs for applications where only mild service coatings are required.

Phenolics

Paints and varnishes made with polymerized phenolic resins tend to be very hard and somewhat brittle, yet they are also extremely resistant to stains, solvents, and acids. Phenolic resins are low in cost and have such splendid electrical insulation properties they are used to coat wire used in windings for electric motors and to coat flux plates inside current rectifiers and voltage transformers. They are not especially good for decorative home or office finishes. The aromatic structure of the molecules causes phenolics to turn yellow or brown in sunlight even when used indoors. However, they have been used abundantly for marine-varnish finishes where film darkening is not considered a drawback. Their water and acid resistance also makes them a good choice for low cost food can linings. Phenolic resins are used to some extent in anticorrosion coatings for metals.

Polyesters

We have seen that alkyds are unsaturated resins; but polyesters are saturated. Since polyesters do not have unsaturated linkages as do alkyd resins, they are called "oil-free" alkyds. Polyesters, except for the lack of carbon-to-carbon double bonds, are otherwise much like the alkyds in chemical structure. This is one reason they, like the alkyds, are used for a wide range of coating needs. And, like alkyds, they are also moderate in cost. Without double bonds they cannot react with oxygen, so polyester enamels do not air-dry. Polyesters are available only as baking finishes in liquid and in powder formulations. Both wet and powder polyesters are tough and durable in outdoor exposure, exhibiting good flexibility and excellent hardness.

Particularly since the early 1980s, polyesters have experienced extensive utilization as powder coatings. Two major polyester powder formulations are being used: one is isocyanate cross-linked and the other is TGIC (triglycidyl isocyanurate) cross-linked. The isocyanate cross-linked powders provide an attractive but slightly less durable finish than the TGIC cross-linked systems. The TGIC powder coatings have outstanding exterior durability. Several years ago, reports appeared that suggested TGIC be outlawed for paint use because in powder paints it initiates health problems. These exaggerated claims have been strongly refuted and TGIC-cross-linked polyester powder continues to be heavily used.

Silicones

Silicone (organo-silicone) resins have superior resistance to water, sunlight, and just about all other normal exterior surface conditions. They do not chalk. It should not be a surprise then to learn they are well above average in cost; many are around US$50–$60/gallon and others are over US$100/gallon. Perhaps their most outstanding attribute, however, is high heat stability, which permits silicone paints to be used wherever conventional organic resin coatings would deteriorate rapidly. Although some are more heat-resistant than others, all purely organic resins will begin to degrade if exposed to sustained heat of 500° F (260° C) or higher. Silicone resin paints are used for applications such as motorcycle mufflers to withstand temperatures of 900–1,200° F (480–648° C). Some can withstand the heat of 2,500° F (1,350° C) and are used to mark identification codes on the heat-tiles of space shuttles. When these very high temperature resistant paints reach their use temperatures, the organic portions of the resins oxidize away. This leaves an essentially ceramic coating that is stable at elevated temperatures. The high-heat resistant silicone paints require long curing times at temperatures of 650° F (340° C) and above.

Silicones are also used at ordinary temperatures when blended with acrylics or alkyds for finishes that have unequalled weatherability and chemical resistance. Alkyd silicones are available in air-dry coatings that are utilized as extended-life marine paints to maximize the time between repainting of transoceanic merchant vessels. Silicone maintenance coatings are used on water towers, steeples, and bridges for which their high cost is justified by the danger, complexity, and expense of painting such structures in situ.

Urethanes

Among the most outstanding resins in paint usage are the urethanes because of their ability to combine gloss and flexibility with superb chemical and stain resistance. **Figure 3-4** compares the properties of urethanes with those of acrylics, alkyds, and epoxies. The

Type	Properties and Characteristics
Acrylics	Excellent weather resistance
	Chemical resistance
	High heat stability
	Above average cost
	Hard, abrasion resistant
	Slightly brittle
Alkyds	Very good overall properties
	Average cost
	Air-dry curing
	Extremely heavy use in paints due to their fine performance
	Formulation widely variable
Epoxies	Outstanding water resistance
	Tough, heat and abrasion resistant
	High alkali resistance
	Above average cost
	Chalking in sunlight
Urethanes	Superior overall properties
	Excellent mar resistance
	Flexible, with high gloss possible
	Outstanding durability
	Relatively high cost

Figure 3-4. Comparison of paint resin properties.

cost of urethanes is two to five times higher than for other paints, but polyurethanes are unique in their amazingly high gloss levels throughout a wide flexibility range. Urethanes make ideal high-solids coatings that require little, if any, additional heat for curing. Urethanes are often the materials of choice for coating heat-sensitive substrates with a low-solvent-content paint. They are exceptionally physically durable, showing excellent water and weather resistance. Most commonly used are the acrylic-urethane types in the two-component formula. For some reason, two-component urethanes are referred to as "2K" urethanes. It is not clear just why this expression is used, but I like it. Perhaps it is the brevity that is so appealing.

A urethane is formed when an isocyanate chemically reacts with an alcohol. At least six types of urethane paints are possible, but two-part urethanes that mix polyol and isocyanate just prior to application have the best film properties. Pre-reacted (one-part) urethanes are sometimes used to form a blend with other resins. The advantage of pre-reacted urethanes is that no potentially toxic isocyanates are present.

Unreacted isocyanates can cause respiratory problems if an individual breathes the vapors for extended periods. Approved charcoal-cartridge filter masks or head shrouds with a separate air supply are recommended for workers applying two-part urethanes. In the United States, neither the Occupational Safety and Health Administration (OSHA) nor the National Institute of Safety and Health (NIOSH) will officially "approve" any filter or cartridge masks, though they will "allow" them if they capture isocyanates. The reason for this is that to gain OSHA or NIOSH approval, the mask wearers would either have to be able to detect any possible malfunction of the device, or a visible warning of mask failure must appear. Because isocyanates have no recognizable odor, olfactory warning is not possible; and since no fail-safe indicator is available to reveal when a cartridge is spent, facemasks cannot be OSHA or NIOSH approved for use when applying isocyanate-containing paints. Another problem with masks, these agencies point out, is that facial hair tends to prevent a tight mask fit; masks may then not adequately protect bearded workers.

Because urethanes have outstanding performance properties, they are used where their flexibility, gloss retention, superior weatherability, mar resistance, and general durability can justify the higher coating cost. The price may run 2–4 times more than the cost of an average paint. Isocyanate-free urethane-type paints and urethane-like coatings introduced

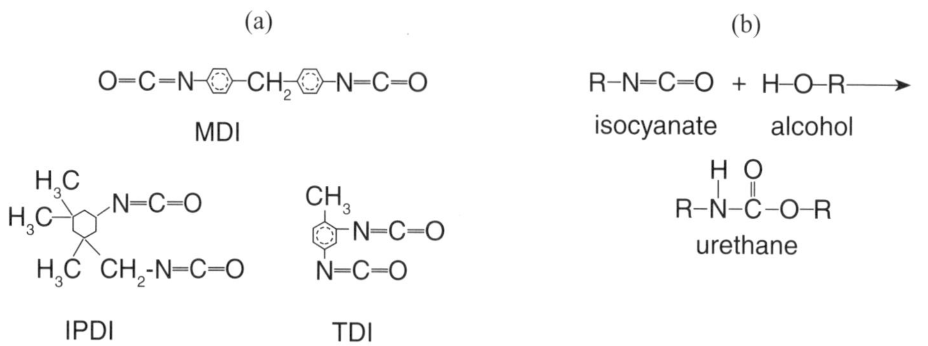

Figure 3-5. Common isocyanates (a), and the formation of polyurethane coating from hydroxyl functional resin and isocyanates (b).

Figure 3-6. Sprayers wearing protective hoods and suits with air-supplied hoods while applying two-part urethane coating.

in the late 1980s, which for some purposes are nearly identical in performance to isocyanate-containing urethane coatings, are considerably safer. A chemical variation of the epoxy linkage is used to cross-link these systems. These paints have not become popular, however, due to inferior across-the-board performance compared to isocyanate-containing urethane.

Vinyls

These resins can inherently be rather rigid, but vinyl resins can be plasticized to varying degrees to achieve greater flexibility when that is required. Vinyls tend to be low in cost yet have extraordinary acid resistances. They possess outstanding water resistance as well, which makes them a frequent choice for maintenance coatings. Their exterior durability is excellent, although halogenated vinyls should not be used in high temperature applications. They will degrade in heat and give off potentially dangerous hydrogen halide acid fumes. For service applications in wet corrosive environments such as offshore oil rigs, vinyls make excellent coatings. See **Figure 3-7**. Polyvinyl chloride (PVC) coatings are used extensively for food and beverage can linings to take advantage of their water and acid resistance. Solution vinyl resins also find frequent use in peel coats for spray booth maintenance.

For structural wood finishes where moisture must be allowed to escape from the building interior, polyvinyl acetate is frequently used because it is porous enough to allow the water vapor to "breathe" in and out through the film. This avoids the paint lifting that frequently occurs on wood siding when a nonporous paint film traps moisture. The water vapor pressure may lift a nonporous paint off the surface and produce blisters and/or ruptures in the paint film.

Resin Mixtures

Mixtures of binder resins, including mixtures both within a single resin family or combinations of two or more different resin types, are frequently utilized in paint formulations. For example, two acrylic type polymer resins such as ethyl acrylate plus butyl methacrylate might be combined in an acrylic paint; or different but related vinyl polymers could be used together in the manufacture of a special vinyl paint. These are intrafamily

Figure 3-7. Air-dry vinyl resins are used in wet environments such as offshore oil rigs.

mixtures. But interfamily resin mixtures are also used to produce cost-effective products or to create coatings that meet special performance requirements. Thus, a coating formulation may contain an epoxy-polyester or a vinyl-alkyd binder. There may be more than just two resin types in a coating; in theory, paint polymer blends may contain any number of compatible binders.

Physical Makeup Categories

Another major way to categorize paint is according to its physical makeup. Sometimes the physical makeup differences are often obvious, yet often they are not. The physical makeup of a two-component paint, for example, is obvious; it comes in two separate containers that require mixing together before coating application. A metallic finish is another paint with an obvious physical makeup characteristic by which it can be identified.

Sometimes the differences in physical makeup are not immediately obvious. Consider the terms "solventborne" and "waterborne." Solventborne paint uses a traditional solvent to disperse the resins, pigments, and additives; a waterborne paint uses primarily water. Note that the terms are "solvent(borne)" and "water(borne)" instead of "solvent(based)" and "water(based)." The term "borne" merely implies "carried or transported by," just as the expression "airborne" means an object held or supported in air. The term "based" refers to the paint's film-forming material. A paint can, for example, be acrylic-based, epoxy-based, or alkyd-based. However, a paint can be silicone-based, or vinyl-based (etc.), yet at the same time be solventborne or waterborne. But it is impossible to have a resin made of water, so in that sense paints are never "water-base" materials, although they might be "waterborne."

Liquid Fluidization Methods

A less obvious physical property of paints is the method whereby they are fluidized by a liquid. The term liquid fluidization refers to how the paint solids are enabled to achieve flow by mixing resins with or dissolving a resin in a liquid solvent or liquid carrier. This liquid fluidization might take place in water or in a solvent. (Liquid fluidization should not be confused with air fluidization employed with powder paints; that subject is covered under powder coatings.)

As shown in **Figure 3-8**, liquid paints can be classified into three major fluidization types, **solutions**, **dispersions**, and **emulsions**.

The divisions are based on the extent to which the resin molecules are dispersed in the liquid. In a true solution, the resin is separated all the way down into individual molecules. Going from solutions to dispersions, and then emulsions, fluidization separates the resin less completely, leaving larger and larger clusters of molecules. **Figure 3-9** compares typical viscosity and percent solid ranges for each.

Solutions

In a solution paint, each molecule of resin (the solute) is dissolved and held apart from other molecules by the liquid medium (the solvent). The individual molecules are essentially floating in a large sea of solvent. As increasing amounts of paint resin are added, the viscosity will also rise proportionally. The viscosity increases because the long thread-like resin molecules physically intertwine and become tangled with each other. The entwined molecules actually resist paint flow (imagine how the entire jumbled pile moves when one of the wires in a piece of steel wool is pulled) and increase paint viscosity. It is desirable to minimize solvent use for cost reasons and to enable lower solvent emissions,

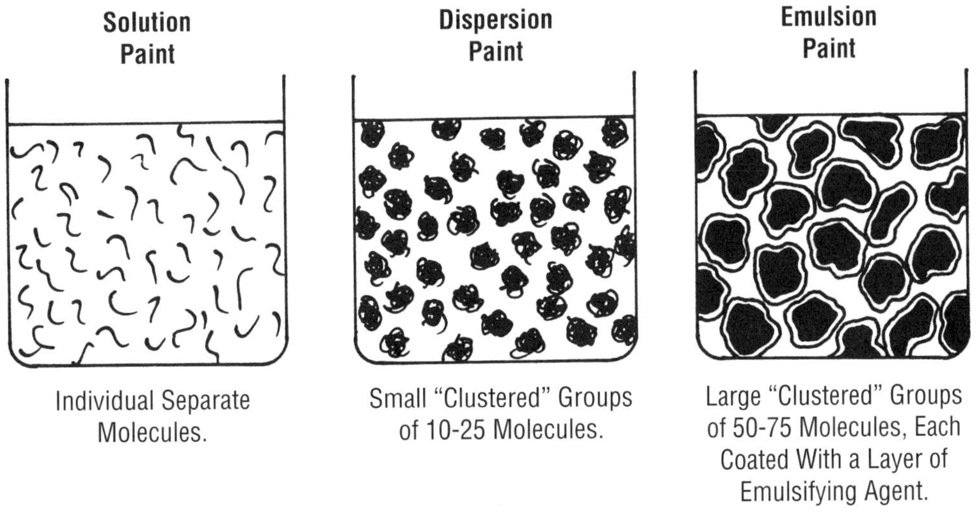

Figure 3-8. Major resin fluidization methods.

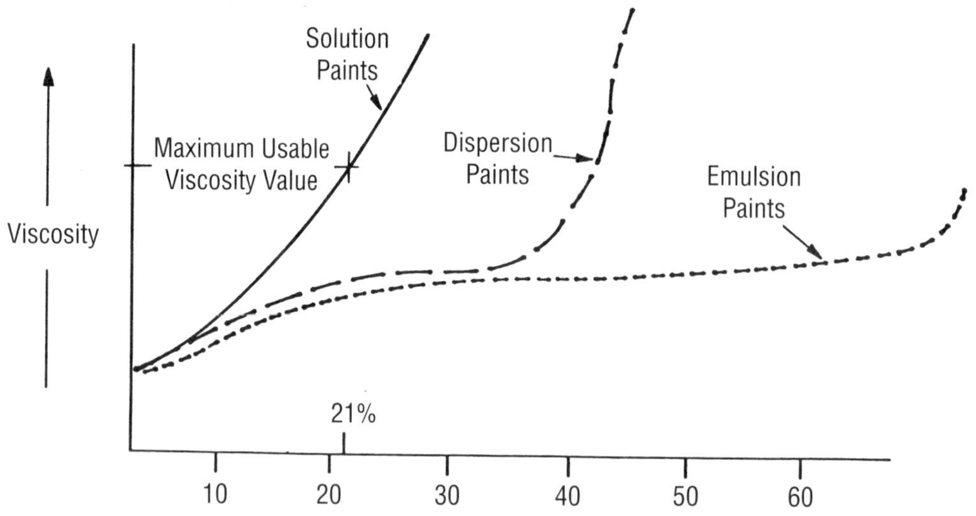

Figure 3-9. Viscosity versus percent resin solids.

but failure to add enough solvent leaves the paint with an unusable high viscosity. You can see, therefore, that to produce a high-solids paint it is not possible simply to withhold some of the solvent; this creates a high solids paint, alright, but one that would be for all practical purposes impossible to apply.

Dispersions

One method to lower solvent use is by forming a resin dispersion rather than a true solution. Dispersion paint is generated using a blend of solvents with lower polarities than those used for solutions. The lower polarity of the solvents cannot fully dissolve the resin, so the resin molecules are forced into a mild clumping or aggregation. Dispersion paints have resin molecules clustered into small nodules containing roughly 10–25 molecules per cluster. The viscosity curve for dispersion paint (see **Figure 3-9**) has a long usable plateau region into the high-solids range. Notice that viscosity rises rapidly to unusable levels when the resin concentration gets so high that the individual nodules of resin contact each other and begin tangling. The advantage of dispersion is the reduced amount of solvent needed and the lower VOC. Dispersions are able to produce paint with the same viscosity as solution paint, but do so using less solvent.

Emulsions

It might seem that producing even larger resin clusters or nodules with perhaps 50–100 or more molecules in each cluster would allow paint formulation with even less solvent. But the difficulty is that dispersion technology cannot be extended quite that far. Resin nodules of that size do not stay mixed; instead the clusters soon begin to agglomerate and the mixture separates into two layers, resin and solvent. The situation is similar to shaking oil and vinegar salad dressing. Vigorous shaking produces only temporary blending. After standing for only a few minutes, all the oil floats back up to the top again. Many decades ago condiment manufacturers discovered permanent blending of salad oil and water could be accomplished if the oil particles were coated with a thin protein layer. The addition of egg to a mixture of salad oil and water followed by vigorous blending (along with some salt and seasoning) creates mayonnaise. The separate oil and water phases of this seemingly single-phase product are visible when examined under a microscope.

The egg protein coating on the salad oil droplets acts as an emulsifier. Although neither egg protein nor any other form of protein is used in the making of emulsion paints, the resin emulsification is accomplished in a similar fashion. In this case, the paint emulsifier is vigorously mixed with resin until it has surrounded all the resin nodules to prevent them from touching each other and being able to form a separate layer. Emulsion technology can produce coatings with even higher solids content than is possible by using resin dispersion technology.

Dispersions and emulsions can be produced in both solventborne and waterborne formulations; however, by far the most common are waterborne emulsions. Both the "wall paint" widely applied in home and office interiors, and "house paint" used extensively for exterior service, are nearly invariably waterborne emulsions. Both are frequently labeled "latex" paints, but here the term latex does not imply that the paint contains natural or synthetic rubber-like resins. It happens that the sap from a rubber tree is a naturally occurring waterborne emulsion and hence it has been named "latex." Technically, latex is an explicit chemical term that identifies an organic polymer emulsified in a water medium. For this reason, household paints are generally labeled and advertised as being latex paints. That term for these waterborne coatings is correct.

The correct plural for latex is "lattices," but "latexes" is used so much that it has almost become equally acceptable. In any case, the most significant thing about latexes or lattices is that they utilize aqueous media to fluidize organic materials that have virtually no water solubility. Most of these organic substances are not especially mobile fluids. A common example of latex is the cream in cow's milk, which is oil (butterfat) solubilized in a watery solution. Milk contains on the average about 11–12% cream. Cream, the source of dairy butter, pours and looks as if it were a single-phase liquid. However, when churned, cream forms clumps of butter and separates from a white watery residue we call buttermilk. During churning, a natural protein emulsifier coating is rubbed off the butter particles, permitting the butterfat globules to join together on contact. They

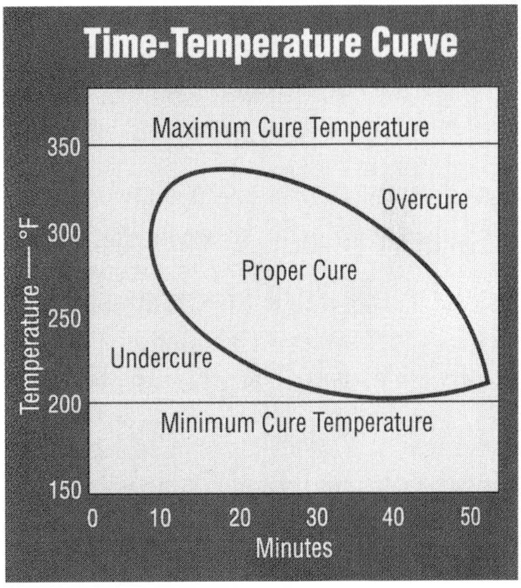

Figure 3-10. Time-line temperature relationship of a typical liquid paint. Adequate cure can be achieved within 20–40 minutes at about 250° F (121° C). Powder coatings are similar, but usually require shorter times and higher temperatures.

soon form larger pieces of butter that have separated from the watery phase. To disperse the cream throughout the milk to the point that butterfat globules will no longer float to the top in a separate layer, the globules are made much smaller, and thus *homogenized* milk is produced.

Dispersions differ from emulsions in that the resin clumps are smaller and therefore do not require any emulsifiers to form a stable fluid mixture. By judicious selection of the solvent blend, vigorous physical mixing alone can be used to create the dispersion. Emulsion paints are by definition only those liquid coatings that require an emulsifier to stabilize the rapidly stirred mixture of water and resin.

Cure Mechanisms

The characteristics of a paint's curing mechanism will, in some cases, determine its suitability for different applications. Lacquers and enamels cure by different methods.

Lacquers

Lacquers are coatings containing binders with large organic molecules that are fully polymerized before they are applied. Powder lacquers simply melt with heat during curing and solidify once again when they cool down. Liquid lacquer paints cure through solvent evaporation alone, although heat may be used to accelerate the evaporation of the solvents. The preparation of lacquer paint requires a considerable amount of solvent in part because the molecules are so large and quite strongly polar. The massive binder molecules necessitate the use of aggressive solvents with strong dissolving properties, typically oxygenated solvents such as ketones, alcohols, and esters. Since lacquers undergo no cross-linking, the molecules must be highly polar to achieve films that are sufficiently hard and physically durable. Dipole-to-dipole (magnet-like) bonds attract and hold lacquer molecules together firmly in the dry paint film. This usually makes lacquers easier than enamels to spray without producing runs or sags.

Even after lacquers are applied and cured, heat or solvent can soften the resin. Application of the right solvent to a cured lacquer film can dissolve the resin. This property is used to facilitate firm bonding between a film being repaired and the newly applied lacquer repair coat. In this way, lacquer paint films can be readily spot-repaired instead of requiring a repaint of an entire panel. The solvents in the repair paint will dissolve the existing paint film enough to achieve strong bonds between the original paint and the repair paint. On the negative side, because lacquers form a harder film, they tend to experience chip damage more readily than enamels.

Heat alone (thermal reflow) in the case of specially designed coatings will reflow the resin in the film and transform minor scratch lines caused by repair sanding into a smooth paint surface. In addition to thermal reflow for repairing scratch lines, solvent reflow has also been used with suitable nonflammable solvents. Hot solvent vapor reflow has been used for lacquers in the past, but like lacquer paints themselves, is now rarely used. This procedure used a methylene chloride vapor gun to achieve a smooth paint surface with minimal hand polishing and no respraying of the sanded item. Reflow was low in cost and fast compared to the labor costs in manual polishing or respraying. These techniques are not used much today because few lacquers are applied industrially.

Solvents cannot dissolve cross-linked enamel films, but they can damage them. For example, if an enamel film is painted with liquid lacquer, often the strong solvents in the lacquer will be absorbed into the enamel paint films, causing severe enamel blistering and wrinkling.

Enamels

Enamel paints—before they are cured—contain resin molecules that are in a few instances totally unpolymerized. Far more often though, the resins in enamels are partially polymerized before application. Enamel molecules must also chemically link together to a much greater degree after the paint has been applied. Fully cured enamels are to varying degrees thermosetting materials due to this additional chemical cross-linking that occurs among the molecules after the coating has been applied. Sufficient cross-linking must take place to provide the requisite hardness and water resistance to the film; however, the extent of cross-linking in enamels is normally only moderate. If too much cross-linking takes place, the paint film becomes overly hard and brittle, causing poor chip resistance and low impact strength.

Enamel curing begins to take place rapidly when the solvent evaporates and heat or radiation activates cross-linking. The resin chemically links together in many places along the molecule chains by irrevocable chemical reactions. If solvent evaporation alone took place with enamel films but no cross-linking, just a soft, gummy film would be formed.

Resin molecule cross-linking can take place by a number of mechanisms, depending on the resin type. Binders may cross-link by reaction with the oxygen in the air; all of the pure alkyd and oil-based coatings cure this way. Other type enamels may react with moisture in the air and undergo subsequent cross-linking reactions. Most frequently the resins undergo self-reaction of chemical groups originally present on the molecules. These reactions occur rapidly only after appreciable amounts of the solvents have evaporated and the molecules are in close proximity to each other. Even if molecules are near each other physically, they must be energetically activated or they will not cross-link. Depending upon their chemical properties, resin molecules may be selected that will be activated by means such as heat from a curing oven or by radiation absorption from infrared emitters, ultraviolet light, or electron beam rays.

The chemical bonds that are formed thus convert the soft, easily deformed resin into a firm, dry cross-linked paint film. The paint formulator can chemically modify resins to produce a soft, flexible film; a hard, highly cross-linked film; or something intermediate, depending on which type of film is most appropriate.

Lower cost hydrocarbon solvents commonly can be used in enamel paints. The short, small enamel resin molecules tend to dissolve readily in comparatively small amounts of solvent. When an existing lacquer film is painted over with enamel paint, the mild solvents in typical enamels rarely cause wrinkles or blisters in the lacquer paint. This is due to their weak solvent power. Enamel films are generally not so extensively cross-linked that they become highly brittle, nor do they become completely able to resist being softened by many solvents. Yet most are sufficiently thermoset so that unlike lacquer paints, they cannot truly be redissolved by solvents or be extensively softened with the application of heat. Not surprisingly then, attempts at producing good results with thermal reflow enamels and vapor reflow enamels have not been successful.

A lasting repair on enamel paint exposed to exterior weather conditions usually must include repainting of an entire panel. Once the original enamel molecules have cross-linked, they can no longer cross-link with the enamel molecules in the repair paint. Adhesion between the original and repair paint molecules is therefore due only to mechanical and not chemical bonds. To obtain a durable enamel paint repair, scuff sanding of the entire panel prior to repainting is an advocated technique. Spot repair of enamels is sometimes performed, even on high-priced items such as autos, but it is not recommended. Enamel spot repair is always inferior to repainting an entire panel.

Final thoughts: HOW COATING CURING HAPPENS

The chemical nature of the resin determines how it will cure. Here are the major processes.

Liquid ***lacquers*** form films by solvent evaporation alone; powder lacquers by melting and then resolidifying upon cooling. A liquid paint that generates a film solely by solvent evaporation contains one or more resins with long-chain polar molecules dissolved or suspended in solvent. As solvent evaporates, polar forces attract the resin molecules together along the length of the molecule chains. No reaction occurs to create chemical bonds among molecules.

Powder and liquid coatings that cure by forming chemical bonds between molecules (cross-linking) constitute a separate class of paints called ***enamels.*** Cross-linking may occur between identical base resin molecules or other molecules designed specifically to cross-link with the base resin. Cross-linking forms a vast polymerized network of chemically attached molecules that generate a viable coating film. Heat is often used to polymerize resins via cross-linking, but other means are also employed.

Cross-linked finishes are classified into five major polymerization categories.

Oxidization cross-linking takes place with unsaturated drying oil resins. These include oleoresinous materials, phenolics, esters, alkyds, and all oil modified resins. As the solvent in an applied liquid coating evaporates, oxygen in the air forms cross-links with the unsaturated chemical bonds in these materials. Powder coatings that contain unsaturation react similarly in air.

Heat-induced cross-linking can transpire with liquid and powder coatings. An example is an acrylic resin combined with a melamine cross-linker. When heated, the two resins interact chemically to form a melamine-cured acrylic coating. Until heated such liquid coatings remain tacky and uncured indefinitely. Applied powder films melt, and then cross-link when heated.

Radiation-cured coatings cross-link when they are activated by UV rays or ionizing electron beams. The UV or electron bombardment triggers free radical reactions among chemical groups that result in rapid film cure. Powder coats must be melted prior to radiation exposure.

Two-component (2K) coatings consist of paired reagents, the coating resin and a cross-linking agent, which react when mixed. An example is a 2K urethane of polyester resin and isocyanate cross-linker. When mixed the two form a polyurethane coating.

Moisture-cure coatings function similarly to oxidation cross-linked coatings except that moisture in the air triggers cross-linking. A single-pack urethane in which the isocyanate first reacts with moisture and then forms a urea-type linkage cures by this mechanism.

Chapter 4

Low-Solids and High-Solids Coatings

The History of Low-Solids Coatings

An understanding of low-solids coatings is helpful as a building block to understanding high-solids coatings. The chemistry of these two coatings systems is considerably different. Low-solids coatings are characterized as having a low percentage of solids, and therefore a high percentage of volatiles (solvent). Percent solids can be expressed both as weight percent or volume percent. Due to density of resins and pigments, weight percent solids is always a higher number than the volume percent solids.

Low-solids coatings have been manufactured and used for a long time. Their formulation and use, however, multiplied greatly after the burgeoning of the petroleum industry in the 19th century. As the pumping of oil increased, numerous oil-derived solvents and resins became available at a reasonable cost. Coatings were formulated for thousands of different products and purposes.

In its early days, paintmaking was hardly a scientific endeavor; the craft evolved primarily by trial and error. At that time, almost anyone could make a paint and prosper (provided one had a flair for marketing). One could say with some accuracy that in the early 20th century, the integrity of many paints was in question. Because early paintmakers were not chemical formulators, their products were often mixtures hastily put together by technicians.

However, as time went by, the chemical reliability and batch-to-batch uniformity of the coatings began to improve. The chemical improvements went hand-in-hand with developments in the industrial revolution and with improvements in the mass educational system. As more chemists became available, paint companies, especially those that foresaw the day when paints would have to measure up as a quality product, began to hire them. Chemists were needed not only to develop new formulations but also to devise ways to maintain consistency from one batch of paint to the next. As a consequence, in about the early 1920s, paint formulation consistency was beginning to be established. This improved quality and performance in low-solids formulations increased at a steady rate until a dramatic shift occurred in the mid-1960s. Patents on coatings formulas (or "formulae" if you prefer) that deviated from the typical low-solids technology were being filed during this period. Yet, since petroleum solvents continued to be so economical, the low-solids paints remained as the most practical choice for most products. Nearly all of the coatings produced were simply minor variations within what are now called low-solids paints.

Although some low-solids coatings made prior to the mid-20th century were still of marginal quality, many more were truly quite good, and some could be called excellent.

Almost everything that was painted was being finished with low-solids coatings—cars, appliances, airplanes, furniture, etc. However, the solids content of these coatings was extremely low, and the solvent content was very high. Lacquers were still in wide use with volume solids content of only 10% and less. By today's standards a solvent content of 90% or more is rather shocking. A number of paints did have less solvent than these lacquers but, still, most of the industrial coatings of the 1960s probably averaged less than 20–30% in percent volume solids. The solvents were an effective, convenient, and inexpensive fluidization medium for the resins, pigments, and additives.

By the late 1950s, total solvent use for industrial paints had soared, and this large-scale use continued into the early 1960s by sheer momentum. However, some paintmakers, seeing the excessive solvent use as environmentally ominous, and fearful of governmental intrusion, were trying to market coatings with higher solids content. The public was being made increasingly aware that solvent emissions were a major contributor in causing poor air quality in a number of geographic areas in the U.S., Japan, and Europe. But industry forces, principally the low cost of solvent, were not ripe for successful marketing. Solvents were just too low in cost to resist using them abundantly. No strong market force was present to serve as a catalyst to change this wasteful technology.

However, in 1966 the first U.S. environmental regulation was adopted in California limiting solvent emissions due to the recognition of their role in the increasingly poor air quality in locations such as the Los Angeles Basin. The coatings industry was only slightly affected by these laws. Then the 1973 Arab oil embargo shook an energy-wasting world into action. This major event and more air quality laws precipitated a crisis in paintmaking. Solvent use began to be restricted through regional and national environmental regulations and, at the same time, solvent costs began to rise sharply due to the oil shortage. In short order, minimization of solvent content became a major criterion in paintmaking. The industry had arrived at a major turning point onto a path of reduced solvent content in paints, a path from which the coatings industry has not swerved and is still following today.

The emerging solvent emission regulations caught the industry almost totally off guard. National seminars began to be held annually in the mid-1970s dealing with high-solids and waterborne coatings. Formulators of paints and those who applied coatings flocked to these meetings to learn how to apply or make coatings that would comply with EPA regulations. Marketing projections for low-solids coatings began to turn downward, and those for compliant coatings began to rise.

For many years, emission regulations have only allowed low-solids coatings to be used in restricted amounts. Limited exceptions may be allowed if the solvent vapors are destroyed by incineration, bioremediation, or if captured through use of devices such as carbon adsorption chambers. The future use of low-solids coatings, if not eliminated entirely, will be limited to companies that can prove the need to use them, and companies large enough to be able to afford solvent capture or destruction systems. Companies too small to be covered by emission regulations, and companies such as those in third-world countries not having restrictions, may also be able to continue with low-solids paints, at least for a time.

The Importance of the Automobile Industry to New Developments in Painting

A sales manager for a resin company in my home state of Georgia is quoted as saying, "As with any paint technology or any coatings products, to build volume you have to link up with the automotive industry. The key to jump-starting any new paint technology is the automotive applications." The veracity of these statements can be seen over and over again in equipment, processes, and materials. Automakers have the financial resources to try innovative items plus the huge production volumes that make entrepreneurial ventures highly attractive to prospective suppliers. For these reasons, you may perhaps notice throughout the following chapters that information is given that mentions one of the automobile manufacturers is using or testing an innovative coating material or process. Often an auto company will "test use" a new method or new product in just one plant before deciding whether to use it in all similar plants. Thus, they can run modifications and gain experience with it, thereby learning how well it works in mass manufacturing and how easily it can be instituted and controlled in full-scale operations. Over the years many new concepts have gone to wide utilization as a result of such initial studies; of course many others have also been quickly dropped when they didn't pan out as well as had been hoped.

High-Solids Coatings

The EPA's limitations on low-solids coatings paved the way for the emergence of high-solids coatings. The term high-solids coatings can broadly include solventbornes, powder coatings, radiation cure coatings, and, under some circumstances, waterborne coatings. In this chapter, however, the term high-solids will refer especially to solventborne high-solids coatings. (Waterborne coatings of all types, whether high or low in solids, will be considered in Chapter 5 because they can also be considered a separate and unique class of coatings. Powder coatings, which are certainly "high-solids" because they are practically 100% solids, will be discussed in detail in Chapter 6.)

Some confusion exists among paint formulators and finishers as to what the solids percentage must be to qualify a given paint formulation as a high-solids coating. Many people making or using coatings arbitrarily consider that if it is applied at about 50% solids or higher, a coating can be classed as high-solids. That is a convenient way to define them even though it is not a strict rule.

Recognizing the difficulty of defining a high-solids coating, some prefer to call a waterborne or solventborne coating of considerable solids as being one having a "higher" solids content. The word "higher" presents another problem: higher than what? It is similarly inexact and its use is not encouraged.

Most high-solids coatings—as applied—typically fall into the range of 50–70% solids. A few 100% solids coatings are used, especially in radiation-cure finishing, but very few paints are used in general manufacturing that are much above 70–75% solids as applied. There are some out there, but they are rare and have limited sales volume.

High-Solids Resins

Simply raising the solids content in a paint formulation cannot make easily applied high-solids coatings. Remember that viscosity rises rapidly as the percent solids increases in solution-type paints. Only by altering the very chemical nature of the resin molecules is it possible to produce a high-solids coating with a viscosity low enough for normal application. The key difference is that low-molecular-weight resins must be used to produce high-solids coatings. Chemically reactive short resin molecules at a low polymerization level are used in the high-solids formulations. The fluidity of resins will increase as the molecule size is reduced. Short molecule chains do not physically intertwine as extensively, nor do they tangle and ensnare each other as much as the long molecule chains used in low-solids binders. Thus, the relative movement and flow throughout the solution is freer and easier with short molecules.

But merely using short molecule resins is still not the complete answer. Small, short molecules would not give the same extent of total system cross-linking during paint curing as would long molecules. Soft and far less water- and abrasion-resistant paint films would result. To get people to buy them, the performance properties of high-solids coatings must at least be equal to those of the low-solids coatings they are designed to replace. So further modification of the small resin molecules is necessary. Increased numbers of reactive sites for cross-linking must be chemically added to these short molecules. This allows more cross-linking so that the overall film properties reach the same performance levels achieved with comparable low-solids coatings.

The short resin molecules in high-solids coatings can sometimes make VOC calculations difficult. Some of the short resin molecules may volatilize when the paint is baked. The EPA in New York found that several manufacturers were certifying their high-solids coatings as being 70% solids by weight. This was based on the assumption that only the solvent portion of the formulation was volatile. In actual tests, the EPA determined the weight solids to be much lower, in one case only 58.4%. This is not a minor discrepancy, nor was it an isolated instance. A number of other paints tested ranged anywhere from 2.0–9.4% lower in weight solids than the information on the paint manufacturer's product data sheets. If resins are not carefully checked, they may contain low-molecular-weight molecules that will contribute to VOC emissions. To prevent volatilization, the resin molecules should be reactive enough to become part of the film or else be of sufficient size not to evaporate at oven temperatures. Manufacturers are now aware of the need to check resin volatility as a component in potential VOC emissions.

The leading resin types used in the manufacture of high-solids coatings are alkyds, polyesters, epoxies, urethanes, and acrylics. However, nearly any resin types can be made into a high-solids formulation by applying the appropriate chemistry to the molecules.

Surface Tension

An unfortunate consequence of increasing the number of cross-linkable reactive sites along the length of these short, high-solids resin molecules is that they produce paints with a relatively high polarity and high surface tension. High polarity is not desirable since it tends to cause excessive Faraday cage problems during electrostatic application and poor adhesion on surfaces having low polarity.

Surface tension for fluids, often expressed in dynes/cm, is defined as the energy needed to generate an area of interfacial surface between the liquid and the air around it. The surface tension of liquid coatings is critical because it must be lower than the surface tension of the substrate it is to cover. To adhere fully and "wet out" various substrates, the paint needs to be as low as possible in surface tension. This is especially true when coating plastics, which as a group are low in surface tension and therefore somewhat difficult to "wet out."

Some adjustment in the formulation of many high-solids coatings is necessary to compensate for the potential surface tension problems. Normally, paint formulators strive for a fairly low surface tension to enable good surface wetting. Paints with high surface tension are characterized by a tendency toward elevated internal cohesiveness. This makes it difficult to spray or otherwise atomize the coating, and in the freshly applied film increases the tendency to form fisheyes and related defects such as blisters and edge pull.

The sprayability of paints is more closely related to their surface tension and cohesiveness than to their viscosity. However, surface tension and cohesiveness are difficult to measure, while viscosity is readily determined with a cup viscometer. Therefore, coating viscosity is commonly used as an indicator of sprayability. Paints are atomized with less energy (lower bell speed or lower air spray pressure) when the surface tension and viscosity are low. Using lower bell speed and lower air pressure wastes less of the paint. Paint transfer efficiency goes down as the energy to atomize paint increases; this is due to droplet "bounce-off" and fine paint droplet "fog."

Each component in a paint formula contributes to the overall surface tension and sprayability of the coating. Solvents have the lowest surface tension of all components, ranging from about 0.00026–0.00053 PSI (17.5–35.0 dynes/cm^2); therefore, solvents tend to improve sprayability. However, high-solids coatings contain less solvent than low-solids coatings, which tends to lower the sprayability of high-solids coatings. In addition, high-solids coatings usually contain more slow-evaporating solvents, which tend to have relatively high surface tensions, further lowering the sprayability of high-solids coatings.

The resins used in high-solids coatings also contribute to the tendency for high-solids coatings to have poor sprayability. High-solids coating resins are more reactive than those used in low-solids coatings and therefore have higher surface tensions. The surface tension of resins generally varies from approximately 0.00053–0.00090 PSI (35–60 dynes/cm^2).

Because paint viscosity tends to decrease with temperature, coatings are often heated prior to application to lower their viscosity and therefore improve sprayability. This had been done since at least the 1950s with low-solids paints; and was not newly introduced just for high-solids coatings. The drop in paint viscosity when heated is, however, far more dramatic for high solids than for low solids. So heating becomes an especially valuable tool in the application of high-solids paints; approximately 85% of all high-solids coatings used industrially are heated for easier spray application.

High-Solids Advantages

The most important advantage of high-solids coatings is the reduced VOC emissions and the resultant improvement in compliance with state and local regulations. Inherent in the lowered VOC emissions is another advantage: solvent usage is reduced and substantial savings can be made, as shown in **Figure 4-1**. Solvents savings can translate additionally into inventory reductions, plant space savings, and fire hazard reduction. **Figure 4-2** illustrates how increased volume solid decreases the overall paint consumption.

Another advantage to having increased solids is the associated reduction in the number of spray application strokes to achieve a given film thickness. Theoretically, this can allow increased conveyor speeds. However, in actual practice this is a minor benefit at best. If rather thin coating films are desired, it may not be an advantage at all but just the opposite. The tendency for film thickness to build rapidly with high-solids paints can make film-thickncss control somewhat difficult. Extraordinarily fine atomization such as that obtained with rotary atomizers is sometimes helpful in overcoming this hurdle, since not all spray guns are able to finely atomize high-solids.

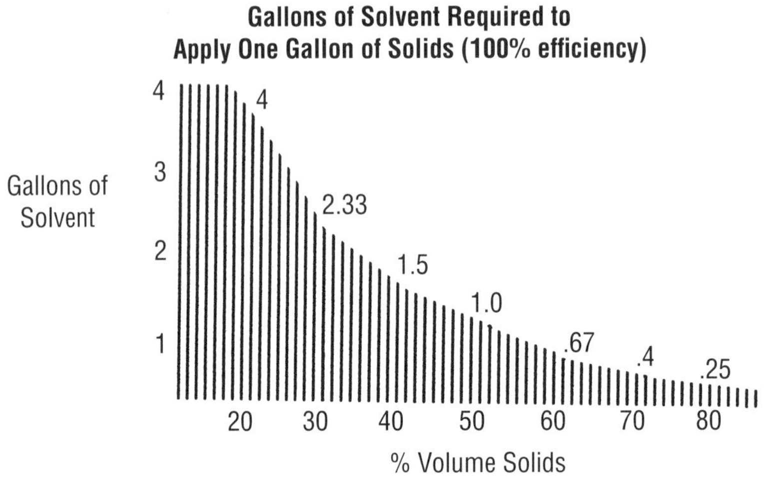

Figure 4-1. Solvent content of paint decreases as paint solids increase.

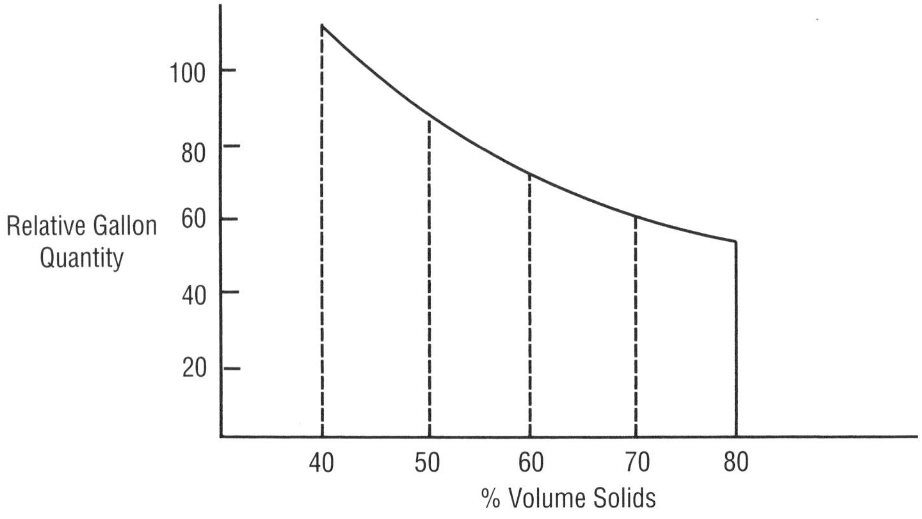

Figure 4-2. The relation of paint sales to changes in volume solids.

High-Solids Disadvantages

Although their advantages are pronounced, high-solids do have a number of significant disadvantages as well. Low-molecular-weight resins needed in the formulation of these paints generally require high cure temperatures. **Figure 4-3** compares the relative cure windows (time and temperature curing cycles) for high- and low-solids coatings. As a result, comparatively few single-component air-drying coatings are available. A 3 mil (75 μm) high-solids air-dry alkyd may be force-dried in 10 minutes at 200° F (93° C), but it may need 10 hours to air-dry to the same degree of hardness at 77° F (25° C). The air-dry or low-temperature-curing high-solids coatings are far more likely to be the two-component types, primarily the two-part epoxies and the two-part urethanes, which cure quite quickly even without heat.

Another disadvantage with high solids is their narrow "time-temperature-cure window." This means that oven times and temperatures need to be controlled closely. The cure window profile is not something a paint user would determine; rather, that information is available from the paint supplier. The cure window, naturally, differs from one particular high-solids paint to another.

These low-molecular-weight resins in high-solids paints are particularly sensitive to inadequate cleaning of substrates. Minor surface oil contamination can cause cratering, blistering, and edge pull. Such flaws can occur with all paints, but they are definitely more pronounced with high-solids. Blisters and craters can result from oil or grease on the surface because these contaminants have low surface tension. When high-solids coatings are applied over traces of oil or grease, the paint solvents at least partially dissolve the

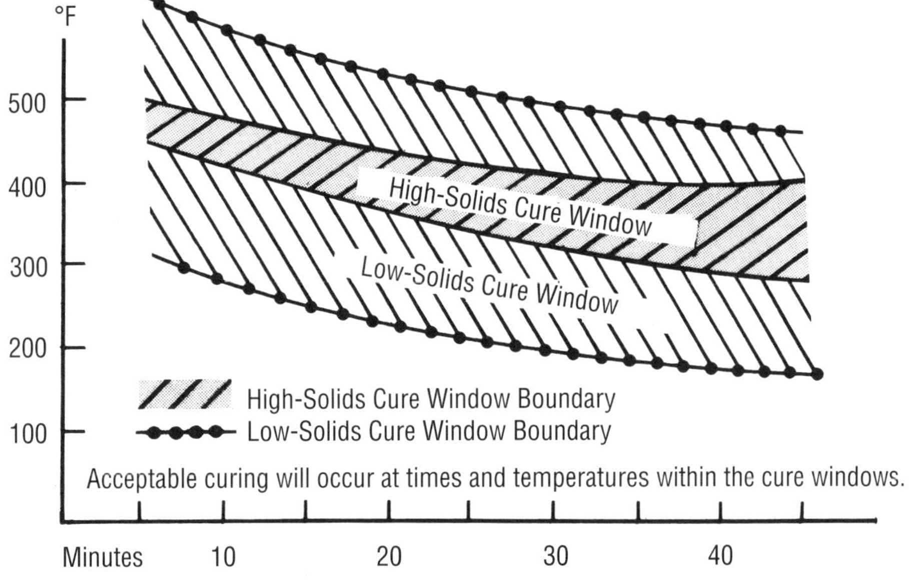

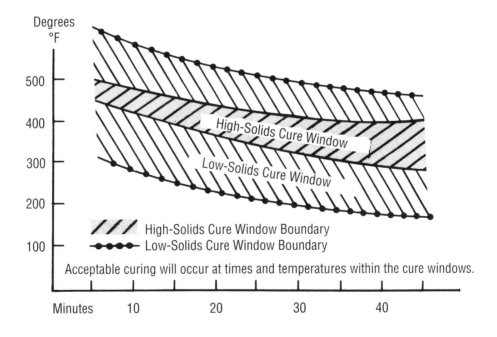

Figure 4-3. Cure windows for high- and low-solids coatings.

oil or grease and incorporate them locally into the wet paint. This creates a region with low surface tension, causing the material to be pulled into the high-surface-tension paint. What is left forms the blister or crater in the finish. Substrate cleaning must be far more thorough with use of high-solids coatings than with low-solids coatings if appearance defects such as these are to be avoided. (Film defects will be discussed in greater detail in Chapter 16.)

Except for two-part coatings, high solids do not normally air-cure and are cured with heat or radiation light. That works well, but the overspray left in the booth is hard to handle because it never hardens—even after months and months of standing. Although the tendency of high-solids overspray to remain gummy forever almost invariably causes cleanup problems, it can occasionally be a small advantage. It may allow overspray of some select high-solids paints to be reclaimed. One of the most readily reclaimed types of paint is high-solids polyester. A limited number of finishers are collecting oversprayed paint and either recycling it themselves or having a paint company do it for them. If the overspray does not become too viscous (and many paints do), overspray can be collected on vertical baffles. It then flows down the baffles into a collection trough and finally into a suitable container. All that is needed to reclaim it is to filter, check and adjust the color, and restore the proper viscosity by solvent addition. Reclaiming overspray has a bonus: much less waste paint needs to be sent for disposal, further reducing operating costs. Currently, this is not practical for most plants; most paints become far too viscous to flow down the collection baffles. Yet, there are significant savings being realized by some companies who have pioneered this cost saving procedure. In the future we can expect to see considerably more reuse of recovered paint materials being attempted.

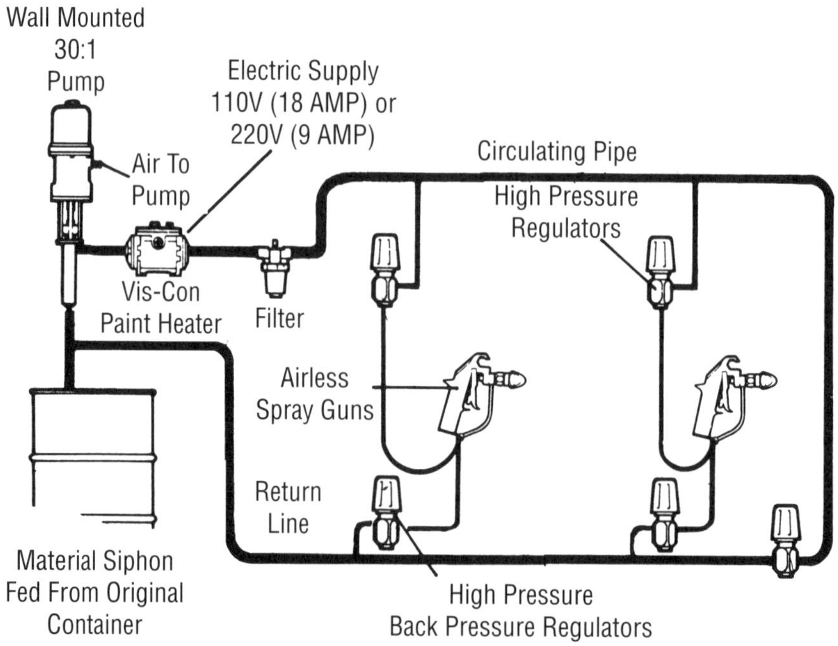

Figure 4-4. Heating the paint supply reduces the viscosity of high-solids paints.

Overspray of high-solids paints, since even after most of the solvent evaporates it tends to remain uncured and gummy, may present an issue during booth cleanup. The tacky overspray can be time-consuming to remove. The sticky nature of the material makes it unpleasant to work around when it accumulates on surfaces, particularly those that people have to contact. Paint accumulations on walking surfaces are fire hazards as well as messy. The coating tends to get tracked over a wide area by people's shoes. In the past, overspray from some high-solid paints was extremely difficult to detackify in waterwash booths. Newer "kill" agents that are better able to reduce high-solids paint sludge stickiness have lessened but not completely eliminated this problem.

Because their viscosity tends to be somewhat higher than that of low-solids coatings, high-solids frequently need heaters to produce a low enough viscosity for proper application and good surface appearance, as shown in **Figure 4-4**. Users do not have much leeway to add more solvent as this would increase the VOC emission levels. The character of high-solids is such that they tend to decrease sharply in viscosity when heated, as shown in **Figure 4-5**. Thus, mild heating can cause significant viscosity reductions. But this marked response to heat can lead to the formation of sags in the paint film during oven curing. Rheology control agents are utilized since they can help prevent this from occurring, but if the coatings are sprayed too wet, oven sagging will still occur. A spray

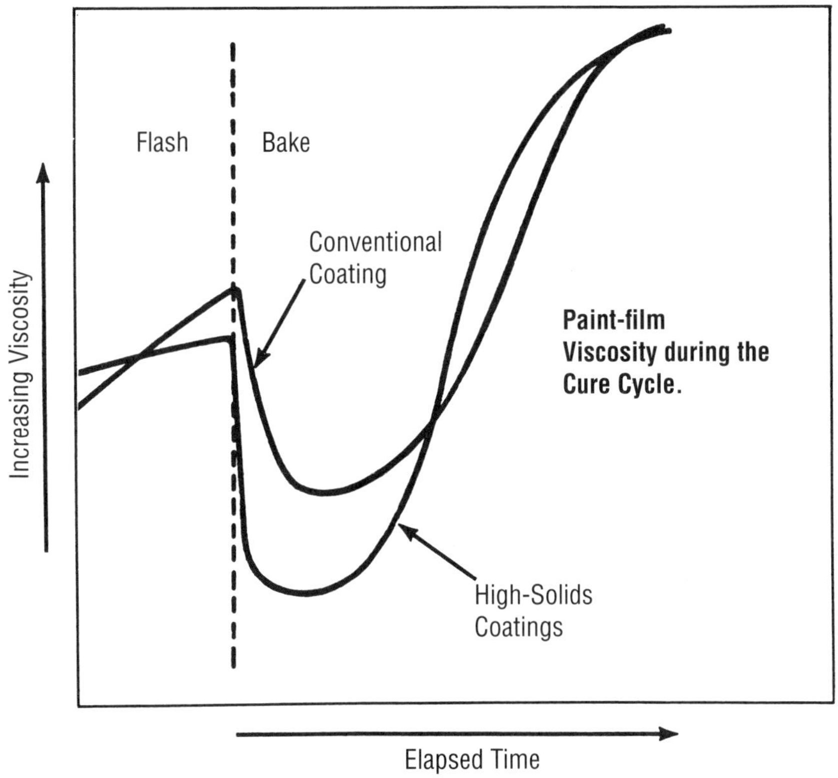

Figure 4-5. Viscosity changes during flash and bake.

operator applying high-solids coatings soon learns to deposit a "dry-looking" finish to prevent such sagging.

In the late 1990s, many people thought that high-solids coatings would be used less and less because waterborne paints would replace them. This belief was based on the assumption that it would ultimately be easier to achieve reduced VOC and HAP emissions with waterbornes than with high-solids. But this has not been the case. Waterbornes are not that easy for many paint shops to use. Instead, the largest shift in paint types has been toward powder coatings. The lower curing UV powders offer a better solution to emissions. Where powder use is possible, it offers a method of near total elimination of paint solvent emissions. Although high-solids paints are beginning to be rejected in favor of newer coating technologies, they will not be displaced very quickly. In fact, their utilization in many applications will continue for at least another decade.

Chapter 5

Waterborne Coatings

Definition of a Waterborne Coating

In solventborne coatings, the fluidizing medium is usually a blend of organic solvents, although some may use just a single organic solvent. To be classed as a waterborne coating, paint needs to have water as an important component of the fluidizing medium. Contrary to popular belief, however, water does not need to be the major fluidizing component, although it often would be. Coatings that qualify as waterborne coatings in most cases contain at least one water-compatible organic solvent called a cosolvent, present in the fluidizing media along with the water. Although these coatings are still correctly classified as waterborne, they may actually contain more organic cosolvent than water. A typical percentage of the cosolvent is less, however, often in the 5–10% range. The cosolvents tend to be low molecular weight oxygenated organic compounds such as ketones, alcohols, and esters because in order to be miscible (fully soluble in water at all proportions) they must be somewhat polar. In contrast, common paint hydrocarbon solvents such as xylenes and mineral spirits are too low in polarity to be more than only slightly soluble in water.

Strictly speaking, electrocoating (ecoat) formulations are waterborne paints because they contain a great deal of water as the fluidizer. However, ecoat is deposited by electrical current and is utilized in a paint application process so individualistic that ecoating will be discussed in detail in its own chapter (see Chapter 10). Waterborne coatings as a category considered here are those coatings containing water that are applied by a conventional method such as spray, dip, or roll coating.

Paint binders (resins) of the solventborne type are essentially hydrophobic, a term that means literally "water fearing." It indicates the material is not soluble in, and not attracted to, water. Many binders of the waterborne coating variety have been chemically treated to render them hydrophilic ("water loving," i.e., water-soluble)—they have been given a strong affinity for, and are soluble in, water. The resin manufacturer makes resin molecules hydrophilic by giving waterborne coating resin molecules a distinctly different chemistry than traditional organic solventborne coatings. Almost any paint resin can be chemically modified for use in a waterborne coating formulation. Some of the common waterborne resins are acrylics, epoxies, vinyls, alkyds, and polyurethanes.

Like their solventborne "sisters," waterborne coatings can be formulated for air-drying, oven baking, or UV curing. And as was true of solventborne coatings, the air-dry waterborne coatings also tend to be somewhat softer, lower in gloss, and markedly lower in water resistance compared to baked versions of these coatings.

Waterborne coatings are categorized into three types according to how the organic resin component in the paint is fluidized. These distinct fluidization types are **solutions**, **emulsions**, and **dispersions**.

Solutions, emulsions, and dispersions are not unique to waterborne coatings; they can be used in all liquid coatings formulas whether waterborne coatings or solventborne coatings. The categories are not mutually exclusive; some waterborne paint formulations actually contain a combination of both solution and emulsion resins.

As with solventborne coatings, the attributes of each of the three types of waterborne coatings will vary somewhat with the resin used. Alkyds and polyesters, for example, are more susceptible to hydrolysis (reaction with water) than acrylic resins, but the latter do not have the film hardness of the others. The home decorating and commercial decorating trade uses acrylic emulsion (latex) paints extensively. Most polyesters require higher bake temperatures or longer bake times than alkyds, although the polyesters may provide harder, tougher films and therefore exhibit better weathering characteristics.

Solution Waterborne Coatings

The chemical definition of a solution is "a completely homogeneous mixture of the atoms or molecules of two or more substances." Examples of solutions are sugar molecules dissolved in water (water molecules), and acrylic resin molecules dissolved in molecules of a solvent such as methyl isobutyl ketone (MIBK). As with sugar molecules in a water solution, each molecule of resin in solution paint is dissolved in the solvent blend. In the case of waterborne coatings, the solvent blend is in most cases mainly water; solution waterbornes are therefore called water-soluble coatings. Many solution waterborne coating resins contain chemically reactive carboxylic acid groups. With the addition of organic amine compounds, the resin molecules form carboxylate anion (ionic and hence polar) groups, thereby making the resin molecules polar enough to dissolve in water. After the paint has been applied, the solubilizing amine and water are volatilized during the cure process, and the soluble carboxylate groups revert back into insoluble carboxylic acid groups once again. The result of this reversion is that the water-soluble polar resin molecules are changed back to their insoluble form. Then once the applied paint has been fully cured, the film is no longer susceptible to being dissolved by water.

If amines should volatilize from a container of water-soluble paint, the pH can drift down to a neutral value (pH of 7). The loss of enough amine solubilizer can make the resin insoluble. When this situation occurs, it causes resin separation, a condition known as resin kickout. Unless it can be reversed, it renders the paint unusable. Poor flowout of solution waterborne coatings can in some cases be traced back to partial resin kickout, which in turn was due to a low amine solubilizer level.

For easier solubility, solution waterborne resins are typically low in molecular weight compared to emulsion and dispersion resins. Their smaller molecular size also brings the advantages of hydrolytic and mechanical stability, minimal agitation requirements, and long shelf life. The resins have a high pH, which reduces the tendency of solution waterborne coatings to rust steel application equipment. On the negative side, their appreciable amine content does present a modest health hazard in the form of potential respiratory problems and skin rashes.

Other characteristics of solution waterborne paints include:

- Excellent freeze/thaw stability
- High gloss capability
- Minimal water-solvent popping
- High film-build capability
- High cost
- Moderate storage limitations
- Excellent appearance
- Application ease (in most cases)
- Minimal gun nozzle clogging
- High cosolvent requirement (causing combustible flash points in some cases).

Emulsion Waterborne Coatings

An emulsion in physical chemistry terminology is a colloidal suspension of one liquid in another liquid. An emulsion consists of two immiscible liquids: one liquid is always present as minute globules dispersed in the continuous phase of the other. The globules or micelles of the dispersed phase can be surrounded by a very thin layer of soap, detergent, or other surface-active substance, and thereby form a stable distribution throughout the other liquid. The surface-active agent surrounding the micelles is called an emulsifier. Its function is to reduce the interfacial tension between the two immiscible liquids.

A well-known example of an emulsion is mayonnaise, which consists mostly of an otherwise immiscible mixture of aqueous lemon juice and vegetable oil. Normally water and oil do not mix. But if the oil particles are coated with a thin emulsifying layer of egg protein, the layers can be seen to mix very intimately. Although it seems well blended, the separate oil and water phases can still be seen through a microscope even after this vigorous blending. In similar fashion, the emulsifier in emulsion paint surrounds the tiny agglomerates of resin molecules. In a waterborne coating, these emulsifier-coated resin micelles are dispersed in a continuous aqueous phase, a phase that is generally mostly water plus some organic cosolvent.

The manufacture and use of water-emulsion paint involve heterogeneous (having unlike qualities) fluids to create paint whose behavior is more complex than that of conventional paints and solution waterborne coatings. Some water-emulsion paints, for example, increase in viscosity when temperature rises, which is just the opposite of what happens in a typical solventborne paint. As a result, some waterborne coating dip tanks require chilling in summer weather so that the heat does not increase paint viscosity. In a way it seems strange to chill a waterborne coating to lower viscosity when solventborne high-solids are heated to lower viscosity.

Waterborne emulsion paints are also called latex coatings. Contrary to what is frequently believed, latex paints contain no natural or synthetic rubber or rubber-like resins. Scientists use the term latex to identify any emulsion of an organic material in water. (The sap of the rubber tree is a natural latex, an emulsion of rubber-producing compounds in water; and hence the name for the material extracted from it.) Most latex paints air dry rapidly. They are the most common types of paint purchased and used by homeowners for both interior and exterior application. But their value is not limited to trade sales paints (sold in retail stores). Latex paints are used extensively for industrial maintenance applications as well.

The chemical nature of emulsion waterborne coatings allows them to be formulated at high volume solids without having unduly high paint viscosities. The viscosity of emulsions is basically that of the continuous phase, in this case water plus cosolvent, over fairly wide ranges of paint solids concentrations. This permits viscosity to be rather constant and not rise significantly from low to moderately high paint solids concentrations. But if solids levels are increased excessively, the material finally reaches a stage at which viscosity will increase sharply. Note that the plateau in the viscosity-solids curve (refer back to **Figure 3-9**) effectively eliminates using viscosity to aid application. With solution paints, using the common viscosity cup method gives a good approximation of the percent solids in a paint. But as you can see in **Figure 3-9**, the correlation between percent solids and viscosity is not linear with aqueous emulsion paints, so viscosity does not indicate approximate solids levels.

The chemical stability of waterborne emulsions is threatened by freezing temperatures. Freezing can permanently separate the paint components, causing insolubilization of the resin—a phenomenon called kickout which, as stated earlier, can also be caused by amine loss from solution waterborne coatings.

Baking-type emulsions and air-dry emulsions both cure in the following sequence, shown in **Figure 5-1.** The organic cosolvent evaporates first, water evaporates next, and the resin particles then coalesce to form a continuous coating film. In the case of an enamel, most cross-linking of the resin molecules occurs following film coalescence.

Waterborne coating emulsion resins are high in molecular weight (100,000–3,000,000). The large molecular size brings about the disadvantages of poor hydrolytic and mechanical stability. Oversized pumps and low paint fluid pressures should be used due to the mechanical shear sensitivity. Shear can result from stirrer blades, gear pumps, reciprocating piston pumps, and flow restrictions in paint lines. But they have advantages also. The low amounts of amine in these resins minimize the health hazard. Because of the low fluidizer concentrations, the resin emulsion paints have a relatively fast air-dry rate, which mini-

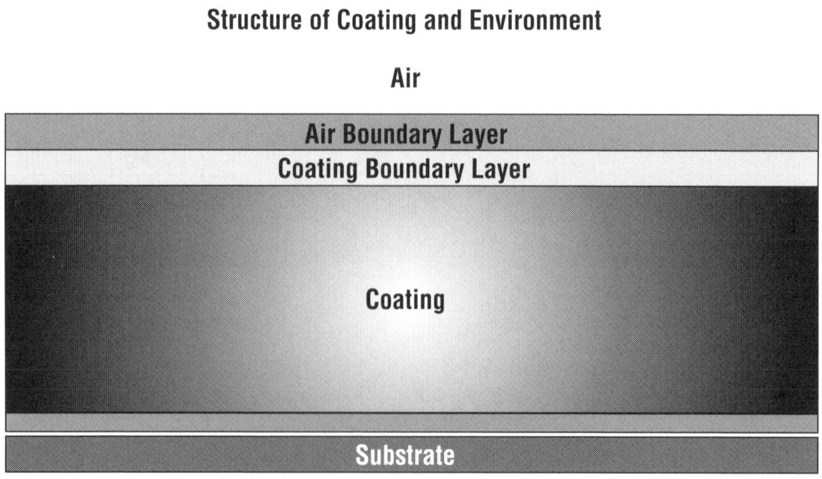

Figure 5-1. The "coating boundary layer" is the surface of the applied waterborne coating. The "air boundary layer" is the very thin layer of air in direct contact with the surface of the coating.

mizes recoat times. The short recoat interval is attractive for shops where multiple coats are required. Unfortunately, the relatively moderate pH of emulsion waterborne coatings tends to encourage flash rusting of steel substrates under the paint.

Other properties of emulsion waterborne coatings include:

- Poor freeze/thaw stability
- Moderate cost
- Low odor
- Minimal VOC emissions
- Medium gloss capability (being improved)
- Rigid storage limitations
- Medium appearance (being improved)
- Ease of application (in most cases)
- High-solids application capability.

Because each of the water-soluble and water-emulsion types has distinct and separate advantages, coatings have been made that are combinations of both resin fluidization types. Some of the combination solution/emulsion waterborne coatings have been utilized in original equipment manufacturer (OEM) automotive paints.

Dispersion Waterborne Coatings

A liquid dispersion paint is a system of dispersed resin particles suspended in a liquid. Dispersions differ from emulsions in that the resin clumps (clusters) are small and do not require added emulsifiers to form a stable fluid mixture. Vigorous physical mixture alone is used to form a dispersion. A resin might be mechanically dispersed or produced from a solution by careful addition of a solvent (or solvent blend) that has a polarity different from the resin, so that resin molecules are forced into a mild clumping amongst themselves. Because many waterborne resins can be thinned using water, they are sometimes referred to by the imprecise terms "water dispersible" or "water reducible."

A graph of viscosity versus solids for a dispersion paint (see **Figure 3-9**) also produces a viscosity curve that has a long plateau region into the high-solids range. The viscosity rises rapidly when the resin concentration gets so high that individual clusters of resin contact each other and begin tangling. If resin agglomerates join together they will eventually separate out of the paint as a second phase.

Waterborne dispersions are slightly different from emulsions but are close enough in properties to be considered as the same general paint type. The properties listed for emulsion waterborne coatings apply almost universally to the dispersion waterborne coatings as well.

Solvent in Waterborne Coatings

During the manufacture of waterborne coatings, limited amounts of organic solvents are added to dissolve the resin. Suspension or solution of paint resins in water alone is frequently difficult if not impossible without the help of organic cosolvents. Because paint resins are essentially oil-like and do not readily mix with water, the chosen solvents must be water-miscible compounds such as alcohols, glycols, ketones, and esters. These cosolvents are of necessity always somewhat polar and soluble in water, but must also contribute resin-dissolving properties to the water/cosolvent blend. The cosolvents also aid considerably in smoothing the freshly applied wet paint film by increasing flowout across the painted surface.

After the solvent and resin have been mixed thoroughly, the mixture may be made more hydrophilic, if the resin has carboxyl groups, by the addition of an amine or chemically similar material. After the addition of pigments and additives plus some additional blending, the entire mixture is thinned down to the desired viscosity by the gradual addition of water.

Two-component (2K) waterbornes have become increasingly popular low-bake paints for plastic substrates, especially for self-texturing and soft touch coatings. Most 2K waterbornes will cure without baking in 7–12 days at room temperature, but more often an oven bake is used to accelerate this to a more reasonable cure time, such as at around 165° F (74° C) for 30 minutes. Compared to the ordinary melamine cross-linking waterborne resins that need a relatively high bake to cure, the 2K paints will cure with far lower bake requirements and still provide equally good or better film properties. As a result of the lower cure temperatures, the 2K coatings can be used on many more heat sensitive substrates.

Surprisingly, the organic solvent portion in some of these coatings can be quite high. The organic cosolvent content of most waterborne coatings when reduced to their "as applied" viscosity typically varies from 2% to 20% of the fluidizing medium with the remainder being water. So most of the coatings at the high organic solvent end of this range may not automatically meet EPA or local VOC emission limits. A number of government and manufacturing representatives have expressed worries that in the early decades of the 21st century, automotive waterborne color basecoat finishes, in particular, may not be able to be applied at low enough pounds of VOC per gallon (1 gal is equal to 3.785 l) of applied coating solids (VOC/GACS) in all plants. Best Available Control Technology (BACT) is permitted in areas where air purity is satisfactory. In ozone attainment areas, BACT mandates a more liberal level of 12.2 lb maximum of Volatile Organic Compounds/Gallon (1.46 kg/l) Applied Coating Solids (VOC/GACS). It is found that on average waterborne basecoats have about 3 lb VOC/gal (360 g VOC/l) minus water. If the normal 50% transfer efficiency is being achieved, calculation can be done to show that the emission rate will be slightly over 10 lb VOC/GACS (1.2 kg/l). This level is therefore in compliance. For ozone *nonattainment* areas, however, Lowest Available Emission Requirements (LAER) must be met. LAER rules specify no more than 6.2 pounds VOC/GACS (740 g/l), a number far lower than most waterborne basecoats are able to achieve. Much work is being done to develop waterborne coatings with zero VOC, and some success in these ventures has already been achieved, but certainly not in all waterborne paints of the types suitable for various product requirements.

The question is often raised, "Why must VOC in waterborne coatings be given *minus water*?" The allowable VOC content of waterbornes is given by EPA rules in units such as "lb VOC/lb applied solids" or "lb VOC/gal minus water." The "minus water" calculation requires that a theoretical gallon of the nonwater portion of a paint be considered and then VOC be calculated for this theoretical gallon and reported for that material as lb VOC/gal minus water. People understand the concept but many question the reasoning behind it. The answer is that by restricting the VOC content reporting that way, no company is tempted to dilute paints with water to lower the VOC per usage gallon and thereby achieve VOC compliance. For example, suppose a waterborne paint has 6.4 lb VOC/gal (766 g VOC/l) and the allowed level is only 3.5 lb VOC/gal (419 g VOC/l). If a gallon of this paint is diluted one-to-one with water, VOC content is cut in half. This would then drop the VOC/gal value to a permissible level of 3.2 lb VOC/gal (383 g VOC/l). But then, of course, twice as much paint, namely two gallons, would be needed to obtain the same dry film thickness as one undiluted gallon. If this were to happen, no net VOC reduction would result. So the necessary "minus water" reporting requirement term was added to EPA enforcement regulations.

Determining the VOC content of a waterborne coating isn't necessarily as easy as simply knowing the amount of cosolvent added by the paint manufacturer. Amine solubilizers and low molecular weight resin components can also contribute to VOC emissions. To comply with VOC regulations, the precise volume percent of organic volatiles must be known. The easiest way to determine total VOC content is by gas-liquid chromatographic analysis, a method that is expensive unless a measuring instrument of this kind is available. Other methods tend to be slow, cumbersome, or inexact. Samples in many instances can be tested for VOC content by fractional distillation of all the volatile organic portions other than water. These organic volatiles usually will have a lower boiling point than water, but there are some exceptions. Another factor can also make simple fractional distillation unusable for VOC determinations. Certain mixtures of solvent and water, however, make distillation and subsequent VOC determination impossible because the mixture becomes azeotropic (having a single boiling point).

The following case history illustrates how azeotropic solvent and water mixtures can distort VOC measurements based on distillation testing. A plant in California received notice of an imposed fine for a VOC violation. They were cited for an illegal waterborne coating because the distillation of all volatiles up to 212° F (100° C) showed "4.37 pounds VOC/gal (523 g VOC/l), an excess of 1.37 lb/gal (164 g/l) VOC over the maximum allowed of 3.00 lb /gal (360 g/l) minus water." The paint manufacturer, however, claimed the paint had a total of only 2.66 lb/gal (320 g/l) minus water. When azeotropic behavior was explained and the regulatory agency was shown that adding water to the paint increased (as measured by distillation) the apparent VOC content, the agency agreed that azeotropic distillation gave an erroneous VOC reading. The fine was promptly canceled and measuring by this fallible method was halted by the EPA.

Whenever waterborne coatings are thinned to achieve the proper paint application viscosity, cosolvents should not be added. One obvious reason is that added solvent will raise the VOC emissions. Because of the nonlinear viscosity relationship with added solvent of any type, all reductions with water must be done in modest increments. Otherwise overreduction is possible. At some regions of the curve, a small amount of added water can cause a sharp drop in the viscosity of the paint, as shown in **Figure 5-2**.

To avoid complications from impurities and dissolved components in water, some plants use only clean deionized water, reverse osmosis water, or distilled water when making additions to waterborne coatings. Exceptionally pure water is only rarely necessary, but use of deionized water may be advisable depending on the quality of the supply water. If in doubt use pure water because the addition of ionic products will often lower water resistance in the cured film. The paint manufacturer, however, is the one most capable of determining whether ordinary tap water does or does not reduce film quality.

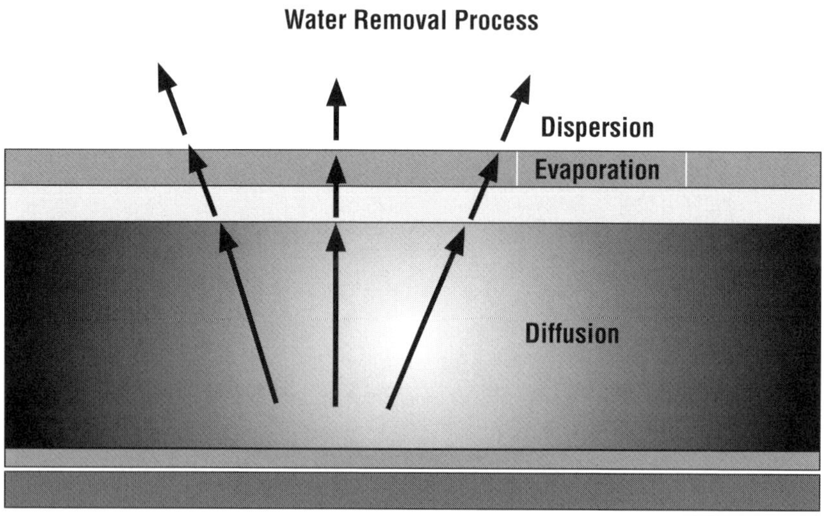

Figure 5-2. Dehydration of a waterborne coating requires the transportation of water molecules from the coating to coating boundary layer, across the interface between the coating boundary layer and the air boundary layer, and from there into the air mass above. This is accomplished by diffusion, evaporation, and dispersion.

Waterborne Wood Finishes

For the past 20 years, formulators have attempted to prepare waterborne materials suitable for the enormous wood product markets in furniture, cabinetry, shelving, paneling, etc. The main difficulty is that water tends to raise the grain of most woods. To a lesser degree there is a problem with heat curing, which can extract the natural moisture from the wood and cause warping. As a result, wood finishes—including fillers, stains, sealers, toners, lacquer varnishes, and enamel varnishes—are still often solventborne types. But EPA regulations and improved wood finishing products are gradually changing this picture. Considerable effort has been expended by wood finish manufacturers to develop waterborne coatings, and they have now achieved good success in developing acceptable waterborne equivalents to many solventborne wood finishes. As many as 15–25% of wood finishers are now using at least some waterborne coatings.

Currently, roll coating and curtain coating give even better results with waterborne coatings than do atomization application methods. Much of this is due to the greater control in roll and curtain coating of film thickness and to the ability to control humidity and temperature more closely in the application, flash off, and curing stages. These variables must be held more tightly with waterborne coatings than with solventbornes. UV curing of waterborne coatings on flat wood panels and boards is an area that has experienced strong growth in just the last several years. The advent of powder coatings for wood, however, has eclipsed the interest in waterbornes for wood applications. Powder coating on wood is still a process subject to severe limitations, especially moisture content in wood and wood products. But understandably, potential rewards for eventual broad use of powder on wood keeps wood finishers and powder suppliers extremely excited.

Spraying of waterborne coatings on furniture and cabinets is finding limited use on the less expensive lines of these wood products. In most instances "hybrid systems" are used; only some of the stain, sealer, and topcoats are waterborne coatings, while other coatings used are solventborne types. Low cost items may receive only 4–5 separate applications, but fine furniture typically will get 8–12 individual coating applications.

In the future, waterborne coatings can be expected to make further inroads into this market due to the continual strengthening of both VOC and HAP regulations. Rule tightening for wood finishing VOC took effect in the summer of 1995, and new HAP standards hit back in 1997. Remember that states may not adopt weaker rules than federal guidelines, but they are free to demand more stringent limits if they so choose. When states do this, however, they run the risk of chasing industries to other less restrictive states. The elimination of VOC and HAP worries is a big factor in the desire of coaters for powders suitable on wood.

Waterborne Coating Considerations

The following are some of the significant factors that waterborne coating users must consider.

- Operation permits
- Pretreatment
- Cost
- Resin availability
- Modifications
- Process commonality
- Electrostatics
- Atomization
- Application Problems
- Agitation
- Odor and cleanup
- Fire hazard
- Storage
- Dip tanks
- Foam.

Operation Permits

The U.S. Clean Air Act (CAA) requires nearly all painting operations to obtain an operating permit. EPA regulated substances include VOC, HAPs, nitrogen oxides (NOx), sulfur oxides, and carbon monoxide, but control of VOC and HAPs emissions are of most pressing concern to painters. Common paint solvents are often both VOCs and HAPs, including toluene, ethyl benzene, xylenes, MEK, and MIBK. (Acetone is neither.)

The CAA provision most opposed was the "potential to emit" condition that classifies plant emission levels based as if they operated nonstop for 24 hours a day, seven days a week. This assumption was applied even though the actual maximum operation might be only 8 hours per day for 5 days a week. Major sources are those painting operations whose potential to emit exceeds any of the following:

100 tons/yr of air pollutants
10 tons/yr of any individual HAP compound or 25 tons/yr total HAPs
50 tons/yr VOC or NOx in "serious" nonattainment areas
25 tons/yr VOC or NOx in "severe" nonattainment areas
10 tons/yr VOC or NOx in "extreme" nonattainment areas
50 tons/yr VOC in ozone transport regions not classed "extreme" or "severe."

Since plants must determine their own need to comply, and operating officers can be held criminally accountable for CAA rule infractions, some people believe this violates their rights against self-incrimination. More changes in this law seem likely to be made. Outcries against "petty functionaries" have become less common recently, fortunately. A lightening of the sometimes ham-handed approach by some regulatory groups has already been observed over the past 5 years in various enforcement areas, under the direction of higher authorities within the government.

Pretreatment

Pretreatment when waterborne coatings are to be applied is far more critical than with solventborne paints; in fact, poor or inadequate pretreatment is the major cause of problems associated with the use of waterborne paints. Because their organic solvent content is low, waterborne coatings are not able to dissolve any spots of oil and grease that might remain on surfaces after cleaning. The dirt and oil sensitivity of waterborne coatings tends to create blisters readily. In all liquid coatings a low surface tension is most desirable to avoid appearance defects. Water has a high surface tension, creating potential edge-pull problems. The choice of cosolvent is crucial in aiding the prevention or reduction of picture framing, craters, blisters, popping, and edge pull. If these terms are unfamiliar, they are discussed in Chapter 16.

Cost

Nearly every waterborne coating will cost more per gallon on an equivalent solids basis than a comparable organic solventborne coating. This reflects the higher cost to manufacture waterbornes. While the price on equivalent mils per square-foot-coverage for waterborne coatings continues to be slightly higher than for organic solventborne paints, the price differential is narrowing. For many plants, the lower VOC emission of waterborne coatings forces them to switch away from solventborne paints despite the cost increase. The continued press for still lower allowed VOC and HAP levels will encourage more plants into conversion to waterborne coatings, especially in smaller painting operations where the cost of solvent incineration is prohibitively high.

Resin Availability

Not as many different types of binder resins are available for waterborne coating formulation as for solventborne coatings. On the plus side, many of the coating polymers used for waterborne finishes, unlike the high-solids resins, can be cured below 200° F (93° C) and thus may be more suitable for wood and plastic substrates than high-solids paints. The inherent high viscosity of waterborne coatings restricts their formulation to relatively low-solids levels in true solution coatings. With waterborne dispersions and emulsions, the percent solids is higher, ranging from roughly 20–35% weight solids.

Modifications

Although conversion from solventborne paints to waterborne paints is still far simpler than switching to powder coating because the handling, equipment, and application processes are similar, converting an existing low-solids solventborne coating line to waterborne use can be complex. Such an existing line may need to have ordinary iron piping, valves, etc., replaced with stainless steel or other nonrusting materials. This is particularly true when light-colored paints are used in which rust is readily visible. Depending on the local climate, air-conditioning or a heated flash tunnel may have to be installed to stabilize ambient temperature and lower the relative humidity for waterborne coating application. Even with heat, the flash-off time may need to be made longer to permit more water evaporation. Oven settings may need to be adjusted to accommodate the high water content in the paint film as it enters the oven to avoid popping. Where possible, oven zoning to establish a lower entry temperature and a compensatory higher temperature in the second half of the oven can be highly effective in avoiding solvent pop problems.

Process Commonality

Conventional application processes can be used with waterborne coatings, including all of the various types of spray guns, rotary disk and bell application, dip coating, electrostatics, and flow coating. This gives waterborne coatings an advantage over high-solids paints. Neither dip coating nor flow coating is possible with the majority of high-solids coatings because of paint drain-off troubles. This limitation for high-solids is due to the inherent elevated viscosity of high-solids solventborne finishes that makes dip or flow application impossible.

Electrostatics

Waterbornes are often applied electrostatically to increase paint transfer efficiency. Because most waterborne coatings are far more conductive than solvent paints, grounded electrostatic systems, such as those used for solvent paints, cannot be used. The conductive waterborne paint would short out the electrical charges through the paint lines and make the system ineffective. To prevent this, waterborne electrostatic application equipment generally uses what is termed an “isolated” system in which the application device, the paint lines, and the paint reservoir are all completely isolated from any electrical ground. (Isolated electrostatic systems should never be used with solventborne paints because the fire danger would be extreme.) Plastic stands and attachment arms are used to isolate all conductive parts of the system from ground. Since the isolated paint equipment components themselves carry a high electrical charge, they are often said to be “hot.” This is a safety hazard. Anything conductive that touches or closely approaches any part of the system will suddenly drain off a large quantity of the electrical charge stored on the ungrounded system or, in other words, “will receive an electric shock.” Since the human body is extremely electrically conductive, a protective cage must enclose an isolated electrostatic paint system for the safety of plant personnel. Appropriate signs warning people about the shock danger also need to be posted. The added cost of isolation items for electrostatic application of waterbornes is moderate. It is more than offset by the large and continued cost saving from reduced overspray waste from using electrostatic application.

Extra safety considerations must be taken if electrostatic equipment is used no matter what paint is used. Any electrically conductive object located within several feet of an electrostatic paint device, even though it is not actually touching it, can accumulate electrical charges unless the object is fully grounded. The more massive the item, the more electrical charge it can store. The ability of objects to retain electrical charges is termed their capacitance. This obviously is more likely to occur with electrostatic application than with nonelectrostatic application, but static charges can actually be generated with any paint application equipment, especially airless spray. Grounding of all objects near painting equipment is a good safety practice. Solvent containers anywhere should be fire safety approved with positive closures. Solvent storage containers in the plant are required by law to be grounded.

Grounded waterborne coatings systems have been used but they have had only limited acceptance. Remote or secondary charging of the paint particles in a grounded system (see **Figure 5-3**) can be achieved in theory by locating the charging electrode about 5 inches (12.5 cm) away from the gun tip or rotary bell head. Secondary charging is generally not very effective. In some cases, an electrostatic charging ring has been used with rotary waterborne coating application. Another remote charging system used for waterborne automotive color basecoats has four secondary electrodes located uniformly around the bell head. The amount of electrical potential that can be reached with remote charging is only about 15,000 volts (V), far less than the 75,000–125,000 V possible with normal charging devices. The ring and electrodes do not act so much to charge the paint, as to direct or push the already charged paint droplets toward the target by electrostatic repulsion. If paint wrap-back onto the equipment occurs, charging rings or "halos" can be installed to reduce or prevent it.

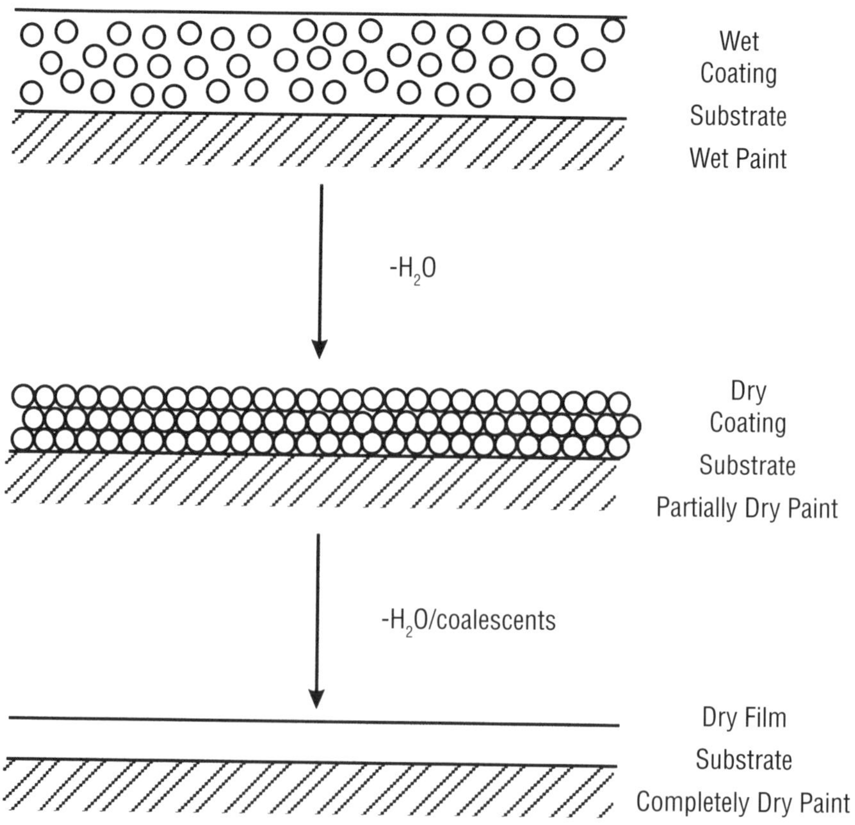

Figure 5-3. Drying of emulsion waterborne paint.

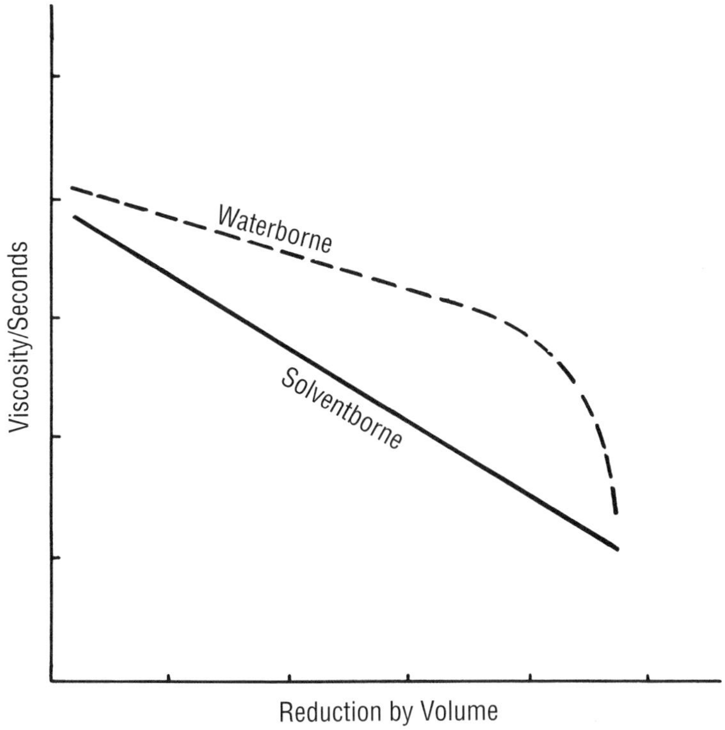

Figure 5-4. Relative difference between waterborne and solventborne finish reduction rates and application viscosities.

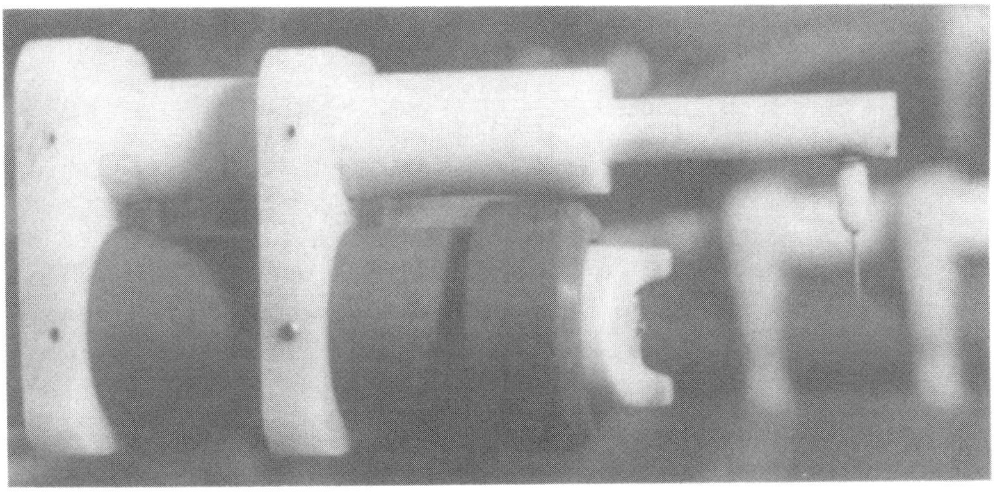

Figure 5-5. Successful electrostatic application in grounded systems with remote electrode charging.

A number of years ago, grounded waterborne coatings systems purportedly would allow electrostatic spraying of waterborne coatings with a fully grounded paint system by electrostatically charging the parts. This required the parts to be hung on insulated hooks. The hooks or hangers had a plastic section that prevented the part from losing its charge to ground. Contact with or even proximity to a charging wire that ran along the length of the spray booth supposedly was able to charge the conductive parts. Reports from plants that had installed this equipment stated they were basically not at all helpful. A similar process used remote "negative" charging of waterborne coatings, plus "positive" charging of large vehicles such as bulldozers that were mounted on isolation stands. The corporation that had installed this system soon abandoned the method of application as ineffective. None of the grounded waterborne coatings systems put on the market has ever had more than lukewarm acceptance; not a single such system could be considered as successful.

For the protection of people and property, all electrostatic systems need a current limiter device to shut down the system if the electrical current draw becomes too high for safe operation. Some plants have found that when waterborne coatings are painted electrostatically, frequent "trip-outs" of the electrostatic current limiter occur from shorting of

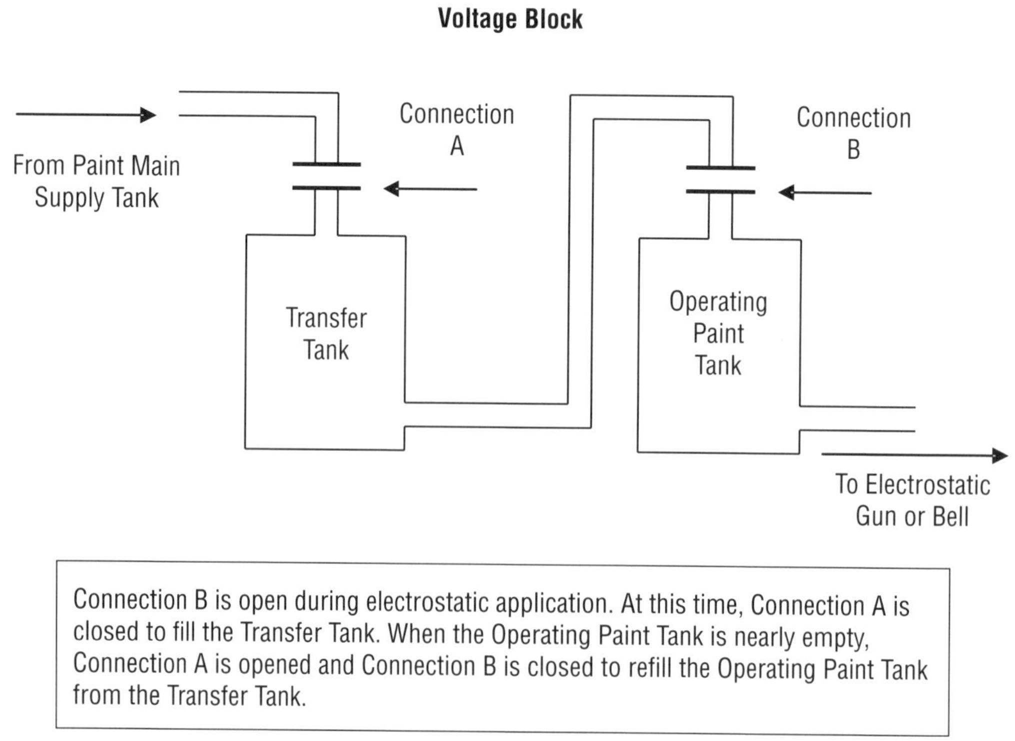

Figure 5-6. Conductive paints can be applied electrostatically by preventing electrostatic charges from shorting to ground. A voltage block system prevents shorting, and also eliminates having to stop paint application for refilling the paint supply tank. Since one connection is always open, shorting to ground cannot occur.

the system to ground. This can be a vexing problem because locating the electrical short frequently consumes a considerable amount of time. Plants have switched back to solventborne coatings for electrostatic application because of continuous trouble from this problem.

Atomization

Waterborne coatings tend to be high in application viscosity and surface tension (highly cohesive). As a result the waterborne paint at the same viscosity does not atomize as readily as solventborne paint. This means having to settle for a low finish quality from poor atomization, or spraying at higher atomizing air pressures. But when higher air pressures are used, it will reduce paint transfer efficiency because more paint fog is created. This is why the transfer efficiency of waterborne paints is generally lower than for solventborne coatings.

Application Problems

One of the major disadvantages with some waterborne coatings can be the difficulty of applying them thickly enough without causing sags, and/or solvent popping (i.e., boiling). The sagging and popping normally do not occur until the parts are in the oven and the paint warms up. Avoiding such application problems while still applying the proper film thickness can be a narrow tightrope to walk. Some plants have resorted to putting on two thin coats because of these problems with one thicker coat. To understand why this situation is more acute with waterborne coatings than solvent paints, let's review what happens when these coatings are sprayed.

When solventborne paint is sprayed, the solids content (and viscosity) of the applied wet paint film increases substantially after exiting the gun tip. Solventborne coatings use a blend of four to eight solvents with varying evaporation rates to avoid a sudden excess of solvent evaporation. The fast-evaporating solvents evaporate quickly from the very large surface area of the fine atomized droplets that are moving toward their target. In the short time after the paint droplets leave the gun until the droplets hit the workpiece, the solids can increase 20–25%. This sharply raises the paint viscosity, which in turn helps prevent sags on freshly coated parts.

With waterborne coatings, however, most of the volatile portion of the sprayed droplets is water. Water has a high boiling point, so virtually none of it evaporates while the waterborne coating droplets travel from the end of the gun to the workpiece. Thus the "as-sprayed" fluidizing media content is nearly identical to that of the paint on the freshly coated part. Because so little of the fluidizing media evaporates from the paint droplets between the application device and the workpiece, it can be difficult for the operator to achieve a sufficient film build, while at the same time avoiding running or sagging the paint. The applicator must walk a fine line between getting an adequate film build and making the paint run.

Where possible, heating the substrate is effective in reducing the tendency to form runs and sags; much more effective than heating the paint, which is rarely done for waterbornes but is common for high-solids. Paint heaters set at 90–105° F (32–41° C) may, in a few cases, help reduce or prevent such sagging. This generally only has a marginal effect, however. Some waterborne coatings have reverse solubility, so increasing the paint temperature can complicate the situation by raising the viscosity instead of helping to alleviate the problem.

High humidity in the spray application area can add to the problem by reducing flash-off to near zero. The application area may need to be air-conditioned to control the humidity. The high cost of air-conditioning may prohibit the use of waterborne coatings. Slow air-dry times will be experienced when humidity is high. Waterborne coatings may be unusable under such conditions. Waterborne coatings need to be sprayed at somewhat higher viscosities to avoid runs and sags, especially when it is necessary to achieve more than 1.0–1.5 mils (25–38 μm) dry-film thickness.

The slow flash-off of water in humid weather can make popping and boiling in the oven a frequent problem, particularly if an appreciable film build is necessary. Heated flash zones are frequently necessary to remove enough of the water so that popping and boiling defects are not experienced in the oven. Zoning of bake ovens may also be necessary to prevent water boiling and popping.

Water's high heat of vaporization can cause problems in curing waterborne coatings in a bake oven. Little of the water evaporates until the temperature of the applied coating reaches 212° F (100° C), then all of the water will suddenly begin to evaporate at the same time. The "outrush" of water vapor can tend to cause blisters and solvent-popping problems. If the organic solvent forms an azeotropic mixture with the water, it lessens the problem because a portion of the water will co-evaporate below 212° F.

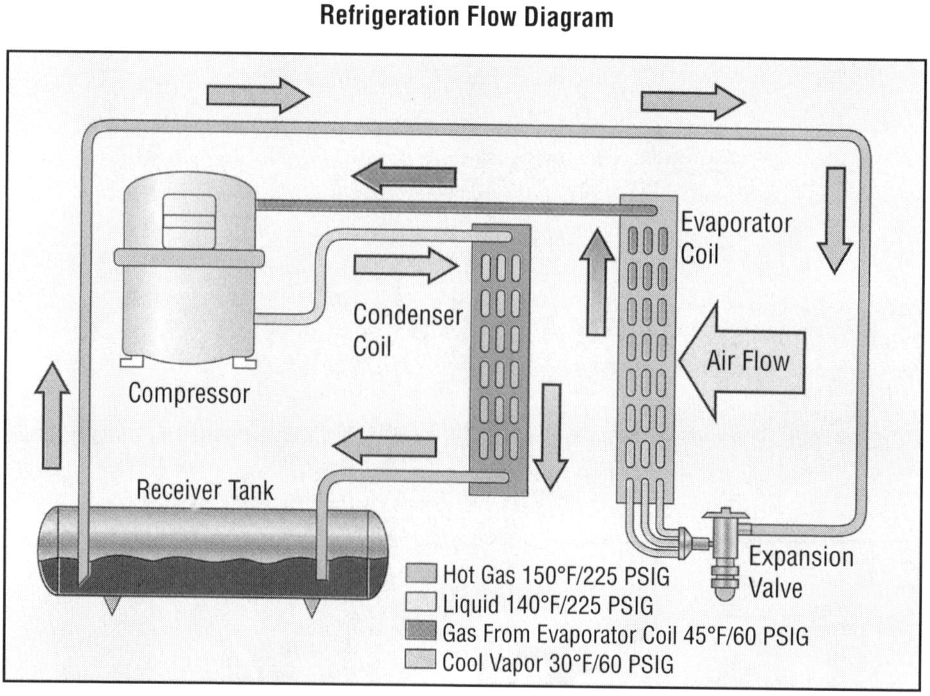

Figure 5-7. Dry air is used to remove water from freshly applied films. Air entering at the right passes over cold coils and the moisture condenses out of the air flow.

Several new dehumidification devices are beginning to be utilized to strip much of the water from freshly applied waterborne paint films without the use of heat. The air surrounding the freshly painted parts is dehumidified and recirculated around the parts. Remote condenser coils remove water from the air and the dry air is then reheated before being run back through the paint dehumidification chamber. This method can also be used to remove water from cleaned parts after they are rinsed, to more rapidly cure air-dry waterborne paints, and to remove moisture from baking finishes prior to oven entry. Water removal helps accelerate the cure process and eliminates or reduces popping due to overly rapid water evaporation during initial oven heating.

Agitation

Some waterborne coating formulations are sensitive to mechanical shear and can be damaged if stirred or pumped too vigorously. This is particularly important since waterborne coatings as a whole tend to settle rapidly and may be subjected to harsh agitation. Even excessive mild agitation can sometimes have a "butter churn" effect and "break" emulsions. Certain filters can produce the same problem. The kicked-out resin then forms an insoluble mass that separates from the ruined paint. It can clog filters and plug fluid passages in spray guns. Sometimes the resin globules are soft enough to be pushed through filters and gun tips. Then they can be ejected as small gelatinous globs onto the workpiece, often causing consternation and bewilderment to those who do not recognize how they might have originated. These visible "bumps" cause rejects and rework; they must be sanded off after curing before the parts can be resprayed.

Odor and Cleanup

One of the great advantages of waterborne coatings (especially latexes) is their low odor due to the type and minimum quantity of cosolvents in them. Easy cleanup is another advantage for many waterborne coatings. But many others form a tenacious "skin" on the inner walls of the paint hoses and piping that resists rinsing with water or aqueous solutions. With the former types, paint lines are easily rinsed clean with water or mild water solutions of cleaning agents. In the case of skinning types, organic solvents must be used to flush out the paint lines to remove the adherent skin, which cannot be removed with aqueous agents alone. This is not always totally successful and in time the paint lines may become so restricted that the lines require periodic replacement.

Since waterborne resins are designed to be compatible with water, it is sometimes difficult to get waterborne coating overspray to float or sink in water-wash booth reservoirs. As a consequence, some waterborne coatings are difficult to detackify and may cause clogging or foaming. Continual improvements are being made in detackifiers or "kill" agents. However, considerable trial and error may be required before a suitable detackifier is identified.

Fire Hazard

Another advantage of waterborne coatings is reduced fire hazards (low flash point or no flash point). This can be a significant plus in many plants. If the organic solvent content of an unapplied coating is high, however, these waterborne paints can still burn. Other waterborne coatings will not ignite even when an open paint container is exposed to the direct application of a blowtorch flame. It must be remembered, however, that after all volatiles including water have evaporated, the resin of a cured waterborne coating film is just as flammable as any cured organic solventborne coating film.

Storage

Waterborne coatings should be stored in a location that prevents them from freezing. Expansion that results from freezing can burst paint containers, and ice crystal formation during freezing can damage the paint, especially by kicking out dispersed and emulsified resins. Solution waterbornes are less sensitive to this problem because they usually have so much organic cosolvent that they will only freeze at extremely cold temperatures.

Dip Tanks

In a waterborne coating dip operation the tank size is critical. The tank should be small enough to get fast paint volume turnover. This will help avoid instability problems inherent with waterborne coatings. Complete turnover of the tank volume in one or two months or faster is considered ideal. With waterborne coating dip coat and flow coat, drain-off continues from the parts for a considerable distance. To recover some of this paint, the drip zone can be rinsed back with a fine water mist or a small amount of cosolvent. One of the most successful recovery methods is to use the paint itself to flush back the coating from the drip zones. This results in a clean and economical operation. Water and cosolvent losses are somewhat reduced when dip tanks are covered during nonuse periods, and it keeps the paint tank cleaner. This practice is more often used with organic solvent dip systems where appreciable solvent evaporation would otherwise occur and increase costs

- No flash-off time between the spray booth and the drying enclosure. This can save on floor space and eliminate additional enclosures.
- Improves the quality of the paint finish. There are fewer pin-holing and other defects than with convection or IR ovens because no heat is added to the process.
- Heat-sensitive parts such as plastic, rubber, and wood are not deformed or damaged because the system is operated at ambient temperatures.
- No heat ramp-up times for heavy parts such as castings or frames. This reduces the overall drying time of the process. The dehumidification system is not affected by the product's mass.
- No cooling tunnel or cooling loop required. Reduces the footprint required for the paint system and saves capital and energy operating costs.
- A dehumidification system is very energy efficient. It operates at approximately one-tenth the energy cost of a convection or IR oven.
- The drying process is controlled throughout the year and the drying time remains the same. There are no production upsets during the humid summer.

Figure 5-8. Possible advantages of drying waterborne coatings using an air dehydration system.

as well as VOC emissions. Nevertheless, covering the waterborne dip tank is a prudent practice if only because it helps keep foreign materials out of the paint.

Foam

Foam generation sometimes can be a problem in agitated waterborne coating tanks, and especially with flow-coat and dip-coat systems. The emulsifiers and similar coating components are surface-active agents that behave to generate foam, because they are similar to soap and detergent molecules. When the paint material is agitated, it can foam up just like any detergent/water mixture would do. The problem with paint foam during dipcoating and flowcoating is that it gets on parts, then when the foam dries it creates appearance defects. Foam can also stall pumps and distort the readings of paint volume monitoring devices as well. Judicious use of antifoamers can reduce the problem, but caution is needed because adding too much will actually cause the antifoamer to create its own foam.

Waterborne Coatings—Advantages and Disadvantages

Waterborne coatings probably have as many advantages as they do disadvantages. One thing is certain: waterborne coating application requires tight process control. It has become apparent that solventborne high-solids technology cannot offer a general range of coatings with VOC levels that will meet future projected VOC restrictions, whereas waterborne coatings seem likely to succeed. Aside from a few 85–100% solids paints, most high-solids paints are not able to go beyond about 70% solids without causing extreme application difficulties. A limited exception would be UV-cure coatings.

The most compelling reason for using waterborne coatings is that their low VOC content usually allows compliance with EPA emission limits. The critical VOC advantage will surely lead to continued waterborne coating growth in the future, especially as VOC limits are tightened. Waterbornes can reduce VOC at the source rather than with additional equipment such as solvent capture and incineration systems. But it is not a one-sided picture. Waterborne paints generally tend to be higher in Hazardous Air Pollutants (HAPs) than solventbornes. Consider also that some solventborne systems with add-on emission controls, although costly, are so low in VOC they have many established EPA standards for emission control. Best Available Control Technology (BACT), Lowest Available Emission Requirements (LAER), and Maximum Available Control Technology (MACT) regulations have been established based on this technology.

Waterbornes, almost everyone will agree, still have considerable room for improvement. Some plants have been lured to switch systems by some of waterborne coating's advantages, but after using them for a while have been so discouraged by their disadvantages that they have abandoned waterbornes for other coating systems. Nonetheless, many plants have happily changed to waterborne coatings and continue to use them with great satisfaction. The environmental pressures will certainly force many companies to look carefully at waterborne coatings, especially for parts that cannot withstand the high cure temperatures needed to cure ordinary powder paints and are not configured to cure well using UV-curing.

Final thoughts: 3K WATERBORNE COATINGS ARE ALSO APPLIED

Most plural component coatings use two materials, a resin and a cross-linking agent. The latter is often incorrectly labeled a catalyst, although it actually chemically reacts with the base resin. A catalyst may be present in small quantities along with one of the two components, but the amount is too small to consider it to be a component in the system.

Three part waterbornes consist of a base resin and cross-linker, plus a third substance—water—that is added to lower the viscosity to application levels. The heaviest users of such coatings at this time are the aerospace coating industry, and, to a much smaller extent, the aircraft manufacturers. As soon as the components contact each other, the cross-linking begins—so once mixed these coatings have a limited pot life. Although the viscosity may not have risen above recommended application levels, do not continue to apply paint that has exceeded the paint supplier's listed pot life, because after this time the film properties will begin to degrade.

Chapter 6

Powdercoating

Introduction to Powdercoating

In the early to mid-1960s, a new painting technology was developed which we now know as powder coating. Instead of a liquid paint, the coating as manufactured and applied is a totally dry, finely ground powder. Its basic constituents are practically identical to those in a wet paint, with *one big exception*: the absence of solvent. Like a liquid paint, a powder coating contains resins, various additives, and pigments. The latter are present in all colorcoats and even in some of the clearcoats. The virtual absence of any VOC emissions makes powder coatings highly desirable on that basis alone, but their exceptional durability is also a powerful inducement for coaters to use powder instead of liquid coatings. There is no question that over the last ten or twenty years the majority of exciting innovations in paint materials, coatings application, and film curing have been in powder coatings rather than in liquid coatings. (Note: The term for this technology can be spelled "powdercoating," "powder coating," or "powder-coating." This author prefers to use the one-word or two-word spellings somewhat interchangeably, but prefers not to use the hyphenated version.)

Today powdercoating has roughly a 20–25% share of the painting market, and that share continues to grow rapidly. Automotive powder primers and antichip coatings are among the recent large, single markets for powder coats. Flat, precut, predrilled, and prepunched sheet metal shapes called "blanks" can be powder coated prior to mechanical bending. For several industries, powdercoating the blanks eliminates the need for postassembly painting. If the piece requires through-holes, these are punched or drilled in the blank before powder coating. In this way the edges of the holes are coated. If punched after being coated, the edges would show bare metal that could rust readily. Of course, blanks can be (and in some instances are) painted with liquid coatings just as well as with powder paints. Consider how the precoated blanks, whether painted with liquid or powder paints, have an advantage over precoated coil stock. If blanks are cut and punched from a painted coil, bare metal is exposed at all cut edges and punched slots and holes. These are susceptible to corrosion. By cutting blanks from unpainted metal and putting in the holes and other required openings before painting, no bare metal edges are exposed.

The nature of powder coating as a technology can best be understood by studying the characteristics of various powdercoating particles. The particles are going to vary dramatically both in size and shape. If an average powder particle were a sphere, it would be about 1.5 mil (38 μm) in diameter. A close scrutiny of the particle will reveal it is a composite of resin, pigment (except for some clearcoats), and various additives. All components are homogeneously mixed together—powdercoatings are not a dry blend of the separate components. Powder coating ingredients are blended and melt-mixed during their manufacture.

In manufacturing powder coatings, exact amounts of resin, pigment, and additives are initially dry-blended. The blend is then heated briefly to the resin's melt temperature, turning the dry blend into a fluid-like mass. The hot melt is quickly thoroughly mixed and immediately extruded into a thin, flat sheet that is quickly cooled and hammered into flakes. Chilled rollers are used to simultaneously flatten the warm mixture into a thin sheet and cool it. The extruded sheet is next broken into small flakes. A hammer mill then pulverizes the flakes into a powder having a consistency similar to that of ordinary wheat flour. Sieves may be used to grade the particles according to size ranges. Extremely fine particles may be melted and added back into another batch; while inordinately coarse particles would be subjected to additional pulverization.

Melting and forming a homogeneous mass of the powdercoating ingredients and then pulverizing the blend results in each tiny powdercoat particle having the same composition. This prevents any possible component separation that might otherwise occur when the powder coatings are shipped or handled. Potential segregation into pigment-rich and resin-rich portions could result from particle size and density differences among the blended components if the powder were not prepared in this way.

Returning to the examination of a particle of powder, the understanding of powdercoating becomes complete when an analysis is made of how the particle becomes a coating. If powdercoat particles are applied onto a part that can withstand heat, and the coated part is heated to the resin's melting temperature, the particles will melt and begin to flow until equilibrium sets in. If a hypothetical sphere particle of powder 1 mil in diameter is heated to the melt temperature of its resin, the sphere will collapse and flow out into a circular disk about 1.6 mils (40 μm) in diameter and 0.25 mil (6.35 μm) thick. The actual diameter and thickness would be functions of the rheology (flow properties) of that particular resin, but this assumption demonstrates the idea of the particle melting into a linearly expanded but reduced-height mass. It's not much different than what happens when an ice cube melts—the cube collapses down and liquid water spreads out over a larger area.

If a surface is covered with a single layer of particles of powder, each a hypothetical sphere 1 mil in diameter, and the particles are heated to the resin's melt temperature, the spheres would flow into a continuous coating film about 0.7 mil (17.8 μm) in thickness. Each melted sphere would merge and become homogeneously united with the neighboring spheres, forming a continuous coating layer.

During an actual powder coating process, however, it would be practically impossible to apply a coating only one powder coating particle thick. The particles almost certainly would pile up at least several particles thick. Therefore, it is easy to see that the minimum thickness of a powder coating would be in the vicinity of 2 mils (50 μm) or so, unless much smaller-sized powdercoat particles were applied.

Nearly all the resins used in wet coatings can also be used in powder coatings, although powder epoxies, acrylics, polyesters, and polyurethanes are the ones most used by far. Also, some resins that are difficult or impossible to make into liquid paints can be readily made into powder coatings. Materials such as nylon, Teflon®, and polypropylene can be used in powder coatings, even though they cannot be dissolved or readily dispersed in liquid systems. **Figure 6-1** compares the relative properties of various powder coatings. Until fairly recently, acrylic powders were totally incompatible with most of the other powders, to the point that many powdercoaters would not even bring acrylics into their shop for fear of causing craters on all their other work. By sad experience, many coaters found that if acrylic powder had been used in a booth, it was impossible to clean it well enough to avoid craters and fisheyes with the next resin used. While this is still a major problem, some of the newer acrylic powders no longer have incompatibility problems with other resins, which is an improvement much appreciated by contract powder coaters.

Resins used in powdercoatings may be either **thermoplastic resins**, which only melt and flow when sufficient heat is applied; or **thermosetting resins** that will melt and flow when enough heat is applied, but then also chemically cross-link. When heated, thermoplastic powder coatings form a paint film by the melting and coalescing of the particles. When they cool, the thermoplastic resins again solidify into a continuous film. Thermoplastic powders fall into the class of lacquer coatings. Lacquer powders are not used much except for special-use coatings; they make up only around 1–2% of all industrial powder coatings. Thermosetting powders, however, are classified as enamels and, as we have seen, make up the huge majority of powder coatings. This is due totally to the superior performance of powder enamels. Thermosetting coatings, when heated, melt and cross-link to form a thermoset paint film, as do liquid enamels. The cross-linking reactions occur between the main resin component and another resin component, designated as a cross-linker or cross-linking agent. Thermoset powders are made of significantly lower

	Weather Resistance	Chalk Resistance	Corrosion Resistance	Chemical Resistance	Heat Resistance	Oven Bake Resistance	Adhesion	Impact Resistance	Flexibility	Abrasion Resistance	Pencil Hardness Range
Acrylics	E	E	G	E	E	E	G	G	F	E	H-4H
Epoxies	E	P	E	E	VG	VG	E	E	G	E	HB-5H
Epoxy Polyester Hybrids	VG	F	VG	VG	G	E	E	VG	VG	E	HB-2H
Fluoro-Polymers	E	E	E	E	F	G	May Require Primer	VG	E	G	HB-2H
Polyamides	VG	VG	E	VG	F	G	May Require Primer	E	VG	VG	70-85 Shore D
Polyesters	VG	E	E	VG	G	VG	E	VG	VG	VG-E	HB-4H
Polyolepins	G	VG	E	Solvents-P Acids And Alkalis-E	F	F	May Require Primer	VG	E	P-F	30-55 Shore D
Vinyls	VG	VG	VG-E	Variable	P	P-F	May Require Primer	E	E	F-G	30-50 Shore D

E = Excellent, VG = Very Good, G = Good, F = Fair, P = Poor

Figure 6-1. Relative properties of powdercoatings.

molecular weight molecules than thermoplastic powders, but after they are applied and become fully cross-linked during curing, the overall molecular weight of the thermoset film molecules is far greater than the molecules in thermoplastic powder films.

Powder coatings are packaged in cartons and drums of various sizes, depending on the amount ordered by the coater. The container weights may range from as little as 25 pounds (11.3 kg) to as heavy as 300 pounds (136 kg) or more. Each powder container is lined with a plastic inner bag which, after being filled, is tightly sealed to keep out moisture and any extraneous contamination.

All applied powder coatings, even UV curing types, require heat to melt and flow into a continuous paint film. Substrates that can be powder coated must be able to withstand at least brief heat ranging from about 200° F to 500° F (93° C to 260° C). Many powders require parts to be heated at 300–450° F (149–232° C) for 15–25 minutes. These durations at high temperatures rule out the use of powder coatings on many heat-sensitive substrates such as woods and plastics, making powder coating primarily a metal-finishing process. However, instead of thermal curing, UV-curing done after the applied powdercoat has been heated enough to melt and flow it out into a film greatly lowers the total amount of heat to which the part is exposed. The UV cure process has permitted a lot more of the moderately heat-sensitive substrates to be coated using powder.

Surface cleanliness requirements for powder coatings are generally the same as for waterborne coatings. The degree of pretreatment needed varies with end-use requirements, both for powders and liquid coatings. End uses with extreme requirements would need maximum cleaning, a good conversion coating, and a quality aqueous sealer rinse. End uses with low requirements might require only minimal pretreatment conversion coating, but the surface must always be well cleaned. Because powdercoating films on average tend to be thicker than wet coating films, powder coatings can usually get by with less rigorous conversion coating than can liquid coatings. Some powder coating end-use requirements permit pretreatment to be limited to blasting clean surfaces with abrasives such as glass beads, aluminum oxide, or steel shot and then no conversion coating at all. The profile in the metal that blasting raises definitely helps achieve stronger film adhesion. The cured film needs to be thick enough to completely cover the peaks in the metal profile.

Powder Application Methods

Successful powder coating application requires depositing the powder uniformly over a surface. Heat is then applied, causing the powder to melt and flow into a paint film. Heat may also be used to cross-link the powder resin molecules. Two general methods are used to apply the powder; and each general method has at least two major variations within that type.

Fluid Bed Application Methods and Powder Application Methods for Hot Parts

In one fluidized bed method, the uncoated parts are heated above the melt temperature of the powder and then immersed into a fluidized powder bed. The powder particles melt and stick to the hot part, then begin to flow slightly although additional heat is normally required for full flowout and curing.

Another fluid bed process named "electrostatic fluidized bed" does not require heated parts. Hot parts are not suited for being coating using this method because the film build gets too high with heated parts. Unheated, but electrically grounded, parts are placed into or above a fluidized bed of electrostatically charged powder, and they become coated by the powder. Parts are nearly always hung above, not into the fluidized bed, again to avoid overly thick film builds. Parts coated this way receive thinner coats than if immersed, and that is the reason this process was developed. Due to limitations of the method, only small items can be uniformly powder coated by this process.

Fluidized Bed for Hot Parts

In nonelectrostatic fluidized bed powder coating, heated parts to be coated are first heated to a point above the powder's melt temperature and then dipped into a nonelectrostatic fluidized bed, as shown in **Figure 6-3**. In some cases, such as with vinyl and nylon powders, a waterborne dip primer may be applied to hot parts first. The primer is baked and parts hot from the primer oven are immersed into the fluidized powder bed. Powder particles contacting the hot part stick immediately and begin to melt (fuse) completely and flow into a continuous coating film. The coating tends to be thick because nearly all powder particles contacting the hot part will melt and adhere to the heated surface. Film thickness may vary from approximately 10–50 mils (250–1270 μm), depending on the temperature of the heated part and how long the hot part is kept in the fluidized powder.

Figure 6-2. A variety of products being powdercoated.

Exact film thickness control is impossible with this type of fluidized bed powder coating, and uniformly thin films cannot be attained.

The powder hopper in nonelectrostatic fluidized bed coating is, in essence, like the fluidized bed container used to supply powder to electrostatic powder guns. Air is passed through a porous plastic diffuser plate that functions as the bottom of the hopper. The fluidizing air must be free of oil and foreign particulate and low in moisture. Preferably the air should have a dew point below 30° F (–1.1° C) and so refrigerated dryers are commonly used to reach this low degree of humidity. Airflow of 5 ft^3/min (at 5–15 PSI) per square foot of diffuser surface (1.52 m^3/min per m^2 of diffuser surface at 0.345–1.030 bar) is typical. When one looks at the top surface of the fluidized powder, it has the appearance of a liquid that has been heated to a full rolling boil. If set correctly, virtually no powder drifts out of the fluidized bed, although an exhaust duct can be built around the top of the hopper to capture possible wandering particles.

The heated parts can be manually or automatically dipped into the fluidized bed. They can be removed after several seconds of immersion since longer immersion is unnecessary. On items such as boat anchors, the very thick coatings from hot dipping are preferred for their greater durability. A thickness of 40–60 mils (1,000 to 1,500 μm) is readily obtained with some powder coatings on heavy mass parts that hold their heat for some time. Except for wire shelving, most parts tend to be manually dipped rather than automatically dipped into the fluidized bed. In some instances, selective areas on parts are preheated to increase the film thickness in hard-to-coat spots on problem parts. As a coated part is raised from a fluidized bed, powder particles in direct contact with the heated part will have flowed out into a coating film. The last powder particles accumulated on the hot part's outer surface may be hot enough to stick but not hot enough to have undergone much flow. At this stage, the newly coated parts tend to look dusty from this partially melted powder. The appearance of this layer has led to it being called the "sugar coating." To fully fuse the powder, parts removed from the fluidized bed are then baked in an oven to complete the flowout of all the powder. With enamel powdercoats, the heat must also be sufficient to cause the resin cross-linking reactions to take place.

Postcure quenching of powdercoating in water is almost never done anymore, although at one time it was used to create smoother finishes with a number of vinyl and nylon powders. Now with improved powder coatings, quenching has become an unnecessary extra step in production.

Powder Spray on Hot Parts and Spraying Hot Powder on Unheated Parts

Instead of dipping hot parts, powder can also be sprayed onto hot surfaces. Hot mold cavities are sprayed with powder where it later fully cures and forms an outer shell for the plastic that is subsequently injected into the mold. This technique is often labeled **powder in-mold coating**. It is basically a variation of the much older liquid gelcoat process for molding plastics, which continues to be more heavily used than powder for this purpose. Flame spraying of powders—the aspirating of powder through a flame to melt it and make it stick to the target—has been done commercially but it never was utilized to any great extent and is still extremely rare.

Electrostatic Fluidized Bed for Unheated Parts

Preheating parts is not the only way to powdercoat with a fluidized bed. For small wire baskets, racks, mesh items, and other modest-sized parts that require a coating thickness of 2–6 mils (50–150 μm), unheated parts can be coated using ***electrostatic*** fluidized beds. The main reason for powder coating this way is to get much thinner film builds than with hot parts. The thickness minimum drops to about 4–6 mils (100–150 μm) with unheated parts, instead of at least 10–15 mils (250–380 μm) with hot parts. Electrodes on roughly 4 inch (100 mm) centers are installed in the air diffuser that serves as the bottom of the hopper. The electrodes project an inch or two up inside the bed. They are charged electrically negative so that they can add electrons to the fluidized powder particles in the hopper. The purpose of this method, remember, is to produce thinner films than is possible with hot parts. This method utilizes the attraction between the negatively charged powder particles and the grounded object being coated. Electrostatic attraction will cause powder particles to be attracted up out of the fluidized bed and onto the electrically grounded parts hanging above the bed. If the grounded parts were immersed down into the fluidized powder cloud itself instead of being supported above it, even though the parts are unheated, overly thick coatings would be the result. Electrostatic fluidized bed technique enables the deposition of a fairly uniform powder layer onto small parts in the desired thickness range of roughly 2.5–6.0 mils (60–150 μm). The parts to be coated must be small because powder cannot be electrostatically attracted up and out of the fluidized bed more than about 6–8 inches (15–20 cm). The thinner coats on small parts can be a major cost advantage, especially when the production volumes are large.

Considerable amounts of coiled metal fencing, wire screening, and similar open-mesh roll stock are coated using electrostatic fluidized powder beds. Screening in coil form is run vertically up between a pair of fluidized beds. The powder that is attracted onto the screening is quickly fused and cured with infrared heat. The coated screen portion is then cooled and recoiled. Open-mesh and expanded metal materials, if made of thin enough gauge, can also be coated in coil form using vertical fluidized bed powder coating. Numerous small parts have been powder coated on horizontal conveyors using an electrostatic fluidized bed. Once applied by either electrostatic or nonelectrostatic means, the powder film can be cured immediately using infrared or convection oven heat, or with just enough heat to melt the powder followed by UV radiation to cross-link the powder molecules.

Electrostatic Application without Fluid Beds

Electrostatic application of fluidized powder with spray guns, disks, or bells onto electrically grounded parts is by far the most popular way to apply powder coatings. Parts can be hot or unheated, but heated parts are rarely coated this way unless thick films are desired. Both negative and positive electrical charging of powder particles are employed.

Powder Spray Guns

Electrostatic gun spraying is the most common method of applying powder coating. **Figure 6-2** shows various products being powder coated. During the coating application, the part must be electrically grounded. Hanging the part onto a properly grounded overhead conveyor as shown in **Figure 6-4** usually affords the needed ground. You will

Figure 6-3. An operator dips a rack into a fluidized bed. Aeration suspends the powder into a fluid-like bath for complete coverage.

Electrostatic Fluidized Bed

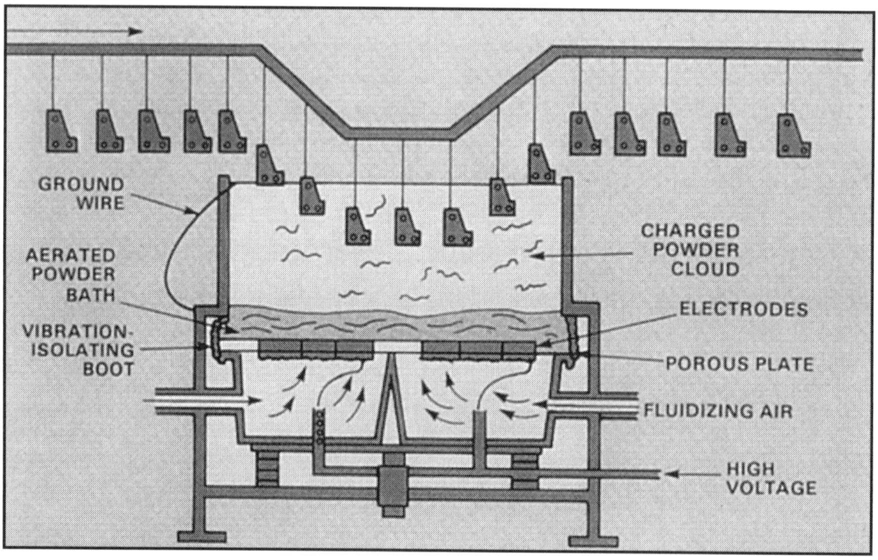

Figure 6-4. Electrostatic fluidized bed.

read that precautions should be taken to ensure proper grounding by checking the actual amount of grounding from part to conveyor, which numerous sources claim should show less than 1.0 microhms resistance. While good grounding is essential, take their advice with a grain of salt. Although it sounds good in theory, I have found in actual test that as long as there is at least some grounding, adequate coating will occur. Even with quite high resistances, full coverage was achieved. Only when there was a total lack of grounding did parts fail to coat.

Powder resting in a container needs to be fluffed up, or fluidized, before it can be pumped to a powder application device such as a powder spray gun. Fluidizing is accomplished by placing a quantity of powder into a special hopper, or bed, with numerous fine openings or perforations in its bottom. Air at a controlled rate is forced up through the hopper bottom, gently lifting the particles into a contained fluid-like powder cloud. For good operation, the air used for any type of powder coating should be relatively dry. A dew point below 40° F (4.5° C) should be attained; this normally requires refrigerated air drying or desiccant columns. Additionally, filtration to less than 9.8 µin. (0.25 µm) size, and removal of oils below 0.1 ppm, is recommended.

A venturi pump provides a means of drawing powder from the fluidized bed to the electrostatic spray gun (or guns). The guns contain a negatively charged electrode at the tip that charges the powder electrostatically. Air delivering the powder from the fluidized powder in the hopper exits the gun and expels (sprays) the powder toward the parts to be painted. **Figure 6-5** demonstrates how an air venturi pump is able to move powder from

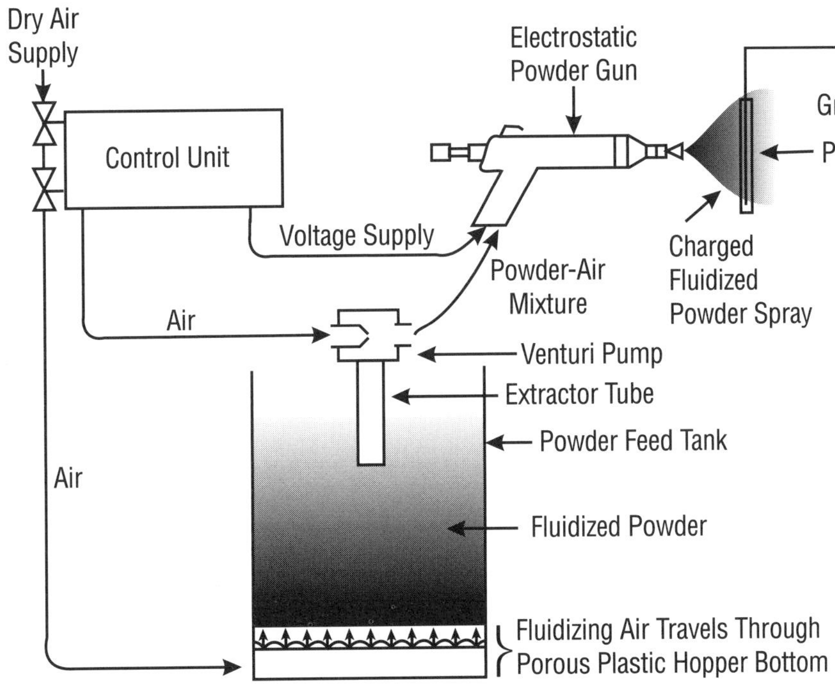

Figure 6-5. Powder electrostatic spray system.

the fluidized bed to the spray gun, where metered air gently sprays the powder out the gun barrel. **Figure 6-6** illustrates examples of a small powder supply tank and venturi pump, filling the tank, and adjusting the fluidizing air.

The spray pattern is adjustable on some guns with a conical deflector located in the spray particle path, and on other more modern design guns using jets of air at the gun tip. Adjusting the deflector distance from the gun tip regulates the size of the spray pattern; moving the deflector increasingly closer to the tip makes the powder pattern wider and wider. Likewise, on air jet pattern adjusting guns, the spray cloud configuration is changed by varying the pressure of air introduced at multiple locations around the end of the gun barrel. **Figure 6-7** shows a drawing of a typical electrostatic powder spray gun.

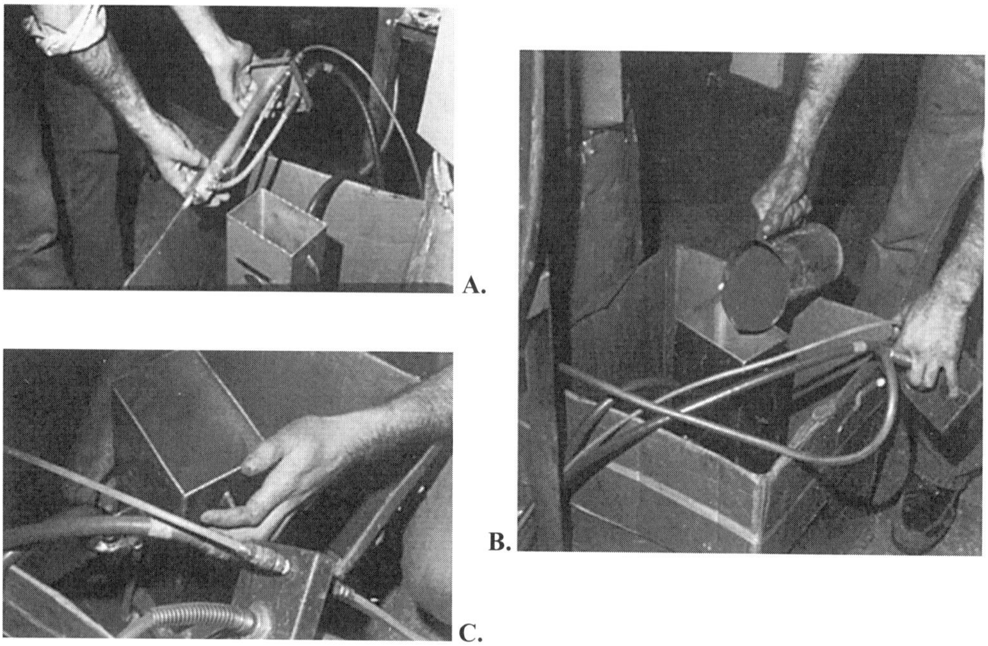

Figure 6-6. (A) A powder supply tank and venturi pump; (B) Tank being filled; and (C) Adjustment of fluid air.

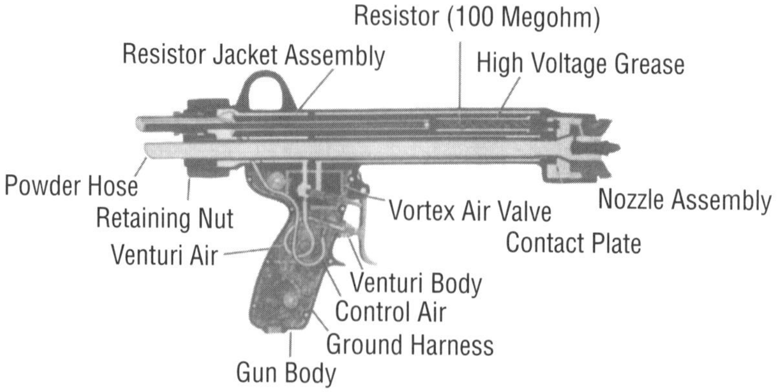

Figure 6-7. Cutaway drawing of electrostatic powder spray gun.

Powder exiting the gun tip is charged electrostatically in a manner similar to that used for negative charging of liquid paint droplets. The powder picks up extra electrons from an electrode (raised to a potential of 75–100 kV) or from the ionized air cloud at the gun tip, and thus becomes negatively charged. The charged powder particles are attracted to the closest ground, which is intended to be the part to be coated. Some people believe that when the negatively charged sprayed particles on the outermost layer of powder contact the underlying powder layers on a grounded workpiece, the high resistivity of these powder layers prevents the negative charges on the outermost particles from discharging to ground. The charge "hangs on" and holds the particle to the part through the electrostatic attraction of charged particles to grounded objects. **Figure 6-8** gives an example of this, showing some small powder coated parts emerging from a spray booth. According to this theory, the attractive forces on the outer layer are sufficient to hold all the powder onto the part until it enters the oven, where heat melts the particles into a fused coating. This retained electrostatic charge theory is widely held, but scientifically hardly plausible. The author has demonstrated how the powder remains on a part even when it is hung for 30 minutes in a highly humid atmosphere that allows all charges to dissipate within seconds.

Another explanation, and one preferred by this author, is that a "packing" effect occurs, which holds powder onto the parts firmly. The effect is not unlike the packing of snow into blocks or snowballs. This is why powder particles hold on well even in high humidity, which drains away all the electrostatic charges very quickly.

Sprayed powder should exit the gun rather gently, forming a powder cloud through which parts are conveyed. Too much airflow would tend to blow powder off parts. **Figure 6-9** shows lawn furniture emerging from a powder booth, the interior of which appears

Figure 6-8. Small powdercoated products emerging from spray booth.

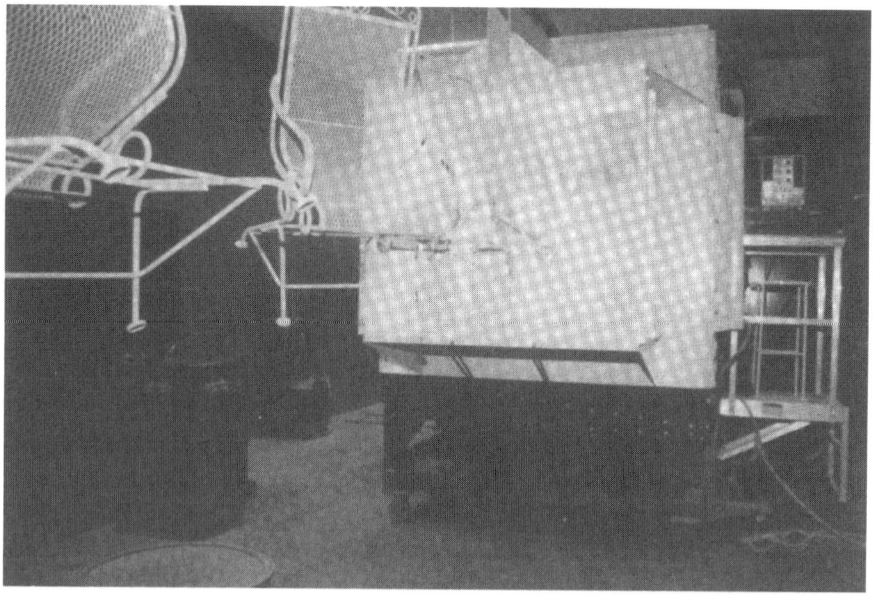

Figure 6-9. Lawn furniture emerging from a powder spray booth.

to be filled with a cloud of powder. The grounded parts moving through the cloud attract the charged particles. The amount of powder attracted to a part depends on the charging voltage and the conveyor speed. A part being conveyed rapidly through a weakly charged cloud might draw only a light coating that would flow to a film thickness of about 1–2 mils (25–50 µm). The same part being conveyed slowly through a highly charged cloud could attract enough powder to build a film thickness of about 3–5 mils (75–125 µm). The voltage and conveyor speeds are not the only factors. The shape of parts being coated, the powder particle sizes, and the polarity of the powder particles also play important roles in determining the film thickness.

For small-volume powdercoat users such as weekend mechanics, auto specialty shops, and custom toolmakers, a relatively new "cup gun" for powder is available. This does not require a separate hopper supply for the powder. Compressed air is used to fluidize the powder in the cup and to deliver powder onto the parts. This is an electrostatic method as well and thus it requires parts to be grounded during application. Wrinkle, metallic, and veining powders can be applied with this style gun. As with all powder coatings, heat must then be used to melt and cure the powder. UV cure powder could possibly be used, of course, but UV curing is not normally appropriate for low-volume powder users.

Rotating Powdercoating Bells and Disks

The design of the powder coating bell is an adaptation of the well-known liquid bell applicator discussed in detail in Chapter 14. An air turbine spins a circular head (a so-called "bell," although in most versions it is far more truncated than the normal liquid bell shape). The bell is electrically charged by a high voltage DC power supply just as in liquid paint bell application. In a liquid application device it is located at the head to create a paint droplet delivery pattern; but in this case the disks and bells produce a cloud-

Figure 6-10. Tool box and racks on this pickup truck are finished with a metallic-look powder coat finish.

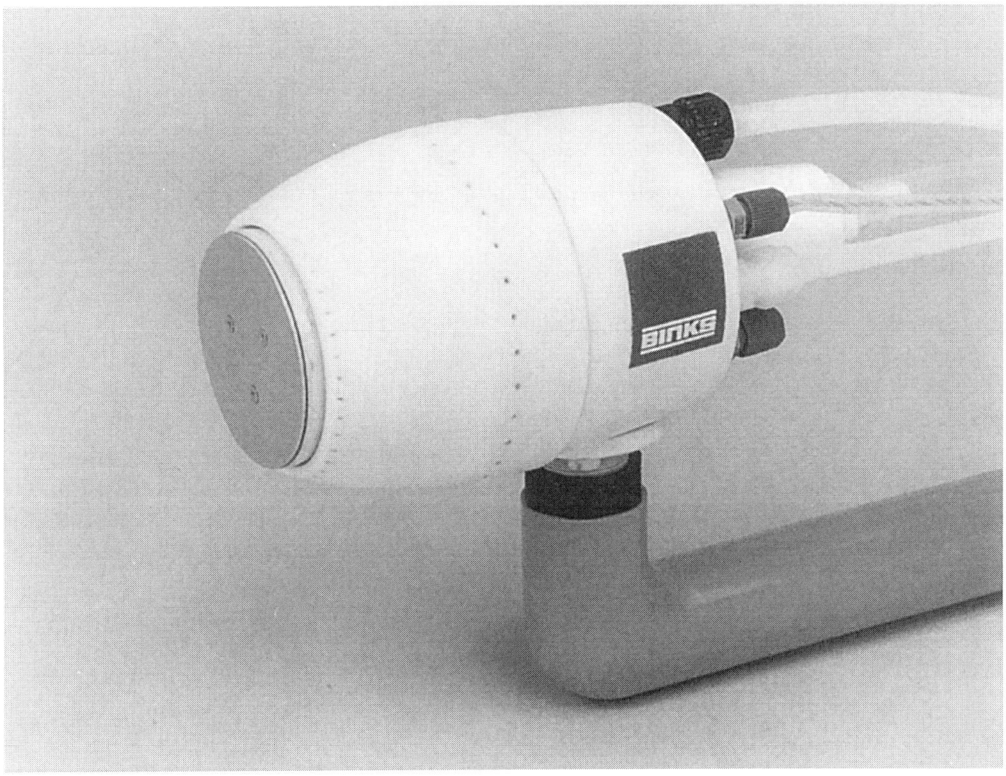

Figure 6-11. Rotating powder bell.

like pattern of powder particles. Fluidized powder is fed through a hose within the turbine housing and then onto the bell surface via a circular ring of holes in the rotating bell head. The powder particles, as they pass along the electrostatically charged bell surface, gain negative electrostatic charges from the bell. The electrostatic principles of powder application are identical to those of powder spray gun application. **Figure 6-11** is a picture of a typical powder bell. Bells are more flexible in use than rotary disks and so bells are more common than disks for both powder and liquid paint application. In liquid application, the rotational force atomizes the paint; clearly that is not necessary for powdercoatings. The rotary motion serves to mix the various size particles that may become stratified as they are carried in an air stream through the powder hoses (sometimes called "roping"). Extremely high rotational speed is not needed so that less expensive electric motors rather than air turbines can be used to spin the powder head. Low speed rotation near 1000 rpm is often preferred; at high speeds the powder pattern becomes smaller.

Full body powder primer for automobiles was first applied on Daimler-Chrysler's model year 2000 Neon car. This primer powder needed less polishing prior to topcoating than wet primers and was more durable, but it also was not easy to control the application to get uniform film builds. Powder primer was applied over an epoxy ecoat to a thickness of 2.5–3.0 mils (60–75 μm) with rotary electrostatic powder bells. Daimler-Chrysler was not the only automaker to introduce full body powder coating. BMW and Volvo started the use of powder clearcoats; having first applied these with electrostatic bells around the 1990–2000 model years. Even so, industrywide outside the auto sector, the use of guns to apply powder is much greater than bells.

Figure 6-12. Powder bells applying antichip paint to an automobile body.

Rotary disk powder application is not unheard of, but must be considered rare. It operates in much the same way as the liquid disks do, as detailed in Chapter 14. Powder is charged by direct contact with the rotary disk head. It was first used in the early to mid-1980s.

Nonrotating Powdercoating Disks and Bells

Some vertical reciprocating powdercoating disk systems are in operation. As with liquid paint application, the disk is charged negatively up to about 100,000 V. The powdercoating disk, however, does not rotate as do disks used to apply wet coatings. The high voltage on the disk charges the powder delivered onto the disk negatively. Since like charges repel, the negative powder particles are electrostatically forced radially outward away from the negative disk edge. Charged powder particles leaving the disk are attracted to the grounded parts that are being conveyed around the disk in an omega loop.

The powder coating disk is claimed to have the output equal to that of six spray guns. It is said to be capable of applying coatings as thin as half a mil with excellent thickness control, at least on parts that are fairly simple in shape. These assertions may almost be true in ideal situations. Normally the disk would reciprocate vertically, but it can also operate in a fixed position or be tilted slightly. Exhaust air requirements are identical to those for powder spray booths. Overspray powder is carried down to the bottom of the booth by a stream of air, and powder can be recovered for reuse with the usual variety of filtering devices.

Rather recently, nonrotating bells for applying powder have been developed, but at best their use has been minimal to date.

Figure 6-13. Interior of a powder application booth.

Tribocharging—An Alternative to Negative Electrode Charging

A method of charging some types of powder without using a charging electrode is termed friction charging or "tribo"charging. Tribocharging cannot generate the much higher charge levels of 15–100 kv that negative electrode-charging can. A tribo gun is equipped with a grounded teflon insert. Powder particles moving along the insert lose some electrons to this ground, and in the process become ***positively*** charged to about 10–15 kV. Small amounts of adsorbed moisture on the particles facilitate the wiping off of electrons. The tribocharge generated in totally dry environments is often too weak to be effective. Tribocharging will not work with all powder coatings. Acrylic powders, for example, charge very poorly in tribo systems.

Tribocharging is often touted as a means of overcoming the Faraday cage tendency for charged particles to avoid depositing in confined areas. Such areas rapidly build up a charge and repel the charged particles. Moderately tribocharged particles can succeed in overcoming the Faraday cage problem, but much of this is due to the weak charge. The same reduced Faraday effect can often be achieved by turning down the voltage in a conventional electrostatic power supply to 10–20 kV, in many instances. Nevertheless, for some parts and with some powders there is no doubt that tribocharging offers superior application ability compared to negative charging. Plants should always test both methods before selecting one over the other. The quantity of powder each individual "tribo-head" can charge positively is far more limited than for negative electrode charging. Powder application head arrays with multiple charge tubes are often employed to overcome this problem. See **Figure 6-14** for an example of that style application device. Powder coated steel coil is often painted using multiple heads to apply a uniform coat across the width of the coil.

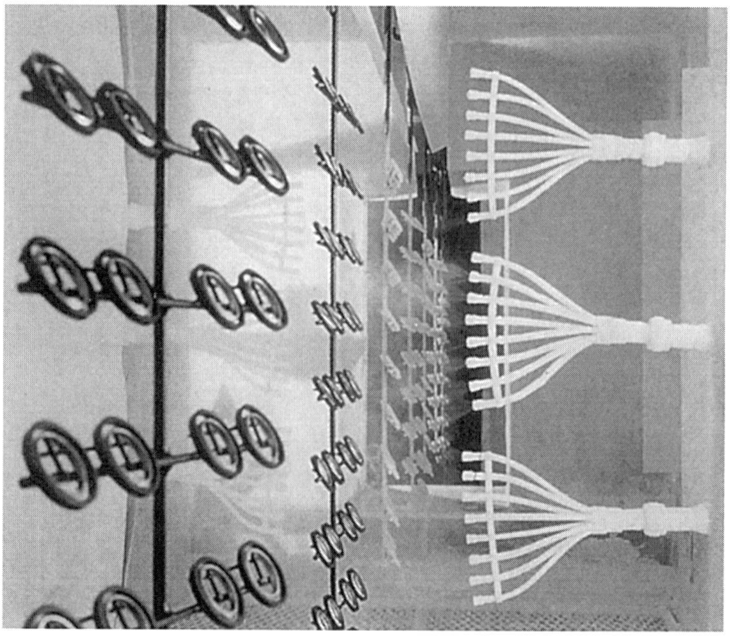

Figure 6-14. Tribo guns in use.

Tribocharging powder with both a nonrotating bell and a nonrotating disk has been successfully adapted for powder application. Neither one, however, has been used to any degree as yet. As we have learned, multiple head spray applicators are usually far better at tribocharging powder. These multihead guns are normally much preferred over using any type of bells or disks for applying powder by tribo-type application. None of the tribocharging devices has much trouble from the so-called back ionization or charge repulsion defects that can be experienced with improper operation using negative (corona) charging.

Specialty Powder Coating Methods

Bulk Powder Coating

Small hard-to-rack parts can be bulk powder coated. Parts preheated on a metal belt fall into and then pass through a vibratory powder coating bowl from which they are fed onto a continuous belt and exposed to additional heat to fuse the powder. Automatic flip-over exposes both sides to curing energy so that no bare spots remain. The minimum film build by this method is 3–5 mils (75–125 µm).

Aqueous Powder Slurry

Attempts have been made to use slurries of powdercoats in water for electrodeposition coating, dip coating, flow coating, and for spraying using ordinary liquid paint application devices. To do this involves extremely fine grinding of powder and then mixing the material in water to form the aqueous powder slurry. Unless the powder is ground extra fine, it separates too rapidly from the water slurry instead of remaining suspended. The economics of aqueous powder suspensions are rarely favorable. It was first examined in the early 1970s and then dropped again until BASF produced a product that was tested by BMW for clearcoat application in the late 1990s. Mercedes-Benz used powder slurry clearcoats for a time starting around the turn of the century on its expensive A-class vehicles in Rastatt, Germany. Powder slurry auto topcoats are also possible and several versions are in their early test phases.

An advantage of powder slurry coating is that films as low as 0.3 mil (7.6 µm) dry film thickness can be achieved. Although the reuse of the powder slurry overspray is rather difficult, it should not be completely ruled out as a potentially useful future technology. For a short time in the 1970s, an ecoat powder slurry was used in a Japanese automotive plant. The finishes were very rough and required excessive sanding. As a result there is virtually no interest in this technology currently.

Powder Coating Booths

To my horror, I have seen powder being sprayed in an open area of a plant without any booth or enclosure. This is foolish, dangerous, and illegal. Almost every powder coating system applies the powder in a confined space designed for powder application. In most systems, powder moving past the part is drawn downward by a gentle downdraft airflow. The airflow must be strong enough to collect overspray but not too vigorous to prevent blowing already-applied powder off the parts. Overspray powder reaching the booth bottom collector duct can be gathered in a hopper or returned to the fluidized bed to be resprayed.

The downdraft air volume is less than the exhaust volume, which creates a negative pressure inside the booth. This prevents escape of powder particles from the booth by bringing in air from outside in through the booth openings. It is recommended that the air in the vicinity of the booth be clean and relatively dry to prevent drawing moisture and contaminants such as dust and lint into the booth, and hence into the reclaimed powder.

The air returning oversprayed powder to the fluidized bed paint supply hopper is exhausted first through the system's primary powder recovery section. These can vary in design, but generally are able to capture around 98% of the powder particles for reuse. The booth exhaust air then passes through very fine (absolute) filters that trap the remaining ultra-fine powder particles. The booth air is then returned to the room. Absolute filters prevent powder "fines" from being blown into the powder coating room atmosphere. The fines do not charge well in electrostatic application, so it is best if they are discarded and not recycled. Various design fabric bag filters, cartridge filters, or "cyclones" can be used for primary recovery of overspray powder. Each has pros and cons that affect which is preferable for each individual powder application situation. **Figure 6-15** shows how a cyclone recovers the powder.

Just as the downdraft airflow in the powder booth needs to be gentle to avoid blowing deposited powder from the part, so too must exhaust and recirculating air movement within a powder-curing oven be moderate. This is true particularly at the oven entry where the powder has not yet melted. As the parts heat up and powder reaches its melt temperature, flow begins. At this stage the requirement for gentle airflow in the oven no longer applies. In a couple of rare instances, highly excessive air movement in a direct-fired powder oven has blown enough powder off parts to cause a dust explosion. The powder ovens are not sealed tightly so minimal damage occurred, but it shows the folly of having too much oven air movement.

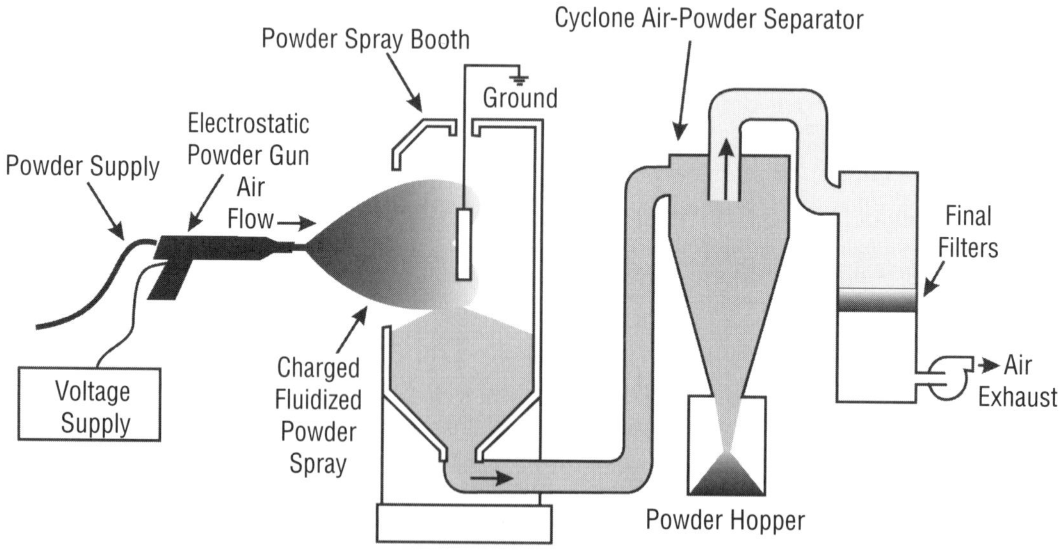

Figure 6-15. Cyclone powder recovery system.

To minimize air circulation before the powder is melted, some ovens incorporate electric infrared heat only at the entry portion of the oven and gas convection for the rest of the oven. This approach is popular in areas where electricity is more costly than natural gas for heating. It is also important for smooth films that the powder be heated quickly to melt and cause it to flow the material before too much cross-linking has taken place. IR is excellent for the rapid heating needed, and in this way smooth films will be produced. Infrared can heat surfaces extremely fast. Another reason why many plants use IR in the oven entrance or even throughout is to shorten the overall oven floor space required. IR may also be able to lessen substrate heating due to shorter oven dwell times. On the other hand, gas infrared is attractive because it is usually less costly to operate, but gas IR temperature control tends to be slow and more difficult. Electric infrared lamps or glow-bars are simpler to install and to focus when supplementary heating is needed.

Powder Coating Advantages

Powder coating's rapid growth is no doubt due to its many advantages beyond the virtual absence of solvent emissions. After 25 years of use, powder coating captured nearly 20% of the industrial finishing market. This percentage is expected to grow steadily until it probably peaks out at about 30–35% of the industrial finishing market. Powder coating advantages include the following.

- Cost
- VOC compliance
- High quality finish
- Ability to capture and reuse overspray
- Ease of application
- Energy savings
- Quick "packageability"
- Wide choice of resins.

Cost

The nature of powder coating allows various traditional equipment or processes on a finishing line to be omitted or minimized, thereby reducing costs. These include the following.

- No solvent flash-off time required. This allows shortening the length of the conveyor line formerly used for flash time. Parts can enter the bake oven immediately after powdercoating application.
- No coating mix room needed. Powder coating lines have no need for a coating mix room. Powders are completely formulated by the powder coating manufacturer for immediate application.
- Shorter oven length required. Powder coating ovens can be short in length. No gradual heat-up is required to drive off solvent slowly to avoid solvent popping. In fact, quick heat-up is recommended for powder to avoid film roughness.
- Lower oven ventilation requirement. The absence of solvent greatly reduces requirements for oven air make-up and exhaust.
- Lower booth ventilation requirement. The absence of solvent greatly reduces requirements for booth exhaust and make-up air.
- Less plant make-up and exhaust air. Booth exhaust air is vented inside the plant, not outside as with solvent-containing paints. In cold weather this saves considerably on

heating costs.

- Floor space economy. A properly designed powder coating line requires from one-fourth to one-third less floor space than needed for wet painting systems.
- Reduced fire insurance rates. The absence of flammable solvents and the elimination of solvent paint mixing and storage hazards can reduce insurance rates. Powder in a bulk form cannot burn, although fine powder dust can possibly explode. However, normal safety precautions can almost eliminate any possibility of explosion. Fire or explosions with powder are extremely rare in powdercoating operations; this is not true of liquid painting.

VOC Compliance

The absence of solvent and the practical elimination of VOCs make powder an unequivocal compliant coating in the eyes of environmental regulatory agencies. The absence of solvents has a number of other advantages.

- No solvent odor. This is highly advantageous inside the plant for employee comfort considerations and outside the plant for eliminating a nuisance factor, which is especially important for plants located in or near residential areas.
- No solvent storage. The elimination of solvents foregoes the need for safety-approved solvent storage rooms, and this saves space and reduces fire hazards.
- No solvent thinning. Since no solvent thinning is needed for coating viscosity control, fire hazard is reduced.
- No solvent health hazards. Powder reduces the potential health hazards to the sprayers. Solvents can irritate the mucus membranes of the eyes. Powder spilled on the skin is not adsorbed into the body and can be removed with a gentle stream of air or with a vacuum cleaner. Cleanup of powder spills on skin using soap and water is recommended, and rapidly accomplished. An appropriate particle filter mask is recommended for safety when spraying powders. A few powdercoatings contain toxic ingredients; special filter masks are required when using these materials.

High Quality Finish

Powder coating has acquired a well-earned reputation for providing a high quality finish. Powder coatings tend to be extremely durable and provide outstanding corrosion resistance. At equal film thickness, powder coatings are almost always superior to corresponding types of wet finishes. One reason for this is the virtual absence of shrinkage in the cured powder coating film, which minimizes stress during curing. When wet films are curing, simultaneous loss of solvent from the coating shrinks the paint film, causing internal stresses on resin molecules.

Powder coatings applied onto grit-blasted surfaces that have not been conversion coated are always far superior to wet paints on this type of substrate, both in adhesion and in corrosion protection. Powder coatings can readily be applied quite thickly in one pass for extra endurance coatings. Edge coverage tends to be excellent for powder, as well.

Powder makers are working to develop more powders for thin film applications, although even these "thin films" are still in the 0.75–1.0 mil (20–25 μm) range. New laser thickness detectors for the uncured powder work well. Thin films have always been harder to achieve in powder than in liquid coatings. High temperature resistance, in contrast, has been better with powders. Silicone-containing powder coatings are heat-stable up to 1,000° F (540° C) for intermittent moderate-length durations with little loss of gloss or color change. This has been useful to manufacturers of heating stoves, cooking grills, and similar products.

Ability to Capture and Reuse Overspray

Most electrostatic powder coating systems capture and reuse the overspray powder. For example, a powder application system may include a cyclone air-powder separator, as shown in **Figure 6-15**, and details of the cartridge collector are shown in **Figure 6-16**. This allows a net utilization of about 95% of the powder, although first-pass transfer efficiency is often less than half that number. The only powder not recovered is the small amount lost during booth cleanup for color changes, and the fines collected in the final air exhaust filters. Some reclaim systems use airflow to pull overspray powder to the floor of the booth, which is a continuous moving fabric belt. At the end of the booth, the powder is vacuumed off the belt and directed to a small cyclone for recovery. Early recovery units used a multibag chamber called a baghouse. The powder was filtered from the air on the surface of the bags, and the powder was periodically knocked off into a collection hopper. The most modern recovery equipment uses self-purging cartridge filters, often with dedicated cartridges reserved for each color. (See **Figures 6-17** and **6-18**.)

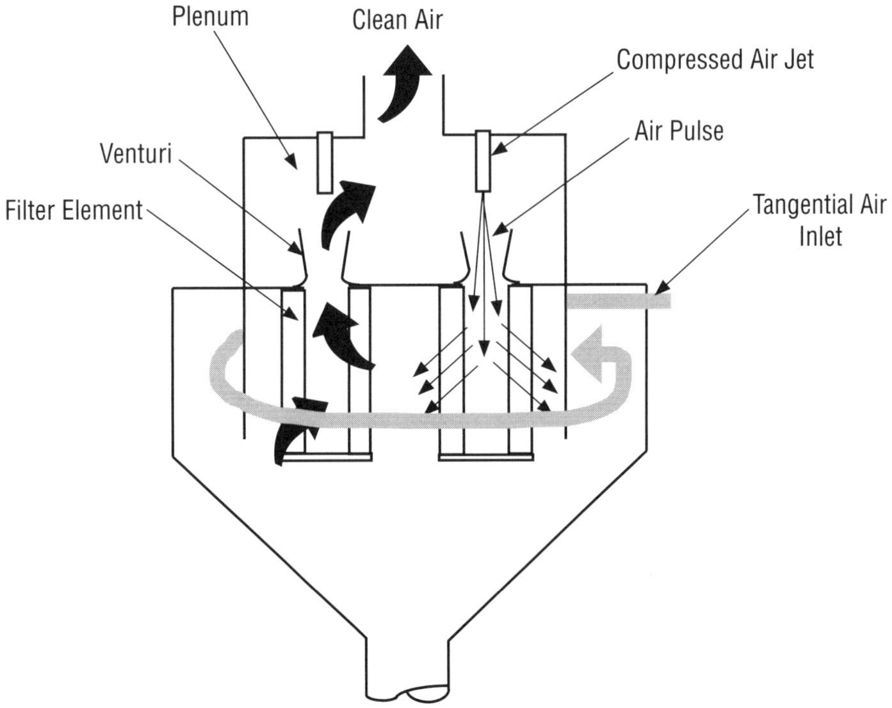

Figure 6-16. Cyclone cartridge collector.

Powder reuse practically eliminates disposal problems, although sometimes the waste powder must be melted into a solid layer or block prior to disposal. Disposal of spilled or contaminated powder is relatively easy compared with disposal of wet paint sludge. But one manufacturer is quoted as saying, "I've been trying to find a practical way to reuse my spent powder for 10 years. Each month we throw away ¾ to 1 ton (680 to 907 kg) of used powder, which gets mixed together." Landfill sites are more likely to refuse to take liquid waste or to charge higher rates for such waste than for dry powders. This is especially true if the paint sludge contains organic solvents. Places for disposal of wet paint wastes containing flammable amounts of solvent are hard to find, which results in high disposal costs. Plants have paid from US$300–$1200 to dispose of each 55-gal (208 l) drum of wet paint sludge! Also, powdercoating systems have no contaminated booth water to treat and no used dry filters from spray booths that require periodic replacement and disposal of used filters.

Ease of Application

Little operator expertise is needed to spray powder. With wet coatings a great deal of spraying practice and finesse is necessary to get uniform coatings and avoid runs and sags. Some plants claim that a person without wet spray experience is often better at powder coating than a skilled wet painter. Manufacturers point out that this allows using relatively untrained labor for powder coating spraying, which permits a lower labor rate and reduces production costs. Runs and sags, while not impossible to produce, are extremely rare with powder coatings. When excess powder is directed at a certain area of a part, the powder simply falls into the recovery system. The electrostatic effect tends to produce a uniform coating as long as sufficient powder is directed into the parts, but point-to-point film thickness is regularly more pronounced with powder than liquid coatings.

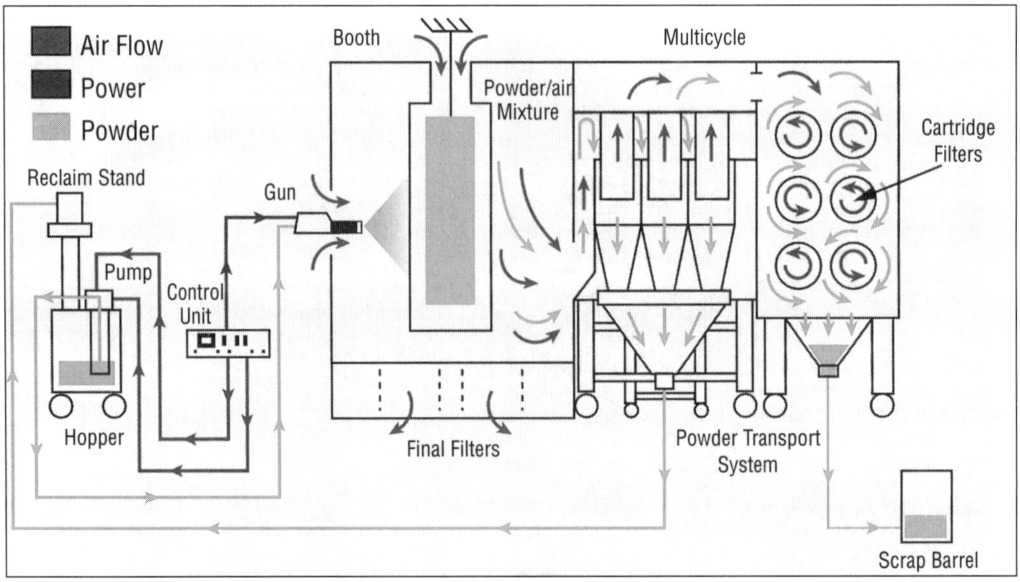

Figure 6-17. Diagram of a powdercoat booth using both cyclones and cartridge filters for powder recovery.

Figure 6-18. Cartridge filters for powder recovery.

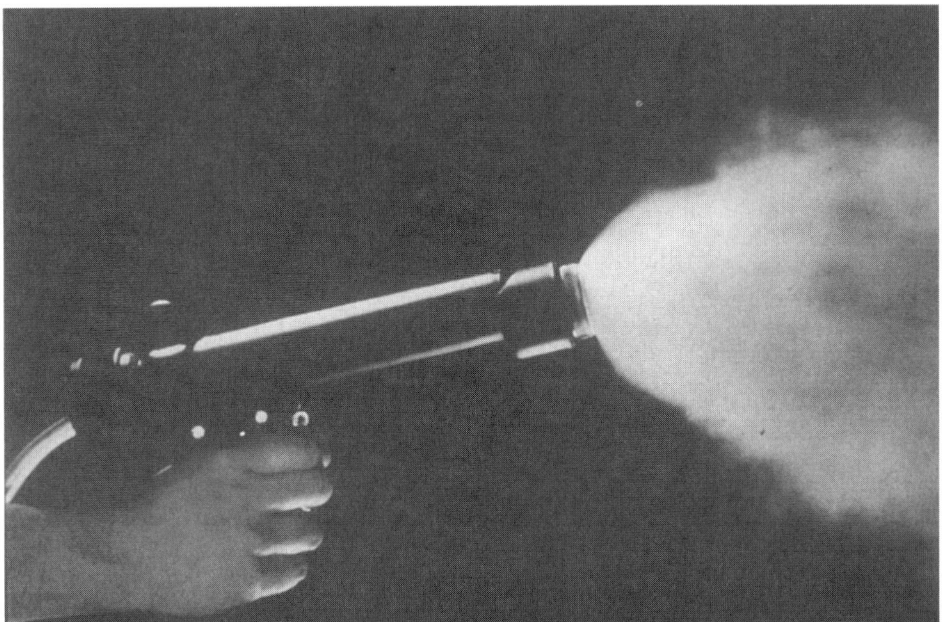

Figure 6-19. Powder spray requires less manual application skill than wet spray.

Coverage on corners, edges, and projections is particularly good with powder coatings because of the added electrostatic attraction (see **Figure 6-20**) and because many powder coatings exhibit only minor edge pull. Powder spray patterns and the electrostatic voltage can be adjusted to produce uniform part coverage, even on parts with complex shapes. Voltage adjustment and feed air control enables adequate management of coating thickness. Spray patterns can be varied from a thin powder stream for covering deeply recessed and concave areas to a wide cloud to coat broad, flat surfaces. Automatic powder spray guns are readily engineered for high-volume production, and multiple-gun systems are abundant.

Powders can be selected that incorporate topcoat and primer properties in one single powder, which can be applied as thick as 12–15 mils (300–380 μm) in just a single coating application. In this way the two separate applications of primer and topcoat with two separate bake steps can be combined into one.

Energy Savings

Although powder coating systems need to use a dry-off oven after parts pretreatment and a powder bake oven after application, powder systems still use less energy than wet paint systems. Utilizing waste heat generated by the bake oven can sometimes conserve energy by directing this heat to the dry-off oven. A powder coating system's greatest energy saving comes from the greatly reduced air-exhaust and air-makeup requirements.

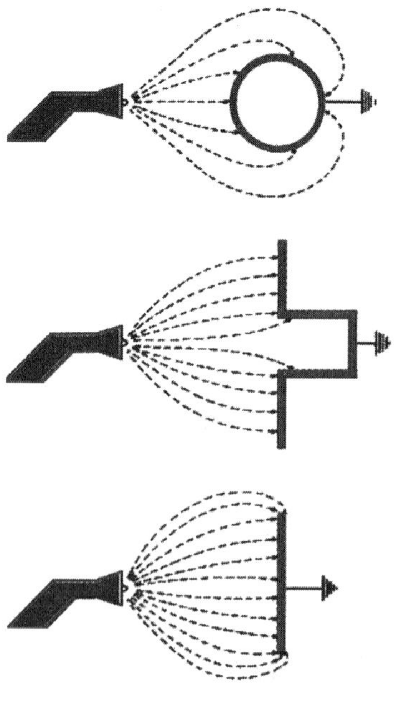

Figure 6-20. During powdercoating applications, the electrostatic field is shaped by the part. The wraparound effect allows fine powder particles to be deposited on the backside or edges of some parts.

Only about 10% as much air is needed for a powder booth as compared with a wet spray booth. Because no solvent vapors form in the powder booth, only enough air is used to recover all the overspray powder. The absence of solvent allows powder spray booth air to be filtered and exhausted inside the plant. In cold climates this saves a lot of heat. The air turnover in a powder oven can be low because no solvent vapors need to be exhausted. Powder systems can often get by with a comparatively smaller oven and this might save energy, yet the powder bake temperatures tend to be higher than for wet paints, so this offsets that advantage.

Quick "Packageability"

When a powder-coated part emerges from an oven, it can be packaged when cool enough to handle. Unlike many wet coatings that continue to cure for days or even weeks, powder coatings develop a full cure during the bake process. As a result, powdercoated parts resist handling abuse immediately out of the oven far better than wet coatings. They are less easily damaged during handling and assembly operations, and they do not require as much care in packaging for shipment. Problems of parts sticking to the packaging (called blocking) because the powder coating is not fully cured do not occur.

Wide Choice of Resins

Certain resins might not be available in solventborne systems due to limited or total insolubility in suitable solvents, but they can still be used in powder coatings. For example, nylon and teflon are virtually unavailable as paints. The same is true for polyethylene and polypropylene. Yet all four of these resins can be made into acceptable powder coatings.

Powder Coating Disadvantages

Although powder coatings have numerous advantages, like everything else they also have a number of disadvantages. Probably their main disadvantage in the past was the requirement for high cure heat (300–500° F, or 150–260° C, depending on the type of resin). These elevated temperatures were needed to fuse powders so that they could flow into a continuous coating film, and virtually restricted powder to being a finish for metals. In a few cases it has been possible to lower the powder cure temperature to near 215° F (101° C), and this has opened up special UV curing powders for heat sensitive parts. While some powders cure lower, for really durable coatings the bake temperatures still need to be around 300° F (150° C) and above.

For electrostatic application, powder coating requires a conductive substrate, which again limits it to metals or else requires a special conductive precoat. Other disadvantages of powder coating include:

- Manufacturing limitations
- Application problems
- Repair difficulties.

Manufacturing Limitations

Because of the way powdercoatings are made (dry mix, melt and mix, extrude and cool, chop into flakes, mill and sieve), it is often not economically feasible to make small amounts of special colors. Powder manufacturers generally liked to make a minimum of 1,000–2,000 pounds (454–907 kg) per batch until recently. One of the problems in manufacturing a new powder formulation relates to color matching when a special shade is needed. Exact powder color matching usually involves trial-and-error procedures, which adds to the "gearing up" costs. A considerable amount of raw materials may sometimes have to be processed before the powdercoat color is exactly matched to the sample chip. Today a few powder producers specialize in preparing small-size orders, although they usually charge a significant premium. With so many colors and shades now available, the need for special order colors is not that great.

Color matching with powder starts at the beginning of the powder manufacturing process. Mixing of powders is not the same as for wet paint. If a can of white and a can of red liquid paints are mixed, two cans of pink paint are produced. If a box of white and a box of red powder are mixed, two boxes of a "white and red" powder will be produced. Even when the powder is applied and cured, the film would be a speckled mixture of red and white. The colors do not mix together with powder as they do with wet systems. Even minute amounts of cross-color powder coating contamination are often visible because extraneous color particles fail to blend.

Other differences from wet paint are found in powder wrinkle and texture finishes. Powders have the advantage. In preparing wrinkle finishes, both wet and powder paints form their wrinkles during curing, but with powders more size variations in the wrinkle pattern are available. With wet paints a "spot" texture is obtained by first painting a continuous film using normal atomizing airspray pressure. After this continuous paint layer

Figure 6-21. Commercial food-mixing paddle powdercoated with a sterilizable nonstick coating (10–14 mils thick, 250–350 μm).

is allowed to firm up a bit, the "spot" texturing is applied using reduced air spray pressure. This produces large incompletely atomized "spots" that dot the film and form the spot texture. Depending on how heavily the spot paint is applied, the density of the spot pattern can be varied. And by adjusting the airspray pressure higher or lower, the size of the spots can be controlled. This cannot be done with powder texture paints; the texture is controlled only by how the manufacturer formulates the powder. But several textures impossible with wet paints are available in powders, such as pebble, sand, and suede-like textures. Veining is possible with both liquid and powders, but much more dramatic multicolor effects are seen in powders. If you have ever seen any "copper and black" or "gold and black" veined powder textures this is already evident to you. There are a host of different color combinations available in these veined texture finishes.

Difficulties exist, however, in preparing metallic powder coatings that match the appearance quality of liquid metallics. Powder coatings that contain mica, metal powder, or metal flake cannot fully duplicate the attractive look and glamour of wet metallic finishes. This has kept powder out of the large automotive topcoat market. In North America at least 70% of all vehicles produced have metallic paint finishes. The lack of "shrink" of the powder coating gives less metallic brilliance in the appearance of the paint film. Shrinkage of wet films as solvents evaporate forces the metal flakes into an orientation predominantly parallel to the substrate surface. This effect is known as metallic "flop" or "travel." Sharp brilliance results from increased reflection because of light striking a greater surface area of metallic flakes. However, recent powder developments have yielded metallic powder coatings that are excellent for many types of parts. One such development encapsulates the metallic particles with an insulating layer of resin before they are added to the manufactured powder.

Application Problems

One of the potential problems with electrostatic powder spray is that the powder recirculating system should create a negative pressure in the booth (unlike the positive pressure of a wet booth) to prevent escape of powder particles. This negative pressure will allow any moisture and contaminants in plant air to be drawn into the booth. Contamination and moisture will then lower the quality of the powder accumulated in the collection system. To prevent contamination and moisture in the powder, the powder application booths are usually installed in a room that is air-conditioned. It does not require a true clean room atmosphere, but the benefits of filtered and dehumidified air for powder application are obvious. Clean and dry powder has better appearance and is far less prone to clumping.

An electrostatic powder bell or gun uses a gentle stream of air to deliver the powder. An excessive airflow within the booth would tend to blow the electrostatically attracted powder away from or off the parts. Similarly, when electrostatic powder guns or disks are reciprocated, the reciprocating speed must be slow to prevent creating undesirable turbulence in the powder cloud.

The required gentle powder cloud from an electrostatic powder gun or bell presents a problem due to an enhanced Faraday cage effect. The air pressure cannot be stepped up to force the charged powder into a confined area. The enhanced Faraday cage phenomenon increases the tendency for powder to deposit at the edges of part recesses with limited amounts within the area, causing wide differences in powder film thickness in these regions.

In fact, all powder coating applications have more inherent film thickness variation than wet paints, due in part to the normal size range of powder particles from about 0.2 to 3.0 mils (10 to 75 μm) diameter. This makes thin film below 1.0–1.5 mils (25–38 μm) difficult to achieve, although improvements continue in this area. With wet systems it is not particularly difficult to achieve uniform films of 0.5 mil (12.5 μm) thickness. Only with the proper powder, careful application, and precise conveyor hanging techniques, it may be possible to satisfactorily apply fairly uniform films at 0.8 mils (20 μm) or slightly less on selected parts. Extremely thick powder coating films can be produced with one pass, but not everyone wants thick coatings. Powder's tendency to yield thick films can be considered an advantage or disadvantage, depending on whether or not the thick coating is desired.

Another minor but recurrent problem with powder coating application is the formation of clumps of powder that can be ejected from the spray gun, causing paint film blemishes and rejects. One cause of clumps is impact fusion, which is the frictional heating and partial melting of powder moving through the fluidized delivery and recovery systems. Impact fusion tends to occur when fast-moving powder particles collide with a hose or pump surface, such as when powder flow makes an abrupt right angle, sending particles crashing into a wall. Good powder circulation design avoids such sharp turns as much as possible. Powder clumping can also be caused by wet or moist powder, powder being

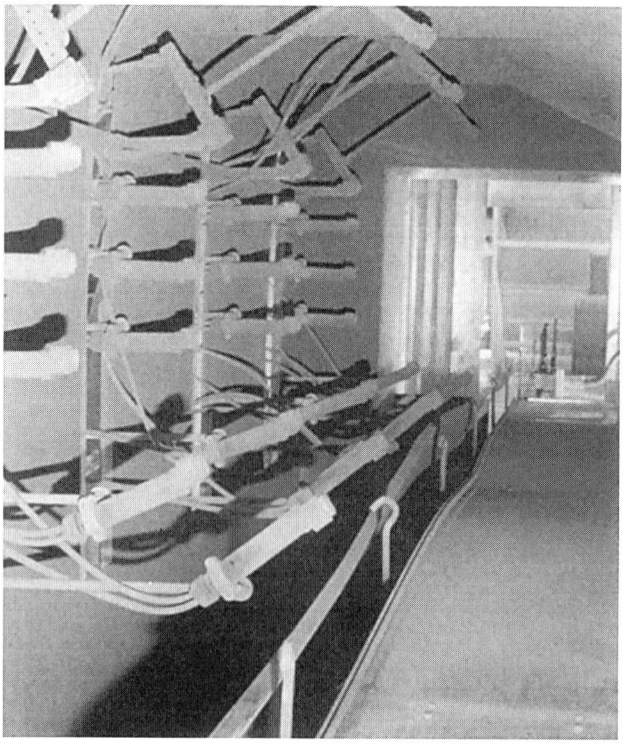

Figure 6-22. Interior of booth showing accumulation of overspray.

stored in areas warmer than the recommended 75–85° F (24–29° C), and powder stored beyond its recommended shelf life.

The difficulty in changing powdercoating colors in a booth is another application disadvantage. If the same booth is to be used, all surfaces of the booth and powder reclaim system must be thoroughly purged of powder before a new color is put into the system. **Figure 6-22** shows the amount of powder that collects in the interior of a booth that has been applying powder. All surfaces of the entire powder recirculation system accumulate powder in the same manner. Before a new color can be applied in the booth, every trace of the previous powder must be removed. Any particles of the previous powder that remain behind will deposit along with the second color, giving a "salt and pepper" effect.

Various quick-color change systems are used for powder. The most common system uses separate booths for each color. This allows a booth to be moved on- and off-line, as its color is needed. Another system has a booth with plastic walls that are rolled up after each color change. **Figure 6-23** shows a powder booth equipped with a continuous fabric collection belt floor and vacuum pickup device (shown in **Figure 6-24**) to somewhat simplify powder color changes.

Some powder coating users make no attempt to recover the sprayed powder—a practice called "spraying to waste." They collect all sprayed powders in a common container and either dispose of it or sell it to someone who can use a mixture of many different colors. Such a mixture can be used on parts where appearance is unimportant, but unless you have a lot of it no buyer will be interested even if you want to give it away.

The percentage of powder attracted to parts moving through a charged powder cloud is about 40–50%, depending on part size and shape. Although this is true first-pass transfer efficiency, it is deceptive because most electrostatic spray powder coating systems capture most of the overspray powder and reapply it. If powder coating transfer efficiency is defined as the total percent of powder used, taking into consideration the reclaiming and recirculating of the powder, then the net figure is often around 90–95% or higher. However, on systems that do not reclaim overspray, the transfer efficiency probably ranges from

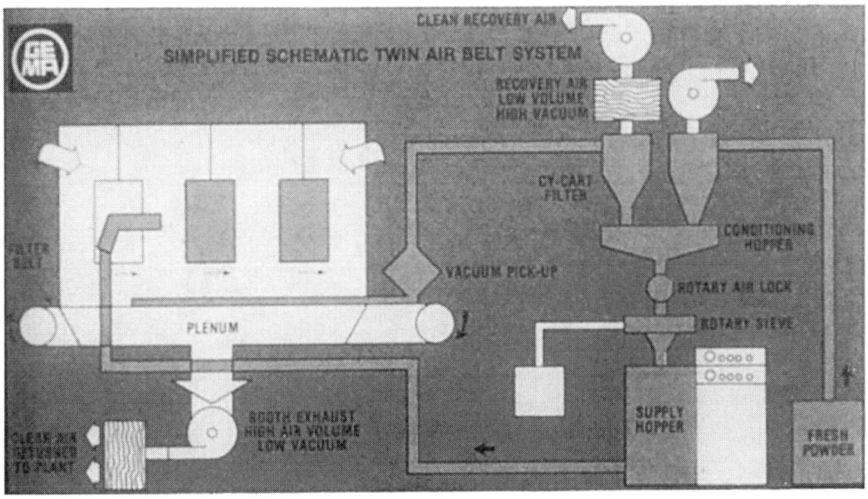

Figure 6-23. Powder spray booth with continuous belt and vacuum collector.

20% for thin, spindly parts to perhaps as high as 55% for parts with large flat surfaces, such as sheet steel panels.

The absence of solvent makes it inherently difficult for powder resins to achieve adequate flowout. This restricted flow tends to cause orange peel and a low distinction of image. For this reason coated parts should be heated rapidly so the powder melts and flows before cross-linking occurs. Additional heating will cause cross-link curing of the coating. Fine grinding of the powder is costly, but in some formulations this can help reduce the visual effects of restricted resin flow rheology.

Powder's requirement to be stored in a cool, dry place can be considered a disadvantage because it adds to capital investment. Powder needs to be kept cool to prevent possible premature cross-linking. Some shipments have had to be transported in refrigerated trucks when especially sensitive powders and unusually hot weather conditions were involved. Powders must be kept dry to prevent clump formation. Powder storage areas should be air-conditioned to maintain cool temperatures and low relative humidity, unless a fast usage rate limits turnover storage to brief periods. Even then air conditioning is a prudent measure.

In the past, powder coating's unique high durability properties would make it troublesome if not impossible to strip with conventional caustic or cold solvent materials. Improved liquid strippers have been developed and now are used effectively for removing even thick layers of powder coatings. For a long while, the most common way of removing powder coatings from hooks and hangers has been by burn-off in controlled combustion high-temperature ovens. This works well providing that the temper of the steel in the

Figure 6-24. Vacuum pickup head collects and conveys powder to small cyclones (at left).

hooks and hangers is not adversely affected. Stripping parts for repainting using burn-off ovens may present similar difficulties. Cold chemical strippers are better for such items.

Repair Difficulties

"Back ionization" is a term that has been used to explain the cause of powder film defects and diminished transfer efficiency that can occur from improper operation during corona electrostatic powder application. Poor transfer efficiency is self-explanatory. The other defects are depressions or craters, even large pinholes in the film. Excessive electrostatic voltage settings and/or positioning the gun too close to the target are major causes for these defects. A much better term for the cause of these defects would be "charge repulsion." The problems are associated almost exclusively with corona charging. With tribocharging equipment only a relatively small positive electrostatic charge can be created, so "charge repulsion defects" are not common using tribo systems. Some people erroneously claim these effects can never happen with positively charged powder particles. But if the positive charges on particles could somehow be raised as high as the negative charges produced from corona equipment, the same defects would definitely be producible.

Unfortunately, a well-meant but totally incorrect explanation for back ionization has been published and, worse, has been unthinkingly accepted by some powdercoat people despite its implausibility. The wrong answers you might read include the use of inaccurate terms and completely false phrases such as "free ion flows" and "creation of positive ions at the part surface." These do not occur. The truth is actually simpler and easier to understand, for it is based on electrostatic and physical science principles involved with electrostatic coating application.

In corona charging, a small cloud of air molecules (mainly nitrogen, oxygen, and normally some water molecules) located around the charge tip will gain extra electrons, which are released by the gun charge tip. This cloud is called the "ionized air cloud" (not correctly because these are not true ions, but we'll ignore that since it's always been called that). In a darkened room this small cloud is visible as a blue "corona" at the gun tip. Since electrons are miniscule negative bits of matter, the extra charges in the air cloud are therefore negative charges. When coating particles pass through the ionized air cloud on the way to the grounded target, the particles pick up some of these negative charges. The coating particles plus their extra negative charges are propelled toward the grounded part by the airflow through the gun. The electrical charges cause the particles to be magnetically attracted to grounded objects, and once they reach the object the particles seek to release their extra electrons. A conductive, grounded part will allow the extra electrons to flow from the paint particles through the part, the hook, the conveyor, and finally through the conveyor supports back into the earth itself.

With freshly applied powder, it may be difficult for the outermost negatively charged particle layers to release their electrons. This is because the powder layer has many air spaces within it and is frequently not a good electrical conductor. Remember that negative charges always repel each other. Either setting the voltage too high or bringing the gun tip too near the grounded part will move the ionized air cloud so close to the charged outer powder layer that three detrimental things can happen. First, the repulsion between the ionized air cloud and charged outer powder layers will force some powder particles off the surface, and so they end up as overspray. Second, if the total charges in the cloud, plus the charges on the outer powder particles, build too high, they will suddenly burst through the powder layer as a miniature spark, scattering the powder and leaving a thin or bare spot. The small sparks will not be visible when they happen, but the resulting powder defects certainly are. Third, newly arriving negatively charged powder particles are repelled, pushed away by the existing negative charges. The arriving powder particles can't transfer to the part so, again, they end up as more overspray.

The so-called back ionization or charge repulsion defects are alleviated if an auxiliary ground is added near parts to drain excess ionized air cloud charges away. Pseudo-scientists call this a "counter electrode," but a more accurate term (albeit less glamorous) is simply "ground." I've seen something similar to this with electrostatic spraying of liquid metallic coatings. It happens when metallic flakes gain excess charges that they cannot quickly transfer to the grounded part. Micro-sparks, again not visible ones, leave a dime-sized spot that is termed a "starburst." The spark aligns metal flakes radially around it to produce this visible pattern defect, but no depression in the film as with powder. Wet coatings conduct charges far better than powders so it probably happens rarely, if at all, with any of the liquid paints other than liquid metallic coatings.

Repairing defect and blemish spots in powder coating films can be a difficult process. If the defective product is spot sanded and put back on the line to be powdercoated again, the coating can become too thick and brittle. If this happens, the part will need to be totally stripped and recoated. A considerable difference among powders is found in their ability to produce a smooth appearance when recoated with powder. The intercoat adhesion of recoated powder coatings, even with the same powder, also varies widely. Before selecting any powder paint, the recoatability of the material should be tested.

Defect areas on powdercoated surfaces can often be recoated with liquid coatings. However, formulating a liquid coating to match the exact appearance of a powder coating can be difficult. Some companies or consumers have rejected the potential use of powder coating because no liquid repair paint could be found that would exactly match the color, gloss, and texture of the powder coating. For example, the State of New York would for a time not purchase powder coated highway lighting poles for this reason, even though the powder finish was demonstrably more durable for this application than liquid paints.

Special Powdercoating Techniques

A somewhat recent innovation has been the use of UV and near-infrared radiation for curing powder as a way to avoid using too much heat. Substrates that are only moderately heat sensitive may be able to be powdercoated this way, but heat is still required. The applied powder is melted first with either IR or convection heat, then the resin is cross-linked very quickly by UV light. Medium density fiberboard can work ideally in this process but it will outgas if it contains too much moisture. The limitations are that UV curing is a line-of-sight method so complex shapes may not cure well, and UV light doesn't penetrate pigmented coatings very well so it is used more for thin or clear finishes on wood.

With UV curing, the parts experience only a moderate amount of heat but the powder must be heated to quickly melt it. Both plastics and woods are candidates for this cure method, but most application to date has been with wood. Not all woods are suitable for powder coating; the moisture content is important. A slight amount of moisture helps with electrostatic grounding of the parts, although a conductive prep coat could be used as well to provide conductivity. Too much moisture makes wood unsuitable for powder; the finish becomes marred because water vapor bubbles the fresh film during the powder melting step.

To overcome the low penetration of UV light through pigmented films, intense near-infrared radiation can be similarly used to cure powder fast, often in less than 8–10 seconds. But remember that IR is also a line-of-sight method. There is a modicum of interest in using near-IR curing for massive parts such as large cast iron valve bodies that can weigh as much as a ton. It takes a long time to heat cure powders on big objects. They heat up very slowly because they are such enormous heat sinks. Once the part is finally brought up to temperature for the required period and powder is fully cured, it then takes another inordinately long time for the massive hot parts to cool down enough for handling. The near-IR cure would be a major help if somehow enough IR light could be directed to all the areas on the part. That may not be easy to achieve.

Flat surfaces, including coil stock, are no problem. Flat sheets and coil can even be powder coated by what is termed "electromagnetic brush technology," a powdercoat process analogous to application of the toners in modern printers.

Coverage in ft^2/lb Related to Specific Gravity
(Multiply by transfer efficiency for actual coverage)

	DFT in Mils (Microns)				
Specific Gravity	**1.0 (25)**	**1.5 (38)**	**2.0 (51)**	**2.5 (63)**	**3.0 (76)**
1.0	192	129	97	77	61
1.2	161	108	80	64	53
1.4	138	92	69	55	46
1.6	121	81	60	48	40
1.8	107	72	54	43	36

Chapter 7

Cleaning the Surface

Cleaning a Variety of Substrate Materials

For proper bonding of paint to any surface, the surface of the material to be painted must be free of contamination. No water, dirt, or other impurities should remain between the surface and the paint because foreign matter, obviously, can reduce bonding integrity and will certainly detract from paint appearance. Substrates to be painted fall into five general categories of materials: cloth, paper, wood, plastic, and metal. Composite materials, although comprised of a combination of these, tend to have predominantly one of the listed materials on the surface.

Cloth and paper are typically manufactured in long sheets that are wound on rolls. They are clean as manufactured and generally need not be cleaned before being painted. Wood, plastic, and metal items are manufactured in all sorts of sizes, styles, and configurations. The paint applicator almost always needs to clean metal items, and often must clean plastic items before painting. In some cases freshly molded or extruded plastic parts can be painted immediately, which is a desirable processing procedure since it eliminates the cleaning step.

Cleaning wood most often involves a mechanical smoothing such as sanding with fine abrasive paper and an air blow-off or wiping with a clean, lint-free cloth. Using fluids to clean wood is usually avoided; liquids such as water and almost every organic solvent will raise the grain, and extra sanding would then be necessary if a smooth coating were required.

For plastic, unless it is a type that absorbs surface moisture, cleaning most often involves simple washing with a low-foaming aqueous detergent followed by rinsing and drying. The wash solution should usually contain a low to moderate concentration of detergent and have just a mildly acidic pH. If any soil on the plastic is particularly heavy or oily, then either solvent or alkaline detergent may be used. Solvent cleaning of plastics, when required, demands great care; many solvents will damage plastics by interfering with their molecular macrostructure. The result can be distortion, swelling, crazing, or even cracking of the plastic parts.

Metal Surface Cleaning

Methods for cleaning metal surfaces before painting vary with the type of metal. Cleaning metal becomes much more involved than cleaning wood or plastic because metal fabricating operations typically use a wide variety of forming and machining oils or greases. These lubricate the metal workpiece as it is pressed, punched, folded, bent, roll formed, or however otherwise manipulated. The fabricated part consequently often becomes oil-spotted and frequently contaminated by bits of metal powder, filings, or chips. The soils to be removed might also include glue, fabrication shop crayon marking, and weld spatter. In addition, for fabricated metals stored outdoors before painting, deposits of oxides will require removal. Even when the item is stored indoors, iron and steel parts can form iron oxide (rust), zinc parts will often collect zinc oxide (white rust), and aluminum will undoubtedly develop a layer of aluminum oxide. These oxides can be expected to interfere with paint adhesion and should be removed or at least treated before painting. Hot rolled steel always has oxide scale that must be removed before painting. Although steel can be purchased scale-free from the mill at added cost as "pickled and oiled," roughly 85% of users buy hot rolled steel without this premium treatment.

The majority of metal surfaces that are painted fall into one of the following categories: ferrous metal (iron alloys including sheet or hot rolled steel and wrought or cast iron), zinc (typically galvanized but including castings), and aluminum. Although all of these metals might have different optimum cleaning chemicals, the actual cleaning procedures themselves are practically identical. These procedures can include:

- Mechanical cleaning
- Solvent cleaning
- Aqueous cleaning
- Acid cleaning
- Alkaline detergent cleaning.

The optimum cleaning method for a particular metal part depends on the type of contaminants to be removed. Methods for cleaning the different types of metal, for example, steel as compared to zinc, may also vary. When a plant paints a blend of these metals, a compromise cleaning process usually becomes necessary.

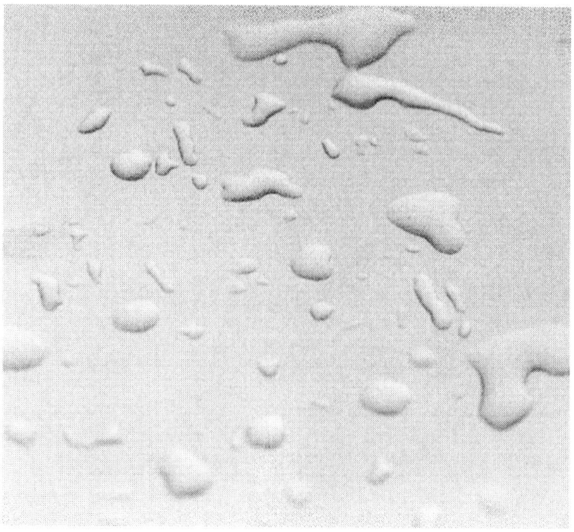

Figure 7-1. Water beading on a surface suggests the presence of oil or grease. Cleaning needs to be improved.

Mechanical Cleaning

Mechanical cleaning can be something as mundane as wiping dust from the surface with a clean cloth. However, when firmly attached metal scale or heavy rust must be removed, more vigorous cleaning methods are essential. Abrasive removal using a sanding belt or disk, or a manually applied abrasive pad, can be effective. Equally beneficial may be the use of a wire brush or wheel to remove tenaciously held rust. Water alone at very high pressure has also been used effectively for abrasive surface cleaning.

When numerous parts or very large areas need to be cleaned, it often is practical to use blasting with sand or other abrasive grit materials propelled in a high-pressure air or water stream. A combination of air and water can also be utilized for this purpose. The grit materials that are used in this technique include sand, steel shot, plastics, glass beads, aluminum oxides, and also softer cutting organic materials such as ground walnut hulls or ground corncobs. Ice, carbon dioxide, and sodium bicarbonate (baking soda) pellets are used as well for their ease in cleanup. Rather than being carried by an air or water blast, grit materials may also be vigorously hurled from a rotating wheel or rapidly moving belt at the object to be cleaned. With grit cleaning or blasting, care must be exercised to avoid leaving grit particles embedded into the surface. These inclusions can cause poor paint adhesion and possible early corrosion under the paint film. When recirculated grit is used, provision should be made for separating and removing accumulated "fines" and debris.

Mechanical sanding-like cleaning may be coupled with liquid cleaners using cushioned plastic wool pads, which have the advantage of not loading up and becoming "blinded" from embedded dirt. As with sanding cloths and papers, the pads are available in a range of physical sizes and grit sizes from coarse to fine. Scoth-Brite®, Brade-Ex®, and Brush-lo® are some common brand names for these three-dimensional abrasive materials.

Solvent Cleaning

The restrictions that are increasingly being placed on most solvent emissions in many of the world's more wealthy nations have produced a severe decline in solvent use for cleaning purposes. Nonetheless, because oils and greases are not reliably removed by mechanical action, for some manufacturing plants suitable solvents can provide a more effective method of contaminant removal than aqueous cleaners. Despite the high labor costs and the possibility of missed spots, manual wipe cleaning is still best for some metal and plastic parts. Solvent cleaning by spray, vapor, or dip may be coupled with wiping using a solvent-saturated sponge or cloth. The wiping material and any solvents used in spraying, dipping, or wiping must be clean

Figure 7-2. A load of parts that have been shot-blast cleaned is being unloaded from the autoblast chamber.

or else dirt, oil, or grease will be left on the part being cleaned. One-way solvent cans for dispensing solvent onto wipe cloths can prevent contamination from working its way back into the solvent container and are highly recommended. Presoaked solvent wipers are also available in "one-at-a-time" dispensing packages, and these convenient wipers likewise avoid solvent supply contamination.

Dip tanks of clean solvent soon become dirty from the contaminants they remove. The resulting problem with dip-type solvent cleaning, then, is that parts may be only partially cleaned. Solvent spray is not commonly used for cleaning, but when it is, recontamination of parts with dirty solvent will also occur unless parts are sprayed only with fresh solvent. Continuous distillation can ensure that the used solvent remains clean. The expense of continuous distillation, however, is not usually practical unless warranted by a sufficiently high cleanliness standard or a large production volume. Solvent cleaning needs to be done in a confined, but well-ventilated, space to keep vapors away from other people and other processes in the vicinity. Flammable solvents should never be necessary, but if for some unknown reason they are used, they have to be used with utmost care for obvious reasons.

Known as vapor degreasing, a once widely used method of solvent cleaning uses boiling solvent vapor to rinse oily and greasy parts clean. The degreasing is enclosed but has an open top for parts entry and exit. Cooling coils at the top condense the solvent vapors to prevent their escape. Pure solvent vapors are continuously boiled (distilled) up from a sump and they condense onto parts being cleaned. The condensed solvent, along with the contaminants it rinses off the parts, drips back down into the boiling solvent. The nonvolatile oil and grease contaminants stay in the boiling liquid, so only clean solvent vapor is able to evaporate and reach the parts for cleaning. This vapor cleaning can only be used safely with completely nonflammable solvents. Solvent vapors would otherwise present totally unacceptable fire and explosion hazards. In the past, various halogenated hydrocarbons termed chlorofluorocarbons (CFCs), including Freon®, methylene chloride, and 1,1,1-trichloroethane (methyl chloroform), were often used in vapor degreasing. The international Montreal Protocol agreements in the 1990s, which phased out both the manufacture and use of CFCs to prevent depletion of the earth's ozone layer, had largely killed this cleaning procedure in the ensuing years. A resurgence in the late 1990s was occasioned by the introduction of several suitable environmentally acceptable safety solvents, but vapor degreasing has never gained its former glory. A few less developed countries that are part of the Montreal Protocol are still permitted some manufacture of vapor-degreasing compounds. In Mexico and Vietnam some manufacture of Ozone Depleting Substances (ODSs) was allowed by the agreement through the year 2006. Still other countries, such as China and India, did not even sign the agreement. During my last consulting work in India (1996), considerable CFC use was evident. It may be assumed they continue the manufacture and use of ODSs for various tasks including liquid solvent cleaning and vapor degreasing.

The basic components of the vapor degreaser shown in **Figure 7-3** are a tank, a source of heat to vaporize the solvent, vapor condensation coils and the ancillary accompanying refrigeration components, water separator, solvent recirculation pump, filter, and a supply of solvent. Vapor-degreasing tanks may range in size from as small as only about 5 gallons to over 20,000 gallons (20 l to over 75,000 l). During operation the heating source maintains the solvent at its boiling temperature, sending solvent vapor up toward the dirty parts. The excess vapors eventually reach the refrigerated condensing coils, where the vapor condenses and flows back to the solvent supply. The water separator catches moisture collecting on the condensing coils, and keeps water out of the solvent. The condensing coils are located about 10–18 inches (25–45 cm) from the top of the tank, leaving a distance above the vapor level called the freeboard. The freeboard length is calculated to prevent solvent vapor escape from the tank due to drafts in the plant and the movement of parts into and out of the degreaser.

A part is immersed for cleaning down into a vapor degreaser only as deep as the vapor zone, not into the boiling solvent. Hot solvent vapors contacting the ambient (room) temperature part will condense and flow off the part, thus solubilizing and carrying away oil, grease, and similar materials. The condensation is an exothermic process and so it heats up the parts being cleaned. Condensing action will continue until the temperature of the part reaches that of the solvent vapor. Withdrawing the part slowly through the freeboard area allows the solvent to drain off or evaporate and return to the vapor zone. Some solvent drag-out is inevitable. The part entry/exit rate must be somewhat gradual to minimize drag-out.

In those limited global locations where halogenated solvents are still being used in prepaint parts cleaning, care must be taken to keep CFC vapors from being drawn into direct-fired ovens. When vapors of halogenated hydrocarbons come into contact with flames,

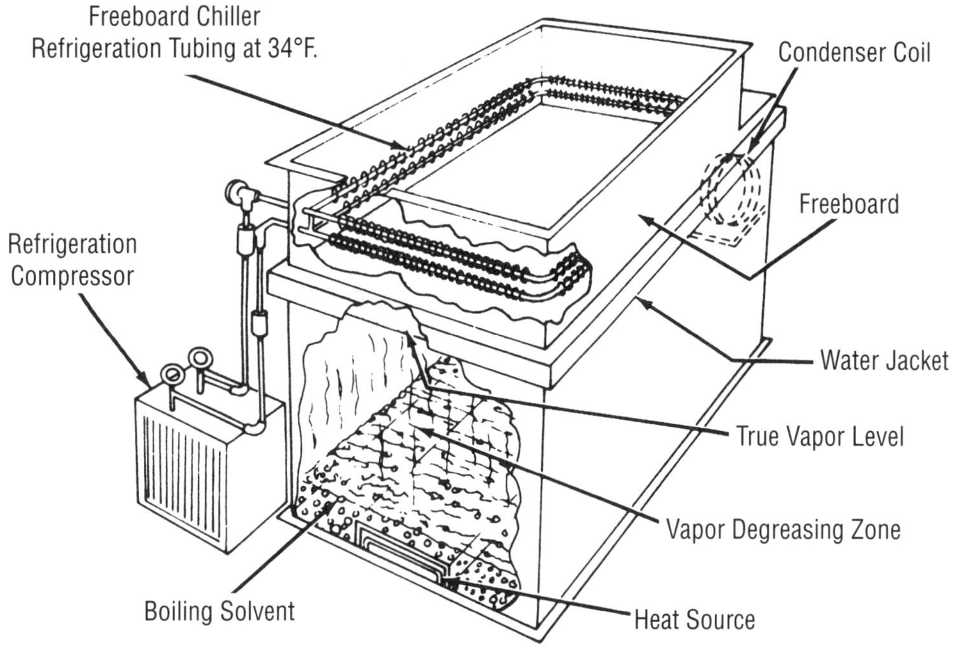

Figure 7-3. Cutaway view of a common type of vapor degreaser.

corrosive gases such as hydrofluoric and hydrochloric acid are formed. These acid vapors can wrinkle the finish on freshly painted parts in an oven; moreover, the acid vapors present a potential health hazard to persons who are sensitive to these irritants.

CFCs do not burn, nor will bacteria break them down, so they persist virtually forever in the environment. Because it is widely believed that halogenated hydrocarbon vapors are a major cause of the atmospheric ozone layer destruction at the earth's South and North Poles, chemical companies have searched for suitable replacement vapor-degreasing solvents that are less environmentally offensive. Fortunately, "drop in" substitute solvents or solvent blends have been developed and are being offered for sale. The return to vapor degreasing has been very slow because so many former users switched long ago to aqueous cleaning equipment. They are satisfied with it and don't intend to spend money to change a system that works well for them.

Yet CFCs are certainly not the only coating and cleaning solvents whose use has been virtually eliminated by a greater awareness of the hazards in using them. Many traditional chemical solvents are no longer acceptable for use, or they are allowed for use only in strictly controlled circumstances. Benzene was eliminated by worries over liver and kidney damage. A number of ketones and glycols were no longer used when it was found they may possibly cause nerve damage, muscle cell damage, or exacerbate certain leukemic cancer dangers. Carbon tetrachloride, for example, was once used in fire extinguishers and as a common cleaning agent. When it was found to be carcinogenic it was no longer acceptable for general industrial or household use.

Other solvents, in addition to the CFCs, are restricted because they are considered Hazardous Air Pollutants (HAPs) and more destructive to the environment than other restricted-use solvents. In the U.S., the Environmental Protection Agency issued final National Emission Standards for HAPs (NESHAP) that required conformance by the end of 1997. Typical paint and cleaning solvent HAPs include toluene, all three xylenes, methyl ethyl ketone (MEK), and methyl isobutyl ketone (MIBK). But HAPs are not limited to solvents—heavy metals such as lead and chromium are also on this list of nearly 200 dangerous materials.

Aqueous Cleaning

Aqueous cleaning systems—water plus detergent and frequently also small amounts of acid or alkali—are the most popular cleaning systems for industrial finishing. Three major advantages of these systems are the absence of solvent emissions, the ready adaptability to meet practically every cleaning requirement, and its safety for use on nearly every type of substrate except wood.

The basic types of aqueous cleaning are spray and dip (immersion). Both can be performed with parts conveyed on a hook or placed in a wire basket to hold them during the operation. Parts may be lowered into and out of the cleaning liquid or carried through the cleaning spray or dip zone. Spray cleaning can be done on parts carried on a mesh belt, or on large parts set on the shop floor. Spray cleaning is most commonly used and is faster because of the scrubbing action due to forceful impingement of cleaning solution. But the spray cannot turn corners to reach into hidden regions of complex part shapes. Dip cleaning is slower, but the solution, except in trapped air pockets, is able to reach into part areas not accessible by direct spray. To offset the lack of impingement and to increase cleaning rates, dip cleaner solutions are often higher in detergent concentration than spray cleaner solutions. They may require longer rinsing as a result. Operations having only a low parts volume commonly use a manual spray or manual dip; while high-volume manufacturing practically demands using automated spray or dip processes. Parts volume is usually the single most important factor in determining if a conveyorized cleaning system is needed. But the choices between manual or automatic, and whether to use spray, dip, or a combination of spray and dip, are affected as well by the nature of the soils on the parts and by the size(s) or shape(s) of the products.

Dip cleaning can be done on small parts by utilizing a wire basket that can be lowered into the cleaner tank, possibly with some mechanical agitation of the basket (preferred) or circulation of the cleaning solution for faster cleaning. Basket cleaning utilizing spray cabinets is possible also, but this is not commonly done.

Large, bulky, relatively low-volume items—such as construction cranes, for example—are often cleaned manually by spraying with aqueous detergent. Most high production items, no matter what their size, are cleaned automatically. Automobile bodies, for example, are automatically cleaned as they move through huge dip tanks at rates of 55–80 per hour. A manufacturer might choose dip cleaning for a complex-shaped product with deep, hard-to-reach areas. A very few manufacturers have tried cleaning systems that incorporate both spray and dip cleaning: impingement spray to clean accessible exterior surfaces plus dip cleaning so that solution can penetrate into hard-to-reach areas. Chrysler did this for years several decades back, but it was never as effective as hoped. Formulating a single cleaner solution that can work well as both a spray and a dip cleaner is not really feasible. Separate cleaning solutions for each method are far more efficient and effective.

Manual spray cleaning (often called wand cleaning) is usually accomplished with a spray nozzle attached at the end of a long wand as shown in **Figure 7-4**. Moderately high fluid pressures of 75–200 PSI (5.2–13.8 bar) are usually used. The wand permits the operator to reach into sizable cavities and yet stay a safe distance from the nozzle in order to avoid getting splashed excessively by hot detergent solution or by rinse water.

In contrast, automatic belt or conveyorized spray cleaning systems use lower fluid pressures and incorporate fixed plumbing components that include headers, risers, and nozzles. Headers, constructed of large-diameter (2–6 in., or 5–15 cm) piping, carry the cleaning solution to the risers, which use smaller diameter (1–2 in., or 2.5 to 5.0 cm) piping. The risers are usually arrayed somewhat U-shaped, containing straight horizontal and slightly curved or angled vertical members. The parts to be cleaned are hung on a conveyor that passes through the U-shaped spray. Nozzles attached to the risers—about 8–12 inches (20–30 cm) apart—provide a 360° zone of spray through which the part to be cleaned passes. Numerous consecutive riser sets can be spaced throughout the length of the cleaner zone, which often ranges from 6–20 feet (1.8–6.1 m) long. The line speed and the length of the spray zone affect cleaning time. Thus, for example, if 1.5 minutes of cleaning are required and the conveyor speed is 5 ft/min (1.52 m/min), the array of risers must extend for 7.5 ft (2.9 m).

The choice of nozzle type is dependent on the stage in which it is used and what chemi-

Figure 7-4. For wand cleaning and pretreatment, solutions are normally run to drain to reduce costs. If this is not feasible, special retention stations can be used to hold and recycle the solutions and rinses.

cals are being sprayed. Whirl-jet nozzles give a flood pattern with low impingement force but lots of fluid volume, while V-jet styles combine high volume and high surface impingement in a flat, wide spray. The K-jet and fan jet nozzles are similarly flat but result in less overspray and allow better containment of chemicals within the spray stage. They are low in both volume and impingement force.

Drainage zones following each of the stages also assist in chemical containment at each stage. The best drainage floor has a slightly off-center "∧" inverted V-shape with the apex located about 60–65% away from the exit end of the active spray zone. Solution tank bottoms should not be flat. Tank bottoms should slope to one end with a drain valve for easier clean out, and also to allow complete draining of the tank's contents. Ceilings can be sloped to both sides from the middle or, better, somewhat off center. It can also be totally sloped to one side. Sloped metal ceilings will have higher construction costs. For composite and stainless steel enclosures, a 15-degree ceiling slope is enough, but for mild steel ceilings the slope must be at least 30 degrees. This is because chemical growths will form on mild steel and act as drip points, with drippage possibly contaminating parts. Slotted top ceilings with the rail outside the oven and pretreatment machine offer a possible advantage in chain life but may allow more steam to escape. Opinions on whether or not they are actually cleaner are sharply divided.

Belt washers most often use a continuous metal mesh belt that travels through the sequential detergent and rinse spray zones. A dry-off oven zone may be included afterwards as well. Parts are normally placed on the belt manually in a random orientation; they can either be taken off at the other end by hand or allowed to fall from the belt into a collection bin.

Large, heavy, and high-volume parts are usually hung on an overhead conveyor and carried through the cleaning and rinsing risers, as shown in **Figure 7-5**. Various configurations of the conveyor and risers are possible to minimize moisture reaching the conveyor parts; moisture contributes to high conveyor maintenance and may allow drippage of conveyor soils onto the parts being cleaned. Cleaner spray and rinses should be directed away from the conveyor so as not to remove lubrication from the chain or trolley wheels.

The cleaner section is usually just the first stage of an automatic multistage machine, but just the same it is most often called a "spray washer." (The rinsing, pretreatment, and sealing that may be done in a spray washer will be covered later in Chapter 8.) In addition to the riser and nozzle sets, automatic spray washers commonly incorporate external reservoir tanks that contain the chemical solutions (see **Figure 7-6**) being sprayed onto the parts to be cleaned. Cleaning solution is held in a tank and a pump delivers it to the risers. The reservoir tanks provide a convenient means of heating the solutions and replenishing the chemical solutions as they become depleted with use. Flame tubes directly inside the main tanks are still used for heating, but separate heat exchangers are now recommended for most solutions because they cause less chemical breakdown, especially in the pretreatment stage.

Water from cleaning or rinsing tanks being dumped, as well as overflowing water rinses, frequently contain oils, detergents, or related chemicals, therefore release of this water to drain may be regulated. Some manufacturing plants are located in areas where absolutely zero process water discharge is allowed. Evaporation of used aqueous solutions can

Figure 7-5. Overhead conveyor carrying parts through an automatic spray washer.

Figure 7-6. Automatic spray washer boxlike protrusions (right) contain chemical solutions and rinses.

be utilized in such instances; it is now economically feasible for some plants at costs of US$0.05/gallon (US$0.05/3.785 l) or less. Naturally, wastewater treatment followed by disposal to drain is economically preferable if this is an available option.

Manufacturing plants are increasingly reusing their wastewater and purifying supply water with reverse osmosis (RO). The technology has advanced to such a degree that it has become cost competitive. As late as the end of the 20th century, RO was not much used, but now its use has soared. Reverse osmosis is able to take a contaminated aqueous solution and separate out most of the water in a fairly pure form. It accomplishes this by "filtering" the dirty solution through a semipermeable membrane filter at pressures of up to several hundred PSI (100 PSI equals 6.9 bar). The large contaminant molecules in the dirty solution cannot pass through the micropores of the membrane, only small molecules such as water pass through. This produces rather clean "RO water" and leaves behind a small volume of highly concentrated dirty solution on the other side of the membrane. The water can never be separated out in a 100% pure form by RO. Some ions and small molecules in the contaminated solution will pass through the membrane, but nearly all the larger molecules are kept back. The reverse osmosis (filtration) process is not unlike ultrafiltration used in electrodeposition coating (Chapter 10), although the pressures against the membranes are far higher in RO.

Acid Cleaning

Rust, scale, and oxides can be removed rapidly from metal before painting using an acid pickling (brightening) solution. Although the oxides react with acid faster than the metal itself, the pickling will also to some degree etch the surface of the metal. The acid may have added inhibitor in it to slow the etching action to a desired rate. The degree of inhibition used depends on the part being cleaned and the kind of soil to be removed. Both hydrochloric acid and sulfuric acid can be used for deoxidizing steel. These are not suitable for aluminum. Aluminum is usually cleaned with nitric acid solutions that are referred to as "brighteners" rather than deoxidizers.

Hydrogen embrittlement can occur readily in acid cleaning because hydrogen gas is produced even if inhibited pickling solutions are used. Hydrogen gas embrittlement is a somewhat complex metallurgical phenomenon that occurs when hydrogen is in contact with metals. Briefly, the tiny molecules of hydrogen penetrate into the crystal lattice formed by the much larger atoms of the metal, thereby making the metal atom array significantly more rigid. This makes the metal more brittle and susceptible to stress fracture. The embrittlement can lower desirable physical properties such as tensile strength and elastic modulus.

Other methods must be used for oxide removal if hydrogen embrittlement of the metal cannot be tolerated. Alkaline derusting of steel is done with very highly caustic (pH = 12–13) aqueous solutions to avoid hydrogen embrittlement. Solutions this high in alkalinity will react so vigorously as to be dangerous with more active metals such as zinc, magnesium, and aluminum. Therefore, alkaline derusting is limited to use on steel alloys only. Alkaline derusting is not common; abrasive derusting tends to be more commonly used since it poses no chemical disposal problems.

Alkaline Detergent Cleaning

Aqueous alkaline detergent solutions are used to clean all types of plastic and metal products. They are now the preferred cleaning agents since they can be formulated to contain little or no phosphate or other materials that are potentially harmful to the environment. On the plus side, the stronger the alkali, the faster the soil is removed and the more thoroughly the soil is dispersed in the cleaning solution. Better soil dispersion results in a longer, cleaner life. But too much alkalinity results in slow and incomplete rinsing (poor rinsability). This is an important consideration because thorough rinsing of alkaline cleaners is crucial for good painting. Enough alkalinity must be introduced into detergent formulations for good soil removal and dispersion, but not so much that rinsing will be long or difficult.

Cleaners may be formulated with various alkaline substances such as sodium hydroxide (caustic soda), sodium carbonate (soda ash), or sodium silicates as builders and may also contain additives such as pyrophosphates, orthophosphates, borates, and chelating agents. Additives are included in detergent formulations to soften water, reduce foaming, increase wetting, aid emulsification of soils, and reduce nubbing (the term used for localized over-etching of zinc alloys).

In addition to a caustic builder, the source of the alkalinity, a typical cleaner formulation contains a low-foaming synthetic detergent rather than soap. Occasionally, alcohols and glycol ether solvents are added to speed cleaning. As a rule, total cleaning effectiveness is increased with nonionic detergents, compared with the more frequently used anionic detergents. The nonionic detergents are more expensive, but they work well even at relatively low temperatures and are particularly low foaming. Since energy costs are sharply higher than they were several years ago, pretreatment stages that are able to operate well at low temperatures are especially advantageous.

Foaming can be a problem in the cleaner stage itself or in the subsequent rinse stage(s), depending upon the formulation of the cleaner surfactants. The *cloud point* of any cleaner is defined as the minimum solution temperature above which foaming is insignificant. A convenient way to restate this is that it is the temperature at which a cleaner (or rinse) solution becomes self-defoaming. For this reason the cleaner stage should be up to operating temperature before the line is started each morning. Turning on the tank heaters well before the shift begins is a prudent practice. If a cleaning stage using a surfactant with a cloud point of 112° F (44° C), for example, is operated before this temperature is reached or if it is allowed to cool down during use below this temperature, excess foam will arise. Since rinses are rarely heated, foam can become a nuisance in these stages as well. Sufficient overflow may avoid foaming in the rinse stage. Cleaner formulations that operate well at reduced temperature levels are preferred as long as they will remove the soils on the parts. Some particular soils, especially die lubricants containing stearic acid or stearate esters, are heat mobilized. They may be very difficult to remove unless higher temperature cleaning is utilized. In general it is least expensive to use the lowest temperature range cleaning that completely removes the particular soils associated with the parts being pretreated.

Emulsion Cleaners

Emulsion cleaners that have kerosene or similar organic solvents dispersed in them have been available for a long time. A few are so high in solvent content that they are flammable. Emulsion cleaners work well for removing some greases, but not for the removal of a wide range of soils and contaminants. They also tend to cost more than ordinary aqueous cleaners, and so they are not used much. Some of the citrus oil-containing cleaners have found good acceptance, however. They clean very well, and the pleasant smell of citrus is abundantly evident when these are used.

Cleaners for Aluminum

Aluminum can be cleaned with various aqueous chemical solutions. The configuration of the parts, the alloy composition, and the desired results strongly influence the cleaner selection. Chemicals used on aluminum include the following.

Silicated alkaline cleaners. Silicated alkaline cleaners, known also as nonetching cleaners, are used at a pH (acid/alkali rating) of 11 to 13 (see **Figure 7-7**) and can remove grease and waxy soils.

Nonsilicated alkaline cleaners. Nonsilicated alkaline cleaners operate at a pH of 8–10 and are mostly used prior to a caustic etch before chromating or anodizing. Nonsilicated cleaners are excellent products and present little problem of cleaning agents drying on parts. Although they are generally considered to be nonetching, nonsilicated cleaners are actually capable of microetching most aluminum alloys if they are operated too hot or at excessively high concentrations.

Organic acid cleaners. Organic acid cleaners use a high concentration of organic surface-active agents, called surfactants, to remove stains, streaks, and related blemishes. They are expensive, but long bath life may justify their use to take off soils not removed by alkaline cleaners.

Acid etching cleaners. Acid etching cleaners, containing phosphoric acid, hydrofluoric acid, wetting agent, and solvents, such as ethylene glycol or propylene glycol, are used to remove light soil and to prepare aluminum for chromating or direct painting. The acid provides a smooth micro-etched surface that is highly suitable for the formation of a chromate conversion coating.

Alkaline etching cleaners. Alkaline etching cleaners are used in hot, strong sodium hydroxide (caustic soda) solutions containing chelates and wetting agents. These cleaners produce a heavily etched surface that has a visible pattern after the black oxides (smut) produced by strong alkali are removed by an acid dip.

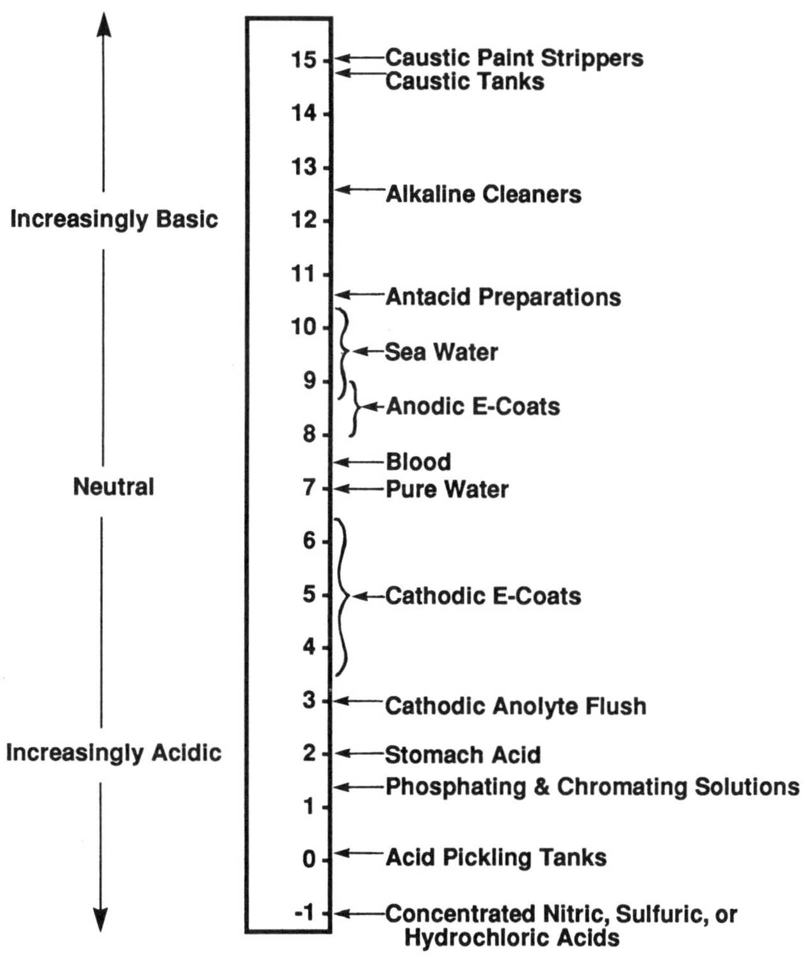

***Figure** 7-7.* pH scale measures alkalinity and acidity.

To ensure an acceptable operating environment, a blanket layer of detergent foam on parts to be cleaned is needed to suppress the alkaline fumes that result from the reaction between aluminum and the hydroxide. If the foam blanket becomes depleted due to oils from the parts, a small amount of high-foaming detergent can be added. Parts should be reasonably clean prior to etching in order to avoid the formation of blotches on the aluminum.

Other Cleaning Considerations

Immersion (dip) cleaning has an advantage over spray because the cleaner solution reaches all exterior surfaces. But remember, since it does not clean as fast as a spray process, it tends to require more concentrated cleaners. These may take longer to rinse and require appreciably more water.

Silicated cleaners can inhibit excessive alkali attack on aluminum and zinc when concentrated solutions must be used in order to clean heavily soiled parts. Although these immersion silicated solutions clean fairly fast, they may passivate (deactivate) the surface and thus cause slower chromating. Silicated cleaners in both spray and dip usage can produce unsightly streaking which in some cases is visible after painting.

Most aqueous cleaners are used at 125–160° F (52–71° C). Cleaners of all types, even so-called "cold cleaners," work better hot than cold; this is because oil and grease are always less viscous and thus easier to break up and remove when hot. Heating the cleaner and rinse stages offers another advantage: it raises the temperature of the metal to be compatible with the temperature of the conversion coating solution that usually follows cleaning (see Chapter 8). The first cleaning stage is normally performed at a high temperature to facilitate cleaning and heating of the metal, and the remaining stages are gradually lowered in temperature to that of the subsequent phosphate or chromate stage. Very few plants use heated rinses; most prefer to save money and use ambient temperature rinse water.

Rinsing After Cleaning

In addition to the items discussed so far, there are other aspects of the cleaning process that are important. Good rinsing after cleaning is crucial because alkali left from the cleaner will neutralize acidic conversion coating chemicals in phosphating and chromating. A portion of the phosphating chemicals is thus wasted because they react with the caustic materials to create insoluble sludge. Additional cost is incurred because the sludge must be separated and disposed. (**Figure 7-8** shows an itemized list of rinsing cost factors.) Poorly rinsed caustic cleaners can also interfere with subsequent zinc phosphating by coarsening the phosphate crystal size and promoting void areas where no protective phosphate coating is formed. Surfactant solutions always need to be rinsed thoroughly from surfaces to be painted. Alkaline solutions are often not easy to rinse, however, and so special care in rinsing must be taken to avoid a reduction in the quality of chromate and phosphate conversion coatings. Periodic checks on the purity of rinse water dripping from parts will establish how well parts are being rinsed free of alkalinity.

Cleaning can be improved using a double-wash system: either a "wash–rinse; wash–

Total Cost of a Rinsing Operation
Water
Drag-out Chemical
Waste Treatment Chemical
Waste Disposal
Waste Treatment Labor
Waste Treatment Equipment

Figure 7-8. Rinsing operation cost factors.

rinse," or a "wash–wash; rinse–rinse" sequence. In double-washing systems, the first wash removes the bulk of the soil, and the second removes any remaining soil. The first wash tank can be replenished by overflowing the second wash into the first. Because of the necessity for very thorough rinsing after cleaning, the "wash–wash; rinse–rinse" system is a better procedure and the one I recommend. Rinsing between the two cleaning stages is unnecessary; double rinsing after cleaning is clearly more effective than just a single rinse.

The water quantity, water quality (purity), and temperature all play a part in determining the efficiency of the rinse. For optimal results and efficiency of chemical use, the total dissolved solids content (TDS) of the water used in each of the aqueous stages, including rinse zones, should be below 300 parts per million (ppm). Chloride and sulfate anions (negatively charged ions) combined should never be higher than 500–600 ppm; the sum of the ammonium, sodium, and potassium cations (positively charged ions) should be below 250 ppm.

Rinses serve to flush away undesirable chemicals from parts that have been cleaned or conversion coated. These chemicals will slowly build up in the recirculating rinse water. Proper overflow of recirculating rinse tanks to drain is essential. This will keep rinse solutions low in chemical contamination from the inevitable drag-in from parts going through the various process stages. When parts are dry, any visible spots and residues found on the parts, especially likely along the bottom edges, are indications of poor and marginal rinsing.

Measuring the conductivity of the final drain-off rinse can provide a simple check of rinse water quality. Rinse water can be collected as it drips from the parts and then be measured for conductivity. A small portion of the drain-off can also be evaporated to see if any visible residue remains. High conductivity and noticeable residue from evaporated rinse water are signs that the rinse water quality probably needs to be improved. The supply water itself may be too high in dissolved solids, or the rinse water may require more frequent changing or a faster rate of overflow to drain.

When rinse water is dried, it leaves behind on the parts nearly all of the dissolved materials that may be present in the rinse water. For this reason the total dissolved solids (TDS) of rinse water should be minimal. In some paint systems the dissolved solids in the rinse water are held as low as 10 ppm or below 30 micromhos/cm conductivity.[1] For critical applications, several paint systems require RO or deionized water that has less than 2 ppm silica and conductivity below 10 micromhos/cm. For less sensitive coating systems, the dissolved solids can be as high as several hundred ppm without causing problems of poor adhesion or weak corrosion resistance. In order to correct for high conductivity in the rinse water, a brief deionized water (DI) or RO rinse may be required after the normal supply water rinse. A recirculating DI/RO followed by a halo or "mist" rinse of fresh DI/RO make-up water will provide superb rinsing. Supply water accomplishes the major rinsing of chemicals; then the supply water on parts is rinsed away by the far cleaner DI or RO water.

In these instances, a recirculating DI/RO rinse of about 5 cc (ml) needs to be supplied for each square foot of work surface. In power washers, 2–4 risers per side are generally adequate. Passing water through two rechargeable ion exchange beds produces deionized water. The first ion exchange bed removes all cations such as sodium, calcium, magnesium, and iron; the second bed takes out all anions such as chloride, sulfate, nitrate, and phosphate. The cation and anion replacements are hydrogen and hydroxide ions. The hydrogen (H^+) and hydroxide (OH^-) ions do not exist for long; they immediately react with each other to form H-O-H (H_2O) molecules; in other words, they form water (see **Figure 7-9**). Thus, pure water without ions is generated, and hence the name **deionized** water.

Wastewater Treatment

Plants must determine and record the type of contaminants present, their concentration, and water quantities for all their effluent water streams. While a detailed coverage of

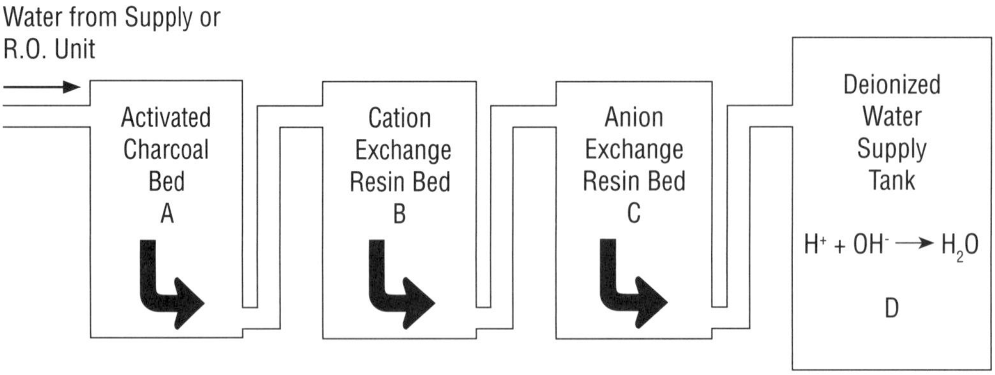

Figure 7-9. Deionized water generation. Bed A removes organic impurities. Bed B replaces iron, calcium, sodium, magnesium, and other positive ions with hydrogen ions (H^+). Bed C replaces carbonate, hydrogen carbonate, chloride, sulfate, nitrate, and other negative ions with hydroxide ions (OH^-). Tank D contains purified water, stripped of ions, and stored for usage.

[1] Conductivity is the measure of the conductance of water to an electric current, and it is usually reported at micromhos/cm, which may also be expressed as µmhos/cm. In most cases, the total dissolved solids content of water is approximately 70% of the conductivity in micromhos/cm.

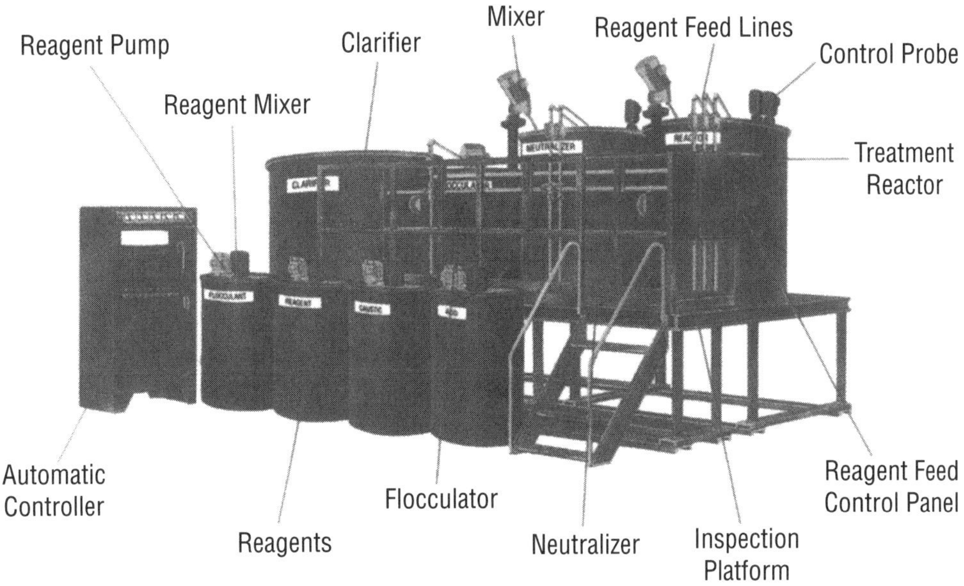

Figure 7-10. Wastewater treatment system.

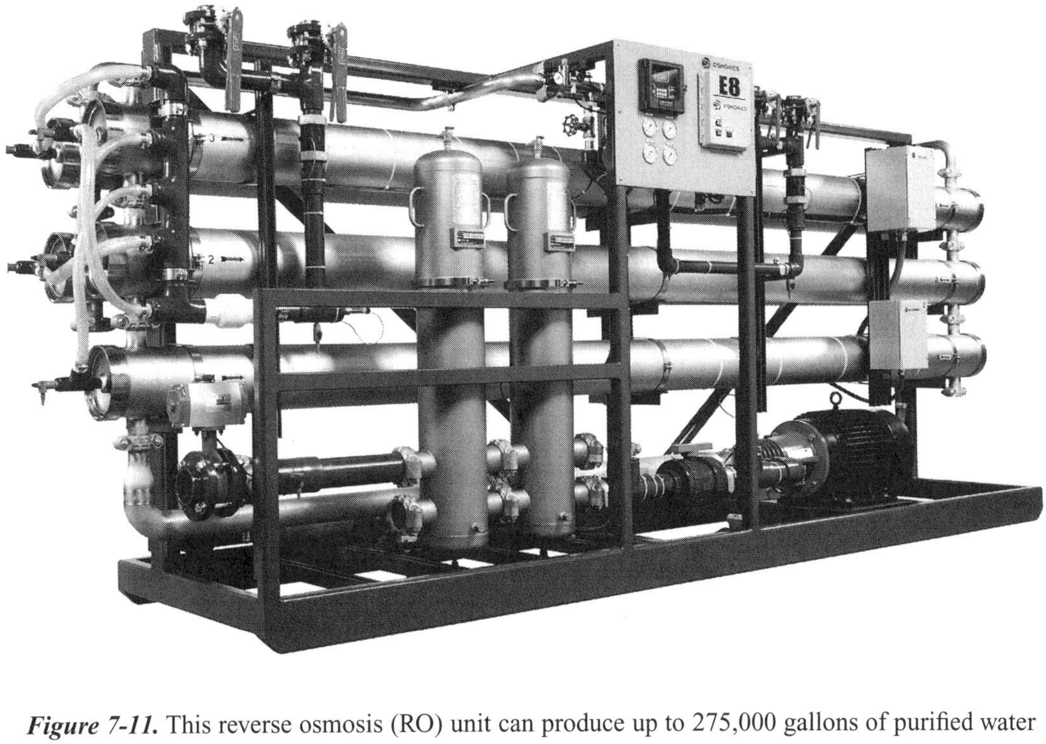

Figure 7-11. This reverse osmosis (RO) unit can produce up to 275,000 gallons of purified water each day.

wastewater treatment is not appropriate here, some typical materials that must be removed before release are heavy metals, suspended solids, soluble and insoluble oils, phosphates, acids, and alkalis. Most wastewater treatments involve flocculation, settling, skimming, precipitation, filtering, and pH adjustment (see **Figure 7-10**). The maximum allowable contamination concentrations in wastewater released to a stream or to a municipal treatment plant can vary widely from location to location. In many instances, plants have chosen to reuse the water after treatment rather than release it to drain. This need not be an added expense to manufacturing operations. More and more plants are finding that wastewater reuse and water-to-drain minimization programs can actually be cost effective. Reverse osmosis (RO) is often an important part of these water reuse programs.

The strong environmental emphasis on reducing water pollution makes the proper management of cleaning solutions and rinses a vital step. Wastewater streams are being monitored more and more by various governmental agencies. Private "watchdog" organizations are also alert to improper waste disposal. If pollution is suspected, non-governmental environmental organizations may get involved to force action through public opinion.

Chapter 8

Conversion Coatings

Conversion Coating Qualifications

The necessity for cleanliness on a surface to be painted can hardly be overemphasized; however, a clean surface may not be sufficient to achieve good paint adhesion. Whether or not additional mechanical or chemical surface treatment is required depends principally on the substrate to be painted. The properties of the paint may also come into play in some instances. In the case of most wood items, a clean and smooth surface is normally all that is required. A natural wood surface is unique because its polar cellulosic composition will provide an excellent adhesion base for nearly all stains and paints.

While clean, smooth wood requires no chemical treatment before painting, the same is only sometimes true for clean, smooth plastic surfaces. In all cases a good bond should be formed between the plastic substrate and the paint film, and both the plastic composition and the paint type determine that bond strength. It is not often possible, but some manufacturers find it advantageous to paint plastic parts within minutes after the parts come out of the molding machine. By painting so quickly, before the plastic parts have a chance to become soiled, manufacturers save washing and drying steps. If painting parts this rapidly is not possible, many plastics can simply be cleaned and then painted. Others, primarily the low-polarity plastics, such as polyethylene and polypropylene, even when clean, do not form good adhesion to paints. These plastics then require an additional treatment to provide a mechanically or chemically modified surface on which paint can find good adhesion.

To improve paint adhesion, surfaces of low-polarity plastic parts can be chemically oxidized. The oxidation can be accomplished by electrostatic corona discharge, by ultraviolet light-initiated decomposition of sprayed or dipped-on activators such as organic peroxides, by controlled direct gas flame treatment, or by special oxidization-causing paints. These oxidative treatments introduce carboxyl (COOH), carbonyl (CO), and hydroxyl (OH) groups into the outermost polymer molecules on the plastic surface. The presence of these groups, which are oxidized forms of the plastic resin itself, can be verified by reflective infrared spectrophotometry. The effect of flame treating lasts only about 10–20 minutes. It is suitable for relatively flat parts or for manual treatment on a limited number of pieces. The corona discharge activation is effective for up to one hour or thereabouts, while the chemical activator method plus UV light allows painting for a number of hours afterwards. The latter method may even be effective as long as 24 hours later, although a shorter interval is obviously advised.

As with plastics, a limited variety of metal parts are also painted immediately after surface cleaning. These metal items typically do not have strict requirements for adhesion or corrosion resistance. Low-cost products that are expected to have a short life cycle, for example, fall into this category. In other cases, you may be surprised to learn, pretreatment other than cleaning may not be used even though the corrosion resistance requirements are high. Items that are quite large, or that may be inconveniently situated for being subjected to chemical pretreatment, are often painted after being cleaned only. Marine vessels, construction equipment, or highway bridges, for example, are frequently sand blasted and then painted soon thereafter. While chemical pretreatment would not, strictly speaking, be impossible in these instances, it would be very awkward and expensive to administer. By using special high-performance coatings, the lack of pretreatment can be overcome so that both excellent adhesion and good finish durability are consistently attained.

Conversion Coatings for Metal

Most metal products that are to be painted receive an insoluble salt-like *conversion coating* after cleaning. The metal on the surface is chemically converted into this insoluble inorganic coating. A conversion coating is a surface layer that helps shield the metal from air, moisture, industrial gases, and the like, thus increasing corrosion resistance and simultaneously providing a roughened surface to improve paint adhesion. The conversion coating has micro-porosity and micro-roughness in the form of pores, fissures, and undercuts. These form an excellent stratum into which the paint can flow and create "fingers" for a good hold onto the surface. In addition, many conversion processes leave a slightly acidic residual pH on the surface, which is more receptive to good paint adhesion than an alkaline pH would be. Alkaline solutions can degrade paint resins, and thus they find use in paint strippers. Generally, alkalinity on surfaces to be painted is undesirable, but there are exceptions for some very mild sealer materials used to rinse conversion coatings.

The thermal expansion and contraction buffer provided by the conversion coating between the metal and the paint enhances long-term paint adhesion. The expansion and contraction coefficient of the paint film is many times that of the underlying metal, while the conversion coating is intermediate in its expansion and contraction characteristics. When sudden hot and cold changes are experienced by the painted metal part, the severe difference in the expansion rates between the paint and the metal would create intense strain on the bond between paint and metal, possibly causing paint cracking. With the conversion coating between them, the distortional strain is greatly lessened.

Around the edges of gouges and scratches that have gone through the paint down to bare metal, conversion coatings sharply reduce the tendency for lateral creep of underfilm corrosion. When bare metal is thus exposed, it will oxidize (rust) from air and moisture contact. A byproduct of this oxidation process is hydroxide ion (OH^-), an alkaline agent that acts to degrade paint films and allow corrosion to creep back under the paint along the edges of the scratch line. The conversion coating, being alkali-resistant, inhibits the undercutting, and the spread of metal oxidation away from the scratch is slowed significantly.

Because their surface profile varies considerably, conversion coating thickness is usually not measured, but rather the amount of conversion coating is given in weight per unit area. Coating weights for a given product are often specified to be within a certain range for optimum cost versus adhesion and corrosion resistance performance. In the United States, the reading is traditionally in mg/ft^2, but multiplying that value by 10.76 gives weight in metric units of mg/m^2. Conversion coatings for metal surfaces to be painted may vary from a few mg/ft^2 to as much as 500 mg/ft^2 (5,380 mg/m^2). Even higher weights are used for some nonpaint related conversion coatings such as for metal lubricant holding coatings. Chemical products used to produce conversion coatings can be formulated by chemical suppliers specifically for a particular type of metal and to yield coatings with certain distinct properties. Thus, conversion coating of the same general type may differ in chemical composition and anticorrosion properties. For example, at least two dozen different iron phosphating formulations are available, each differing in some way from others in this same family. A lot depends on what metal or variety of metals is used in the plant, and how much corrosion resistance is required for the product.

Metals used most commonly in manufacturing and of primary interest in industrial painting are hot-rolled steel, cold-rolled steel, stainless steel, clad and metal-coated steel, hot dipped galvanized steel, electro-galvanized steel, galvannealed steel, galvalume steel, diecast zinc, and sheet or cast aluminum. Passing heated steel through a bath of molten zinc forms hot-dipped galvanizing. Air knives (steam may also be utilized) directed at the sides of the steel control the adherent zinc layer thickness. Hot-dipped galvanneal results from a postgalvanizing annealing process that produces an iron-zinc alloy surface composition instead of pure zinc only. Galvalume is a hot-dipped steel that has been given an aluminum-zinc coating containing more than 50% aluminum. It has better heat reflectivity but is considerably less ductile than hot-dipped galvanized steel. This is due to a thicker alloy layer formed between the substrate steel and the coating bath, primarily as a result of the elevated bath temperature and the increased amount of aluminum present. The smoothest and most uniform zinc layer is deposited by electroplating zinc onto steel, forming what is called electro-galvanized steel.

Conversion Coatings for Steel

Conversion coatings for steel are usually either iron phosphates or zinc phosphates. Somewhat different chemicals are used in iron phosphating than in zinc phosphating, but the overall processes are similar. They both include use of a dilute aqueous solution of phosphoric acid that contains a soluble dihydrogen phosphate salt. The chemicals react with the metal to oxidize its surface atoms, and convert them into insoluble metal salts onto the metal surface. The mixture of insoluble salts is what comprises the conversion coating.

An important prerequisite for applying a phosphate coating is a thoroughly clean metal surface. Cleaning may be done as a preparatory step, but in some iron phosphating operations, cleaning and phosphate coating deposition are accomplished simultaneously with the same solution. If a separate cleaning stage is used, all traces of the cleaning solution must be rinsed off the metal. Unrinsed alkaline cleaner is particularly undesirable since it will create excess sludge and reduce the quality of the phosphate conversion coating.

During the phosphating process, phosphoric acid attacks the metal surface, oxidizing iron atoms and solubilizing them into the coating bath. This simultaneously lowers the phosphating solution acid strength at the interface between the metal and the phosphating solution. In areas on the metal surface where the acid strength has been lowered, the type 1 (primary phosphate salt) converts to type 2 (a monohydrogen phosphate salt) and then immediately to a type 3 (tertiary) phosphate. Although it is not necessary to understand the complete chemistry of this process (somewhat simplified here), it is beneficial to know the principal chemical steps that take place as explained in the previous sentences.

$$H_2PO_4^{-1} \rightarrow HPO_4^{-2} \rightarrow PO_4^{-3}$$

$$\text{Type 1} \rightarrow \text{Type 2} \rightarrow \text{Type 3}$$

In iron phosphating, type 3 ferrous and ferric iron phosphate, $FePO_4$ and $Fe_3(PO_4)_2$, are formed; in zinc phosphating, type 3 zinc phosphate, or $Zn_3(PO_4)$, is formed. Both type 3 iron and zinc phosphate salts are insoluble. During phosphating they crystallize rapidly out of the phosphate solution, bonding tightly to the metal as they grow onto and into the surface irregularities and etch pits created by the acid attack.

Additional chemicals such as accelerators, additives, and oxidizers are often present in the phosphating solution. They include nitrites, nitrobenzene sulfonates, chlorates and bromates, fluorides, manganates, nickel ion, cobalt ion, and others. Accelerators are added to keep the many chemical reactions occurring at an acceptable speed. Additives are utilized for such purposes as reducing sludge formation and preventing "seedy" deposits in the phosphate layer, particularly in zinc phosphate coatings. More recently there is interest in using solutions containing no ions of heavy metals such as molybdenum, cobalt, nickel, titanium, and others to minimize potential water pollution.

When ferrous metals are phosphated, the oxidizers convert the iron ion dissolved by the attack of the phosphoric acid, from the +2 (ferrous iron) to the +3 (ferric iron) oxidation state. At the same time, oxidizers also prevent the formation of large hydrogen bubbles on the surface of the metal. Hydrogen is formed by the reaction of acid with metals. So that hydrogen gas bubbles do not block the access of the acidic phosphate solution to the metal, oxidizing agents are added to the formulation to quickly convert any hydrogen that is generated into hydrogen oxide, or more simply H_2O, water molecules.

Most iron phosphates and zinc phosphates have a clear to a dull gray appearance; however, the addition of molybdic acid or a molybdate salt added as a corrosion inhibitor will give a distinctly purple cast to some iron phosphate coatings. These types of iron phosphate have been used in the past when no postphosphate sealer rinse was included in the conversion coating process.

Process time, temperature, and chemical concentrations are the three basic factors that affect the chemical reaction rate of a phosphate process. The process time is frequently dictated by the speed at which the conveyor line must run. Temperature and concentration of the solutions are the conditions that can be regulated and should be monitored during phosphating. Some zinc phosphating systems require extremely tight temperature control—far more so than for iron phosphate. This is done to create extra-fine zinc phosphate crystals, which have been shown to provide superior adhesion and corrosion performance compared to larger-size crystals.

The temperature of the parts and of the phosphating solution can in some instances dramatically affect the speed of conversion coating deposition. This in turn may alter the coating composition. If the temperature of the preceding stage—such as a rinse stage—is lower than that of the phosphate stage, problems could develop that would eventually diminish phosphate quality and increase solution maintenance. With low cleaner and rinse stage temperatures, the heating burden shifts to the phosphate stage. The increased demand for heat in the phosphate stage results in formation of excessive amounts of sludge, and this tends to plug risers and spray nozzles. It will also reduce heating efficiency by building scale inside the heat exchanger, and may drop the temperature of the phosphating bath below specifications. For the phosphate to coat properly under these conditions, a high level of accelerator is required, which will generate even more sludge. This is not a huge worry for many phosphating operations, but they must at least be aware such situations can indeed arise.

Increasingly, we see the use of plastic headers, risers, and nozzles in cleaning and phosphating spray systems. Not only are these lighter in weight and faster to remove and install, they are considerably less prone to collect sludge deposits (insoluble iron phosphate compounds), which often plug metal pipe risers and spray tips. Plumbing connections are simpler as well. Plastics are not immune to scale; the system must still be checked periodically for build-up, especially in the spray nozzles.

Iron Phosphating

Iron phosphate coatings are used mainly on steel alloys, but also to some extent on zinc and aluminum. Coating weights on zinc and aluminum are always going to be low compared to steel. Coatings formed on steel alloys are sometimes called *amorphous* (or noncrystalline) iron phosphate. In actual fact, they are finely crystalline and therefore are not truly amorphous. These coatings are mostly a mixture of iron (ferrous and ferric forms) oxides and iron phosphates. Some studies reveal that normally about two-thirds of the "iron phosphate" coating is actually comprised of iron oxides, while only about 35% is mixed ferrous and ferric phosphates. Despite that fact, the name "iron phosphate" has been and will continue to be used for these conversion coatings. The typical industrial coating weight of iron phosphate ranges from 15–60 mg/ft^2 (160–645 mg/m^2). Compared to zinc phosphate coatings, iron phosphate provides only moderate corrosion resistance. Iron phosphate systems do have a number of desirable characteristics, however, including relatively low operating costs, simple equipment requirements, low sludge formation, and moderate parameter control requirements. The main variables to regulate and monitor are the phosphating time requirements, solution temperatures, and solution concentrations. Iron phosphating can be done with ambient (room) temperature solutions; but for better quality phosphate, most plants operate at between 120° F and 140° F (49° C and 60° C).

The coating color may be, and usually is, varied across the parts. But this does not necessarily denote any process problems, coating thickness differences, or other inadequacies in the phosphate coating. Iron phosphate color varies generally with thickness and/or coating weight. Typical colors are correlated to phosphate coating weights in **Figure 8-1**; but the specific phosphating solution formulation, the operating parameters, and also the particular metal substrate alloy will possibly affect iron phosphate color.

Iron phosphating and cleaning are often done in the same bath, for it is possible to clean and iron phosphate simultaneously in a single solution that contains a suitable cleaner

Iron Phosphate Weight	Approximate Color
10–15 mg/ft^2 (105–160 mg/m^2)	Colorless to faint gray or blue-gray
15–30 mg/ft^2 (160–320 mg/m^2)	Blue-gray or gray
30–50 mg/ft^2 (320–535 mg/m^2)	Blue or gray plus slight gold tone
50–80 mg/ft^2 (535–860 mg/m^2)	Blue or gray plus light red-gold tone
80–125 mg/ft^2 (860–1,345 mg/m^2)	Blue-purple plus red-gold tone

Figure 8-1. Phosphate weight/color chart.

and phosphating chemicals. Such a combined solution, however, almost always provides inferior cleaning and phosphating compared to when these two steps are done separately. One reason is that phosphating solutions must be acidic to etch metal and thus deposit a coating. If a cleaning agent is combined with phosphating chemicals, the cleaner must be an acid-type cleaner. Acid cleaners, unfortunately, are less able to remove most soils than are alkaline cleaners. On the plus side, soils do not generate nearly as much foam in acid solutions as they do in alkaline cleaners. Acid cleaners remove smut and minor traces of metal oxidation (red rust and "white rust") better and rinse off more easily than alkaline cleaners. The combination "cleaner-plus-phosphate" solutions are usually satisfactory only if the parts have fairly light soil levels and no tenacious contaminants. Instances of tough-to-clean soils are very common, however, and for these applications combined cleaning and iron phosphating is not recommended. But remember, it can be a totally satisfactory way to accomplish cleaning and conversion coating in just a single treatment stage. This obviously can save money because a smaller pretreatment machine is required, and operating cost is reduced with two fewer stages and the elimination of the rinse stage between cleaning and phosphating.

Be careful. If one is ever in doubt about which system to install, it is definitely advisable to install separate cleaning and iron phosphating stages with an intervening water rinse stage. This could later prove to be a prudent measure. The extra capital cost for two added stages is not excessive if installed while the original process machine is built. More than a few plants that have installed combination cleaner/phosphate systems have later contacted me for help after they start receiving paint adhesion complaints from angry customers. Combination cleaner plus phosphate baths may work well for a given set of products, but later can prove to be inadequate when new production items are introduced, or when new manufacturing steps are added. Only light soils may be present on products for which the combination cleaner/phosphate bath was first designed, but subsequent changes (such as a different alloy, a new metal supplier, a different manufacturing process which may require heavier oils) can result in parts not fully cleanable in a combination bath. Then expensive retrofitting must be done to add a separate cleaning stage plus another rinse stage, which often will cost more than the original pretreatment equipment. Another problem often arises in that although two extra stages are needed, there isn't enough floor space available to add them without moving a lot of machinery. Separate cleaning and phosphating stages are almost always more desirable because of the flexibility they afford. It is foolish to install a combined cleaner/phosphate system when maximum corrosion protection is needed for the products. On the other hand, this does not mean combination systems should never be considered. They are still very common; they can perform satisfactorily and do indeed work well, *under the proper circumstances*.

One situation where combination cleaner/phosphate baths are practical is manual spray wand pretreatment, where products are too large or too few in numbers to be practical for conveyorized pretreatment systems. Heavy industrial machinery and large construction equipment, for example, are pretreated this way. In manual spray wand cleaning and conversion coating, a steam generator provides the necessary heat and motive power to a combined cleaner and iron phosphate solution. The solution is aspiration-fed into the hot water/steam spray. The chemicals are rather low in cost and are nontoxic so they are often just allowed to run into the drain rather than being reused, although equipment is sold for collecting and recycling the solution. The manual spray wand is used for water rinses as well, however, iron phosphate/cleaner formulations that require no rinsing are also available. In actual practice, postphosphate sealers are almost never used in wand phosphating; yet with the availability of nontoxic sealer rinses, there is no reason other than economics for not using them if added corrosion resistance is needed.

Solutions that form dense iron oxides—some with coloring agents to make iron and steel parts blue, brown, or black—have found only infrequent use as prepaint treatments. They may be coated or used as the only protection on metals but are not sold or recommended for use under paints. By themselves, the oxides offer extremely limited rust protection. Most often these colored oxides are coated with drying oils or waxes to provide a low-cost rust preventative, which is a function they do well under mild nonabrasive indoor conditions. They are not intended for severe service or for outdoor situations, nor are they considered to be decorative coatings. An exception may be "bluing" on firearms to reduce light reflection. In other niche applications, metal darkening does find use as an aid to holding heavy oil and wax lubricants, such as used on small hand tools and adjustable automobile seat rails.

Zinc Phosphating

Zinc phosphate comes in a large number of formulations. Zinc phosphate coating weights on steel parts to be painted are typically in the range of 100–350 mg/ft^2 (1,076–3,764 mg/m^2). (For other purposes such as adhesive bonding, the coating weights may be much higher. Even 2,000–2,500 mg/ft^2, or 21,500–26,900 mg/m^2, is not uncommon.) Little increase in paint adhesion or corrosion resistance is gained from coating weights heavier than 400–500 mg/ft^2 (4,200–5,380 mg/m^2). In fact, heavier coatings may become somewhat friable and thereby cause a slight decrease in paint adhesion.

The formation of a fine, dense crystal pattern of zinc phosphate (see **Figure 8-2**) noticeably improves corrosion resistance and paint adhesion compared with coarse zinc phosphate crystals (see **Figure 8-3**). One of the most common ways to produce a fine, dense crystal pattern is to add crystal grain refining agents, such as titanium phosphate, either to the zinc phosphating solution or preferably to the rinse preceding the zinc phosphating stage. These *titanated rinses* neutralize trapped cleaning agents, orient the crystal growth in a configuration more closely parallel to the surface, and, most importantly, initiate the growth of finer, denser zinc phosphate crystals. Phosphate crystal growth under ordinary conditions occurs only at metal grain boundaries, but titanium ions act as numerous additional nucleation sites for growth of zinc phosphate crystals. This nucleation occurs wherever minute amounts of titanium compounds have been adsorbed onto the steel surface during the rinse stage immediately preceding phosphating. The end result is many more phosphate crystals but all of them much smaller in size.

Pure zinc phosphate, $Zn_3(PO_4)_2$, is called hopeite, but this is not the preferred chemical form of zinc phosphate. Incorporating ferrous iron ions (Fe+2) into zinc phosphate coatings as they are forming will produce an iron-rich zinc phosphate known as phosphophyllite (pronounced fahs-FAHF-oh-lite, and not fahs-foh-FILL-ite), $Zn_3Fe(PO_4)_2$, which is more alkali resistant than pure zinc phosphate. Increased alkali resistance diminishes the rate of creepback from areas where corrosion has already begun at a paint scratch or a stone chip. The strong tensile properties of phosphophyllite also result in reduced crystal fracture when paint films are under mechanical or thermal stress. Similarly, manganese ions and nickel ions incorporated into zinc phosphate coatings also significantly increase the corrosion resistance. Compounds of these metals are usually included (individually and jointly) in immersion phosphate formulations intended for automobile body sheet metal pretreatment and other metal parts requiring strong corrosion resistance.

Although zinc phosphating baths can be operated effectively at temperatures as low as 85° F (30° C) and as high as 180° F (82° C), most are maintained at some point between

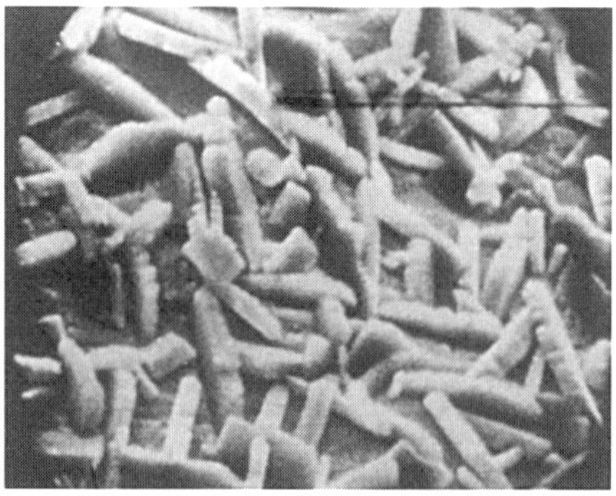

Figure 8-2. Fine, dense crystal pattern of zinc phosphate.

Figure 8-3. Coarse zinc phosphate crystals.

120–140° F (49–60° C). The lower temperature zinc phosphate systems work fairly well, but they require extra clean parts for good operation. If a temperature of 135° F (57° C) is selected, let us say, then the operating temperature should be held within ±2 to 3° F (±0.6 to 1° C) of this value. By operating the phosphate bath within the midrange temperatures, chemical concentration control becomes somewhat less critical than it would be using higher or lower operating temperatures. Because the concentration of phosphating solutions must be maintained within a far narrower range when the operating temperatures are lower or higher than midrange, more frequent titration to maintain proper chemical levels becomes absolutely necessary to maintain good phosphate coating quality. As you can see, it is generally easier to operate somewhere in the 120–140° F range. The criticality of these variables varies, however, from one formulation to another. So for some plants it may be preferable to run the phosphate at 160° F (71° C), or some other suitable temperature.

In contrast to iron phosphating, combination cleaning and zinc phosphating is not technologically feasible. Because of the chemistry involved, it is simply impossible to do a good job of both cleaning and zinc phosphating by utilizing a single solution. Zinc phosphating reactions are disrupted by the presence of both cleaning agents and soils; separate stages for cleaning and rinsing are always required prior to zinc phosphating. It is rarely done (but it is indeed possible) to use manual wand phosphatizing to produce zinc phosphate conversion coatings. The zinc phosphate quality is not very good when done this way because temperatures cannot be readily controlled. This method of zinc phosphating was used years ago on fieldwork when repainting large immobile structures, such as bridges and towers or large refinery equipment. The concern over heavy metals in wastewater now all but eliminates use of zinc phosphating in such cases. Where the solution and rinse runoff can be captured for metal ion and phosphate removal in a wastewater treatment process, it would still be possible, but in all likelihood would also be economically prohibitive.

Conversion Coatings for Zinc

Many modern finishing systems regularly clean, conversion coat, and paint parts with zinc surfaces. Zinc phosphate is the best conversion coating for iron and all steel alloys; it is also the best for zinc and zinc alloys. This is true for zinc itself and for zinc-coated metals such as galvanized steel. Because zinc reacts rapidly with zinc phosphate solutions, other metal ions should be added to slow the rate of coating deposition and to limit the coating weight. Nickel and ferric iron (Fe^{+3}) are widely used for this purpose. In contrast, some hot-dipped galvanized surfaces containing aluminum react slowly and are difficult to phosphate. Fluoride ion activation can be used to initiate the coating reactions. Fluoride ion added to an acid solution forms hydrogen fluoride, which strips aluminum oxide off the surface and precipitates the dissolved aluminum ions that otherwise would tend to "poison" the phosphating solution (i.e., disrupt the chemical reactions in the bath). The fluoride and aluminum ions combine to form insoluble aluminum fluoride, AlF_3, which separates out as part of the sludge.

A typical zinc phosphate coating formulated for zinc can also be used for steel. Such formulations generally use nitrite ion—typically added as sodium nitrite—both to accelerate phosphate coating formation and to control the iron content in solution. As an oxidizing agent, nitrite readily oxidizes ferrous iron ions to ferric ions.

$$\text{ferrous ion, } Fe^{+2} \rightarrow \text{ ferric ion, } Fe^{+3}$$

With hot-dipped galvanized steels, a nickel- or cobalt-activated bath can cause white pinpoint deposits or "seeds" that can sometimes be seen through the paint. These appearance defects are eliminated if fluoride ion is also present in the bath. Oxalic acid $(COOH)_2$ systems have been used, although in practice they are now extremely rare. Normally a single zinc phosphate formulation will be designed to work better on one type metal than another. So one formulation might be better for zinc and another formulation might be better for steel. When parts having these two metals need to be run on the same phosphate line, some compromise in phosphate formulation will become necessary. Stainless steel is rarely phosphated; however, when it is, it is treated somewhat like zinc. As with aluminum, the phosphate solution should contain fluoride ion.

Suppose a plant intends to zinc phosphate some zinc parts and also some steel parts on their pretreatment line. In this case it is preferable to mix the various metal parts together on the conveyor, rather than run a block of steel parts for a several hours, and then later run a block of zinc parts. Mixing the parts keeps the bath metal ion concentrations more balanced than would occur when large blocks of an individual metal substrate are run. The larger the single metal blocks, the more out of balance the phosphate solution will tend to drift.

The majority of iron phosphating formulations do not produce any appreciable amounts of coating on zinc, whether electrogalvanized steel, galvanneal steel, hot-dipped galvanized steel, or die-cast zinc. However, iron phosphating will sometimes remove thin layers of zinc oxide (white rust) from the surface and in that way improve paint adhesion. That may be good enough for some products. If strong corrosion resistance is not essential, iron phosphating may be completely satisfactory as a prepaint treatment on zinc. There are iron phosphate formulations, however, that can actually produce crystalline deposits on zinc substrates. The typical coating weights are low, only in the 5–20 mg/ft^2 (50–215 mg/m^2) range for such systems, which is not a problem since coating weights greater than these are neither readily achievable nor desirable.

Conversion Coatings for Aluminum

When aluminum is cut or abraded it very quickly reacts with oxygen in air to form a weak oxide layer that can fail cohesively. The oxide therefore provides only poor to fair adhesion for paint. For this reason aluminum should be conversion coated prior to painting unless it will be exposed only to mild service conditions. Like any other metal surface, aluminum needs to be cleaned completely of all soils and oxides before being conversion coated. Aluminum should be cleaned carefully and properly to avoid forming black aluminum oxides that are called "smut." Smut is formed easily when strongly alkaline cleaning solutions are used. When concentrated alkaline cleaners are unavoidable due to tenacious soils, the smut can be readily removed by dipping or spraying parts with a mild nitric acid brightener solution. Dipping is preferred for safety reasons since spraying acid is always potentially more hazardous. The acid brightener converts insoluble black aluminum oxides to soluble aluminum nitrate, which is colorless in solution, and can be easily rinsed away.

$$Al_2O_3 + 6\,HNO_3 \rightarrow 2\,Al(NO_3)_3 + 3\,H_2O$$

aluminum oxide + nitric acid → aluminum nitrate + water

The proper conversion coating type for a particular aluminum product depends in large measure on the amount of corrosion resistance desired. In the past, the absolute best way to obtain maximum corrosion resistance was with chromium-containing pretreatments. Although this situation is changing as plants desire to avoid the toxicity associated with chromium compounds, far more chromium-containing conversion coatings are still being utilized for aluminum than those without chromium. This is especially true whenever corrosion resistance is a top requirement. This pretreatment situation has until now been changing very slowly, but it will inevitably continue to shift toward completely chromium-free and toxic component-free materials.

Selection of the most appropriate conversion coating for a particular aluminum product also depends on whether the aluminum is being pretreated alone or along with other metals. When only aluminum is involved, chromating is usually the method of choice because chromate and chromium phosphate pretreatments are superior to phosphating. If zinc, or zinc and steel, are processed along with aluminum, then a phosphate pretreatment would be more suitable, depending on the relative percentages of each metal to be pretreated and painted.

Chromate is the real workhorse of aluminum conversion coatings. It includes a number of similar chromate deposition processes that are known generically as *chrome oxide* or *amorphous chromate*. The trade names of two proprietary chromating systems have also become practically synonymous with chromating: Alodine® from Henkel and Iridite® from Allied-Kelite. (This is so pervasive that more than one person has told me, "At our plant we alodine with iridite.") Chromate does indeed provide outstanding paint adhesion and corrosion resistance, especially when sealed afterwards with a chromic acid rinse. Chromate processes are most frequently used for aluminum pretreatment, but can also be used to provide corrosion resistance and paint adhesion on cadmium, copper, magnesium, zinc, or titanium. Chromate can be used on iron and steel too, but it is less effective than iron phosphate or zinc phosphate.

Chromate coating weights normally range from 10–80 mg/ft^2 (107–869 mg/m^2); most plants apply weights of 25–35 mg/ft^2 (270–375 mg/m^2). Higher coating weights, as expected, increase the corrosion protection. Heavy coating weights in the range of 65–85 mg/ft^2 (700–915 mg/m^2) are common for aluminum parts used underwater, such as the outboard components of boat motors and drives.

Below 10 mg/ft^2 (107 mg/m^2), chromate is colorless. This low coating weight is sometimes used when clearcoats are applied to aluminum, such as on styled aluminum auto wheels. This invisible chromate coating provides some extra adhesion and corrosion resistance without discoloring the aluminum. At 10 mg/ft^2, the coating contains a very low concentration of hexavalent chromium ion, which colors the coating at higher weight levels. At about 15 mg/ft^2 (160 mg/m^2), chromate coatings begin to exhibit an iridescent light yellow-golden color. As additional coating weight develops, however, it forms an increasingly golden appearance and then a darker metallic tan or brownish-bronze color. The exact shade is determined by several factors including the pH, chemical formulation, concentration of the bath, duration the bath has been in use, spray pressures, and aluminum surface roughness.

The various chromating processes include those with accelerated and nonaccelerated coating baths. One of the oldest and best (but highly toxic) accelerators is the ferricyanide ion. A coating that uses an iron ferricyanide accelerator has the composition of $CrFe(CN)_6 * 6\ Cr(OH)_3 * H_2CrO_4 * 4Al_2O_3 * 8H_2O$. Safer accelerators, utilizing ions such as molybdate and molybdate/zinc, are now more common.

In general, chromate conversion coatings produced by nonaccelerated baths are quite durable, but the processes are slow. Nonaccelerated baths have fairly short working lives as well, and require frequent bath replenishment. The nonaccelerated chromate coating typically has the representative chemical composition of $Cr(OH)_2 * H_2CrO_4 * Al_2O_3 * 2H_2O$.

Chromating solutions contain chromic acid and hydrofluoric acid at a pH of 1.5–2.0 and a fairly high concentration of hexavalent chromium ion (Cr^{+6}). While amorphous chromate can be applied by immersion or by spraying, immersion is far more frequently selected because of the inherent dangers in spraying toxic chromium solutions.

A freshly chromated coating is subject to two types of potential damage: 1) the fresh coating is soft and easily abraded and 2) the unpainted chromate is heat-sensitive. A maximum skin temperature for chromated parts during dryoff (usually with hot air) prior to painting is 140° F (60° C). Even 10 minutes at 300° F (149° C) will cause noticeable chromate decomposition. Once the parts are painted, however, heat no longer causes significant deterioration, and paint oven temperatures can be as high as required without degrading the chromate. The difference is presumed to be due to the exclusion of oxygen plus the effective sealing of the water of hydration in the crystals by the paint layer. Too much heat before being painted destroys chromate effectiveness because it alters the chemical composition of the film, expelling some or all of the water of hydration in the crystal structure. A color change to a powdery bluish-purple is apparent when overheating has occurred. Loss of the water of hydration shrinks chromate coating crystals, and this leaves bare metal areas devoid of conversion coating. Note in the chemical formulas given previously that the nonaccelerated coatings contain only two "waters of hydration" versus eight for the accelerated versions. The lower water content makes the nonaccelerated coatings somewhat less susceptible to heat deterioration.

Phosphates for Aluminum

Iron, zinc, and chromium phosphate coatings can be used on aluminum with various levels of effectiveness. (Chromium phosphate was not covered earlier when iron and zinc phosphates were discussed, but it will be explained shortly in a subsequent paragraph.) Iron phosphating solutions usually deposit little or no coating on aluminum, but they rather effectively clean the surface of weakly adherent aluminum oxide just as they do on zinc. Aluminum is a rather forgiving metal to paint. Many aluminum alloys exhibit good paint adhesion with only this moderate acid treatment. Some iron phosphate systems are formulated to deposit a small amount of coating on the aluminum, but this is probably a mixture of iron and aluminum oxides rather than a true phosphate. It is an adherent layer, however, and does enhance paint adhesion. The coating composition will vary widely with the dissolved iron concentration in the bath. When production involves mixed metals with steel plus a small amount of aluminum, the preferred pretreatment is likely to be an iron phosphate rather than chromating.

Zinc phosphating usually would be used on aluminum only if steel and/or zinc were also being surface-converted in the same system. Either fluoroborate (BF_4^{-1}) ion or fluoride (F^{-1}) ion as an additive to the bath is helpful in etching the aluminum, which is a necessary step in the deposition reactions. The hydrogen fluoride that is formed will remove aluminum oxide from the surface and allow acid attack on the metal. The fluoride ion also acts to precipitate excess aluminum ions as insoluble aluminum fluoride (AlF_3) to avoid poisoning the bath by aluminum ions.

Zinc phosphate produces a clear-to-light-gray coating on aluminum, much the same in appearance as the zinc phosphate coatings on steel and zinc. The immersion-produced zinc phosphate on aluminum gives finer crystal sizes and better corrosion resistance than spray-applied zinc phosphate, just as is true on steel and zinc.

Most architectural aluminum siding and panels are pretreated with chromium phosphate. The aluminum is first cleaned in a separate stage, usually by spray. The chromium phosphate bath contains a mixture of hydrofluoric, phosphoric, and chromic acids. The weights of coatings produced range from about 25–200 mg/ft^2 (270–2,150 mg/m^2); the normal amount is approximately 40–60 mg/ft^2 (430–645 mg/m^2). At around 20 mg/ft^2 (215 mg/m^2), chromium phosphate is a rather pale green with a slight iridescence. The green color becomes more evident as the coating weight increases. It has a medium to deep green color with coating weights between 150–200 mg/ft^2 (1,600–2,150 mg/m^2). Finishing operations that paint only small amounts of aluminum may also use chromium phosphate pretreatment, but chromium phosphate is not used widely except in certain types of industry.

A typical chromium phosphate coating has the formula $Al_2O_3 * 2CrPO_4 * 8H_2O$. Notice that it contains no toxic hexavalent chromium ion (Cr^{+6}), but only the relatively nontoxic trivalent form of chromium (Cr^{+3}). For this reason, the U.S. Food and Drug Administration has approved the use of chromium phosphate conversion coating inside food and beverage cans. A subsequently applied chromic acid sealer will considerably increase the corrosion resistance, but the coated metal cannot then be used for food container applications. As we have learned earlier, chromic acid contains toxic Cr^{+6}, hexavalent chromium. Additionally, a number of Cr^{+6} compounds are known to be carcinogens (cancer-causing agents).

Dried-in-place, no-rinse-required aluminum conversion coatings are sold by at least two pretreatment chemical suppliers. The coating produced is said to be based on tannic acid, tannates, zirconates, or related anions. Full details are not available on their chemistry. A third company has a similar dried-in-place, no-rinse-needed pretreatment for aluminum that is based on either titanium or zirconium chemistry. The principal advantage of no-rinse systems is the total absence of toxic wastewater, which can be very appealing to any company. Nothing goes down the drain, so no wastewater treatment is necessary;

no discharge permits are needed; and no water pollution violations can occur. Although this sounds great, not every plant will want to use no-rinse aluminum pretreatment. The drawback is that no-rinse coatings are low in coating weight and offer only limited corrosion resistance. Thus, their usage is limited to parts where corrosion resistance is not a major consideration. Most exterior parts need greater protection than no-rinse conversion coatings can offer, but they are ideal for many items used indoors.

Some aluminum parts receive clear protective coating so the attractive metal color and polished or textured mechanical finish will remain visible. An example is widespread use of styled aluminum wheels for motor vehicles. Unprotected aluminum would show signs of weathering and staining if used in outdoor exposures. Clearcoats on aluminum can be used for protection when bare aluminum appearance is desired, but the clearcoat adhesion on aluminum will be weak without a conversion coat under it. The golden-tan iridescent appearance of most chromate coatings makes them unsuitable here; a completely colorless form of conversion coating is obviously preferred for such applications. Low weight (<10 mg/ft^2, or <108 mg/m^2) colorless chromates have been used for this purpose, as well as heavier weight chromates. With the latter nearly all the hexavalent chromium is then leached out with water, and this turns the coating colorless. New proprietary chromium-free colorless conversion processes that provide both outstanding corrosion resistance and excellent paint adhesion are now being used in this market. Suppliers of these novel systems have revealed only partial chemical information about the chemical reactions that form the coatings, but it is believed that they are based largely on transition metal chemistry.

All conversion coatings, even iron and zinc phosphates, are harmed by excessive heat, but the amorphous chromate is by far the most heat-sensitive of them all. Iron and zinc phosphates are the least susceptible to thermal damage; while chromium phosphate coatings are only slightly more heat tolerant than chromate coatings. Where heating of conversion-coated aluminum parts before painting cannot be avoided because of a particular manufacturing process, one of the chrome phosphate conversion coatings would be a wise choice.

Anodizing of Aluminum

Another type of conversion coating for aluminum is called *anodizing*. This gets its name from the process whereby a tightly adherent aluminum oxide coating forms on the surface when aluminum is made the anode in an electrochemical cell. However, it is not an exceptionally effective prepaint coating in most cases. Anodizing has been preferred for pretreatment in the aerospace industry because it causes the least fatigue stress to be induced in the aluminum portions of spacecraft. Anodized parts are frequently used without further painting for exterior parts in many applications. Use of unpainted anodized aluminum on aircraft may be considered because of the important weight savings that can be realized. Two of the major virtues of anodized coatings are their corrosion resistance and hardness. This provides weatherability and abrasion resistance. Both factors are important in the aerospace and aircraft industries. Aircraft encounter considerable abrasion from sand and various other airborne particulates, as well as raindrops, sleet, hail, snow, and general cold-weather icing. For protection of the thin aluminum skin, the anodizing is far more important than a layer of paint. In contrast to the 1,000 mg/ft^2 (10,760 mg/m^2) aluminum oxide coating weights for aircraft, much heavier anodized coatings of up to 5,000 mg/ft^2 (53,800 mg/m^2) are mandated by military specifications for aluminum used in underwater applications.

Anodizing is usually done in a dilute acid, such as chromic, oxalic, or sulfuric acids. The recent trend is to use sulfuric acid processes, however. While immersed in the acid medium, the aluminum part is connected to a direct current electrical power source so that it becomes an anode in an electrochemical cell. The electrical flow forces the oxidation reaction of the aluminum with the solution. The resulting oxide forms a tight uniform layer about 0.25 mil (6.35 μm) thick that weighs 500–2,000 mg/ft^2 (5,380–21,520 mg/m^2). The process is very slow, and to form 1,000 mg of oxide may require 15 or even 20 minutes. Initially, the oxide film is soft and somewhat gelatinous, but on drying it becomes very durable. Anodizing done in cold solutions produces a much harder oxide layer than in warm solutions. Anodizing bath temperatures as low as 20° F (–7° C) have been used. Pores in the newly formed oxide coating can be sealed effectively by immersion in hot water or in aqueous solutions of corrosion-inhibiting substances. Also available is a unique process that simultaneously hydrates and impregnates the anodized coating with a polyurethane resin.

Color anodizing of aluminum is not a painting operation; it involves placing the freshly anodized part in a liquid dye solution that fills the pores. When the pores are sealed, the color remains trapped in the oxide layer, giving color-anodized parts their distinctive though often slightly garish appearance.

Specialty and Mixed Metal Phosphates

For years the appliance industry has used a microcrystalline calcium-modified zinc phosphate that gives good corrosion resistance, but which needs a relatively high process temperature close to 160° F (71° C) for consistent coating. This material is an extremely heavy sludge producer. If the phosphate stage of a washer is to use this high-temperature solution, the method by which the solutions are heated may have to be modified. This is because of thermally inverse solubility problems with some of the chemicals, in which the compounds become less soluble rather than more soluble as the solution temperature increases. Heaters for the solution should have a large surface area and only a moderate temperature differential of less than 12° F (3.6° C), rather than running full bore at high temperatures with a small heat tube surface area. Use of a heat exchanger is indicated rather than a less expensive flame tube. This will help to prevent localized hot spots that might force substances with thermally inverse solubility out of solution as sludge. Temperature control often has to be within ±2° F (±0.6° C) for good operation. In these systems the concentration of the crystal grain-refiner is extremely crucial.

Many closely related process formulations fall under the general headings of iron phosphating or zinc phosphating. Industrial finishing magazines and directories list dozens of manufacturers who formulate and sell the basic phosphatizing chemicals along with the numerous proprietary ingredients used. Particularly common in recent years are the zinc phosphates that also contain manganese, which is added to increase corrosion resistance.

Spray and Dip Processes

Phosphating solutions can be applied to metals either by spray, immersion, or combinations of the two. Spray application (see **Figures 8-4** and **8-5**) promotes fast crystal growth because fresh solution constantly impinges on the surface at pressures of roughly 15–25 PSI (1.0–1.7 bar). However, recessed areas of complex-shaped parts do not receive much coating because of the difficulty in spraying solution into these somewhat hidden areas.

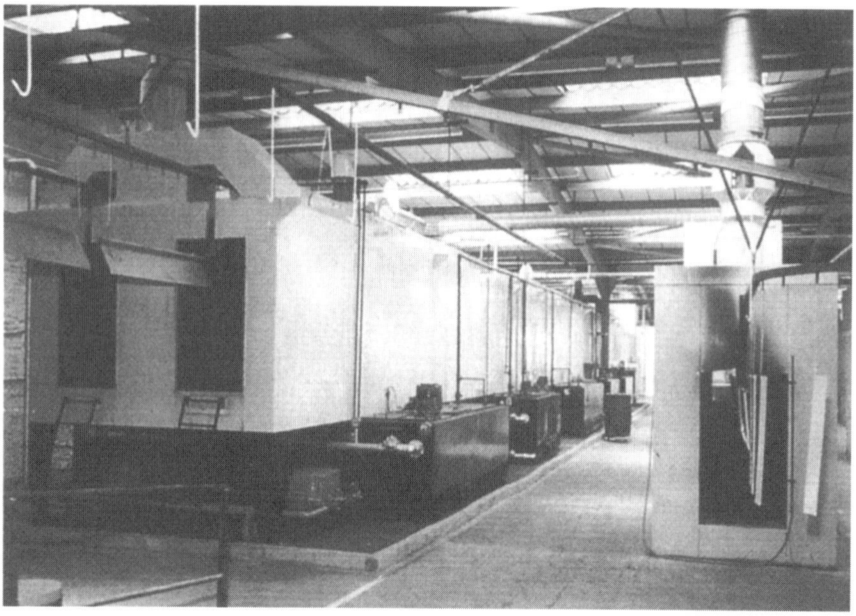

Figure 8-4. Automatic spray line (left) and dry-off oven (right).

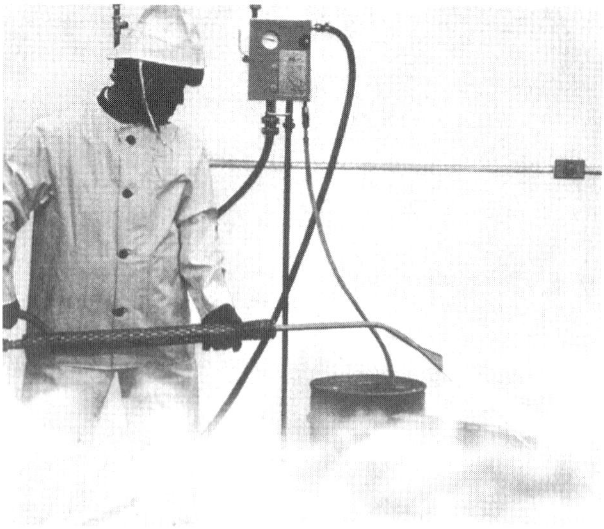

Figure 8-5. Applying phosphate with manual spray wand.

To get solution into those areas, the immersion (dip) process (see **Figure 8-6**) can be used. It has a number of distinct advantages, although immersion conversion coating is slow and requires a large bath of solution, especially in conveyorized lines. If a 4 foot (1.2 m) tall part on the line travels at 8 ft/min (2.4 m/min) and the phosphate stage requires full immersion for 90 seconds, the phosphating tank must be at least 20–24 feet (6–7 m) in length. It needs 12 feet (3.65 m) for the process (8 ft/min × 1.5 min) plus extra length to allow for part entrance and exit from the bath. The advantages of immersion conversion coating, however, include the following.

- Higher coating weights can be achieved.
- The immersion zinc phosphate coating on steel is a high-quality iron-rich zinc phosphate.
- A high quality and small size crystalline structure is produced.
- It is relatively easy to maintain the bath because there are no spray nozzles that have to be cleaned periodically. The problem of clogged nozzles that interfere with the overall spray phosphate quality is a common one, but avoided with immersion.
- Neither heat nor water is lost from continuously spraying hot aqueous solutions into the air during operation.
- Sludge formation is minimal with immersion but significant with spray.

New automotive painting facilities and retrofit systems installed in recent years use full immersion cleaning and phosphating (and rinsing) to increase corrosion resistance. It is interesting that many years ago Chrysler used a combination spray + dip process known as the "slipper dip" method. The lower half only of car bodies was immersed in an at-

Figure 8-6. Car body immersed in phosphate dip tank.

tempt to increase rust protection in these more corrosion-prone areas, while the upper half was sprayed with phosphatizing solution. This proved far less effective than anticipated because phosphate baths work best when they are formulated either for spray or dip operation, but not both. Trying to do both spray and dip phosphating with the same solution does not work well and is therefore not recommended.

A typical system for phosphating either iron or zinc parts may have five (and often six) stages:

1. Aqueous alkaline clean
2. Water rinse (single or double), usually with city supply water
3. Phosphate
4. City water rinse
5. Aqueous postphosphate sealer rinse, and
6. (Not always required) City water rinse, often followed by a brief rinse with deionized or reverse osmosis water.

The sludge from zinc phosphate pretreatment of steel includes salts of zinc and iron (principally iron oxides and phosphates), spent accelerator, and oxidizer. Some of whatever metal is pretreated will invariably be found in the sludge because acid dissolution of the metal is a necessary step in all conversion coating processes. Phosphate sludge may be allowed to settle and then be removed, or sludge can be continuously removed by centrifuging or filtering it out. Excessive sludge may be produced when very hard water is used to make up pretreatment solutions. Utilization of deionized, reverse osmosis, or even softened water for making up conversion coating baths may turn out to be economical for this reason. Remember, though, that softened water still has a high TDS content.

Phosphate Troubleshooting

To do troubleshooting on problems in phosphate coatings requires a considerable knowledge of cleaning, rinsing, and phosphating technology. The successful troubleshooting process generally follows four steps:

1. Determine the problem(s)
2. List the probable causes of the problem(s),
3. Review the possible remedy or remedies to correct the problem(s), and
4. Institute the remedy or remedies and check their effectiveness in correcting the problem(s).

An examination of four categories of phosphating problems follows. Each problem is first stated and discussed. Then, possible causes are listed and the potential remedies are explained.

PROBLEM 1: Streaky, Blotchy, Thin, or Discontinuous Phosphate Coatings

Streaky phosphate coatings might not reduce corrosion resistance, but blotchy, thin, or discontinuous coatings will. It is good practice to take corrective action to obtain phosphate with a uniform, light- to medium-gray, and tight, fine crystal structure. Possible causes and remedies for these conditions are provided in **Troubleshooting Chart 8-1.**

Troubleshooting Chart 8-1. Possible Causes and Remedies for Streaky, Blotchy, Thin, or Discontinuous Phosphate Coatings.

Possible Causes	Possible Remedies
Excessively dirty metal entering the cleaning stage. Failure to remove contaminants such as metal seam sealant, dried drawing compound, or heavy oil deposits. Inadequate solvent precleaning. Dirty solvent, rags, or sponges used in cleaning. Poor rinsing after phosphating or a contaminated rinse stage.	Ensure that equipment is properly maintained and that chemical concentrations, cleanliness, and temperature requirements are met. If parts are manufactured using metal sealant, check if the sealant is applied only where specified. Remove all excess sealant prior to part cleaning stages. Excess sealant can contaminate chemicals, plug nozzles, and alter the correct solution spray pattern onto the parts. Review the prewash process for residual soil and/or cleaning agents.
Weak cleaner concentration. Low temperature and pressure in cleaner stage. Inadequate rinse after the cleaner stage. Drying of parts between stages due to the cleaner bath being too hot.	Review phosphate process procedures. Make necessary changes to temperature, spray pressure, and chemical levels.
Contaminated cleaner. Plugged or misaligned nozzles.	Make daily checks of nozzles to correct misalignment and replace plugged nozzles. High spray pressures can indicate that some nozzles are plugged. Remove accumulated dirt from the cleaner. Ensure that the sludge removal system works properly. Drain, clean, and recharge the tank at necessary intervals to keep it operating effectively.
Low accelerator or high free-acid concentrations. Other chemical imbalances. Poor bath agitation.	Make chemical additions as needed. Low accelerator or high free-acid concentrations will lower the ratio of total acid (TA) to free acid (FA).
Stained, discolored, or oxidized metal surface.	Remove defect(s) by wiping with acid and/or abrasive pads before parts enter the cleaning stage. Do not allow acid to dry on parts. Rinse thoroughly with water when acid is used.
Drips of chemicals from washer roof or conveyor.	Adjust nozzles to avoid spraying the roof or the conveyor. Install or adjust drip pans to catch drips. Reconstruct the enclosure to have a domed roof and/or move the conveyor outside the washer enclosure.

PROBLEM 2: Loose, Powdery, and Nonadherent Phosphate Coatings

Loose, powdery, and nonadherent phosphate will not provide adequate coating adhesion to metal and will reduce corrosion resistance. The crystalline structure will not be tight or dense, and crystals will be larger than normal. This phosphate coating can be easily wiped off. Possible causes and remedies for these conditions are provided in **Troubleshooting Chart 8-2.**

Troubleshooting Chart 8-2. Possible Causes and Remedies for Loose, Powdery, and Nonadherent Phosphate Coatings.

Possible Causes	Possible Remedies
High total-acid/free-acid (TA/FA) ratio in the phosphate stage.	Correct the TA/FA ratio by adding phosphate makeup material per instructions.
High accelerator concentration.	Add fresh water to reduce the accelerator content. Recheck the total-acid and free-acid concentrations and add phosphate makeup as required. Excess accelerator will neutralize the free-acid content and will cause a high TA/FA ratio. A high TA/FA ratio will yield powdery, nonadherent coatings.
Poor cleaning.	Check to ensure that the cleaner concentration, pressure, and spray impingement are adequate. If the cleaners are excessively dirty, dump the cleaner stage, clean the tank, and recharge. Inspect all solvent-wipe operations to ensure properly cleaned exterior surfaces. If cleaning is inadequate, the quality of the phosphate coating will be poor.
Slow line speeds or frequent line stops.	If line speeds are operating below design capacity and if line stoppages are frequent, it may be necessary to reduce the phosphate chemical concentrations. Check with the phosphate supplier for recommended concentration reductions.

PROBLEM 3: Salt Deposits, Acid Spots, and Water Spots

If phosphate salt deposits, acid spots, or water spots remain on the metal surface, they will act as contaminants. They may have adequate adhesion to the metal but will usually be the starting point for humidity blistering or scab corrosion. Since they lessen the overall corrosion protection, they must be corrected immediately. Possible causes and remedies for these conditions are provided in **Troubleshooting Chart 8-3.**

Troubleshooting Chart 8-3. Possible Causes and Remedies for Salt Deposits, Acid Spots, and Water Spots.

Possible Causes	Possible Remedies
Overhead conveyor dripping onto parts due to chemical buildup and condensation.	Adjust nozzles and/or pump pressure to avoid spraying the conveyor. Adjust overhead drip pans to catch dripping.
Inadequate water rinsing after phosphate stage or after sealer rinse stage.	Check nozzles in the water rinse stages for inoperative or plugged conditions. Determine if rinse stages provide sufficient water for thorough removal of residual chemicals. Large-capacity rinse tanks and/or pumps, additional risers, and/or extension of rinse risers may be required.
Parts drying due to line stops.	Ensure that water rinses are activated during all line stops.
Deionized rinse water is too high in solids content and/or conductivity.	Check the deionized water if used. If necessary, recharge the deionizer. When recharging, flush all chemicals from water lines prior to rinsing. Check quality of RO water if used. Replace membranes as required.

PROBLEM 4: Flash Rust

If the metal is only slightly oxidized or rusted, corrosion resistance may or may not be affected. The severity of rust or oxidation will vary, depending on the cause. Corrective action is required in all cases so that flash rust is no longer observed. If in doubt, remove the oxidation from all parts. Possible causes and remedies for these conditions are provided in **Troubleshooting Chart 8-4.**

Troubleshooting Chart 8-4. Possible Causes and Remedies for Flash Rust.

Possible Causes	Possible Remedies
Low accelerator concentration.	Adjust the accelerator concentration. A low accelerator concentration will imbalance the TA/FA ratio, resulting in a thin coating susceptible to flash rusting.
Incorrect TA/FA ratio.	Correct the ratio. A low TA value or excessive FA value will result in thin coatings that may allow flash rust.
Drying between stages.	Avoid overheating any stages. Install misting nozzles between stages to keep parts wet. Drying of parts between stages may occur from high solution temperatures, line stoppages, or insufficient rinsing.
A weak or insufficient spray pattern in the phosphate stage.	Check for plugged or misaligned nozzles and for low spray pressures—these may alter the phosphate spray pattern, resulting in discontinuous and/or thin coatings. These conditions will usually permit rust formation. Correct spray pressure and nozzle alignment as required.
Rust is present on the metal prior to phosphating.	Rusted parts entering the phosphate systems will result in rusted parts exiting the phosphate system. Remove all rust prior to phosphate by mechanical abrasion or acid strippers. Acid must not be allowed to dry on the parts. When acid is used, thoroughly rinse with water. Be aware that freshly derusted steel is prone to rust again when wet.
Stripping of phosphate coating by a chromic acid sealer.	Minimize line stoppages. Check the chromic acid concentration. If a phosphated part stops in a chromic acid sealer rinse or the acidity of the chromic acid is too strong, stripping of the phosphate coating can occur and allow flash rusting.

Sealing Phosphate Coatings

Even when phosphating is done perfectly, one or two percent to as much as five percent of a surface will not be phosphate-coated. This results from hydrogen gas bubbles forming pinholes and crevices that prevent crystalline growth due to interference with the phosphoric acid attack on the metal.

Improved corrosion resistance can be attained when these open areas are sealed with an inhibiting material. In the past, a chromic acid rinse or a chrome-free sealer rinse was most often used. Even more effective was an acidic mixture of trivalent (Cr^3) and hexavalent (Cr^6) chromium compounds, a blend known as "reduced chromic acid sealer."

Typically, the chromic acid rinse accomplished the following.

- It removed possible traces of unrinsed phosphate solution. Phosphate chemicals are hygroscopic, meaning they are able to pull water to themselves out of the air. If they are not thoroughly removed by reaction or rinsing, the result may be early blistering of the paint film.
- It removed loosely attached conversion coating deposits.
- It greatly enhanced corrosion resistance by sealing bare metal areas and pores in the phosphate coating, and by leaving behind residual amounts of the chromate ion anti-rust agent. This is the principal purpose of any postconversion coat sealer.

Today few industrial plants still use chromium in the sealer rinse because chromium is a toxic heavy metal. If an extremely dilute aqueous chromate or similar solution is used as a postphosphate sealer, the parts can often proceed directly to a dry-off oven and then be painted. Better corrosion resistance is frequently obtained when a more concentrated sealer rinse is used, but this requires that the sealer be rinsed afterwards. Deionized or reverse osmosis water spray rinse is usually used for this rinse. Unfortunately, the chemicals deposited by the sealer are considerably more soluble than the phosphate coatings. If a thorough rinse were to be used on the sealer, all the newly applied material would be rinsed away. For this reason, a brief "compromise" rinse is used of about 20 to 30 seconds duration. This removes the excess sealer compounds without removing all of the deposited sealer chemical. The rinse will also remove residual dissolved solids possibly present in plant water.

Because of the extreme toxicity of chromium compounds, a wide variety of chrome-free sealer materials have become available. Silanes work very well on iron and zinc phosphate. Other sealers use phosphates, tannins, titanates, molybdates, zirconates, thioglyconates, and aluminates. Depending largely on the particular paint films that are applied, chrome-free sealers may rival chromium-containing materials in anticorrosion performance. Virtually no new installations use chromic rinses unless tests show they are absolutely needed to get the best corrosion resistance on metal parts.

As water (and solids disposal) restrictions are being tightened, chrome-free nontoxic materials that can run directly to drain after possible pH adjustment and dilution are increasingly being used in industry. As **Figure 8-7** indicates, the cost of dumping a toxic solution includes many items beyond the solution cost.

Most of the nontoxic sealers are acidic, but a few are slightly alkaline. While they are in many cases as effective or almost as effective as reduced chromic acid sealer rinses, their performance cannot be expected to exceed that of chromium. Sealer supplier claims to the contrary find little industry acceptance.

Other Conversion Coatings

In addition to those previously discussed, other conversion coatings are available. They are generally intended for special appearance properties, corrosion resistance, or to hold lubricants, but are not intended as a base for paint. Various bluing, browning, and blackening oxide coatings provide some aesthetic value as well as minimal corrosion resistance when used on iron and steel. Some of the blackening done in hot aqueous solutions has better corrosion protection properties than ambient temperature processes, however.

Total Cost of Dumping a Process Solution
Wasted Chemicals
Labor
Waste Treatment of Chemicals
Waste Disposal
Waste Treatment Equipment
Lost Production
Quality of Product

Figure 8-7. Process solution disposal costs.

Heavy manganese phosphate coatings (as much as 2,500 mg/ft^2, or 26,900 mg/m^2) are applied on metal to hold die lubricants in place during deep-drawing operations and are applied on friction surfaces of gears and bearings to stop scoring during break-in periods. It also may be used as a base for the glue in adhesive bonding of metals. In rare instances, lead phosphate is used as a conversion coating. Neither of these is used under paint.

While not a true conversion coating, hydrochloric acid and table salt are employed in artificially "aging" or oxidizing copper, producing the variegated green "verdigris" finish associated with long-term weathering. (The letter "s" in verdigris should be pronounced; it is not silent although many people mistakenly think so.) Some paint finishes try to duplicate this attractive appearance on other substrates, such as stone and wood.

Dry-off Ovens

Parts cleaned or pretreated—whether they are plastic, wood, composite, or metal—still need to be dried before being painted (unless they are metal and are to be immediately ecoated). As much water as possible should be removed before parts enter the dry-off oven. Compressed air can be costly but may be necessary for some items. Manual compressed air vacuums can be used as well to suck water out of part cavities. Vacuum removal is preferred because blowing the water out can blast contaminated water onto other nearby parts. The cost of oven drying can be reduced if parts are designed with water drainage holes, and are hung on the conveyor in a way that will facilitate drainage. A high degree of air turbulence in the oven will speed drying; oven air seals or designs utilizing bottom oven entry and exit will further lower total energy consumption.

Conversion Coating Chemical Waste Treatment

Conversion coating solutions and rinses need to be treated before discharge to drain if they do not conform to local, state, and federal discharge codes. Discharge regulations almost always severely limit the concentrations of hexavalent chromium and nearly all heavy metals. Harsh fines can be expected to be levied against companies for allowing release of wastewater containing amounts of regulated substances beyond specified concentrations. Limits on phosphate ion concentrations in wastewater have also been established for many locations even though phosphate itself is not inherently a toxic substance. Iron ion is often present in well water and is not considered a dangerous or toxic metal; nonetheless you may need to remove excess iron in wastewater if local regulations impose limits.

It is likely that in the future social and legislative pressures may force many manufacturing companies into increased voluntary and mandatory material recycling, perhaps of paint sludge or of other chemicals. Societal pressures for such legislation are increasing rapidly since many communities have long established voluntary recycling of paper, metals, and plastics. A number of plants are already completely reusing all of their water (except from toilets), sending zero water to drain, using water supply only to make up for evaporative and process losses and for sanitary uses. Landfill restrictions also continue to proliferate, but for now in most cases it is still cheaper simply to buy new paints and chemicals than to recycle them. The cost of conversion coating chemicals is far too low to consider trying to directly recover and recycle chemicals from sludge and spent pretreatment baths. At least this is true at present, and I dare speculate that we won't see it happen in the next decade either.

Electrical Values versus Water Purity (TDS/PPM)
(Values are approximate for ionic impurities)

Resistance (megohms)	Conductivity (micromhos)	Impurity Levels
10,000	0.1	0.05
5,000	0.2	0.1
1,000	1.0	0.4
500	2.0	0.8
250	4.0	1.6
166	6.0	2.5
125	8.0	3.2
100	10.0	4.0
50	20.0	8.0
33	30.0	14
25	40.0	19
20	50.0	24
16.7	60.0	28
14.3	70.0	33
12.5	80.0	38
11.1	90.0	43
10.0	100	50
5.0	200	100

Chapter 9

Paint Application I: Traditional Methods

Historical Applications

Although no known records exist about early paint application, the most probable method of application was by some form of spreading. The methods used by cavemen in painting their caves almost certainly included the use of their fingers, sticks, and perhaps straw-like spreading devices. The application of paint by means of a device held in the hand evolved over the years; even now a brush is one of the most common methods of applying paint, either for art or industry.

The essential feature of paintbrush construction is primitive—merely tying or binding thin reed-like fibers together. The quality differentiation of brushes is linked to three parameters: the nature of the bristles, the quantity of bristles, and also how well they are bound together.

That painting with a brush is slow, labor-intensive, and messy quickly becomes apparent to anyone who has ever done much of it. If the paint is too thick, it doesn't spread easily; if it is too thin, it will spread easily, but the paint will run and the application will be extremely messy. In the past, many paints commonly contained toxic pigments made of substances such as lead, chromium, and other heavy metals. Painting could obviously be hazardous to health not only from heavy metal poisoning: years ago, more dangerous solvents were used and these fumes generally ended up being breathed by the painter.

Modern Application Methods

Several alternative methods have emerged for applying paint, and each has its own advantages and disadvantages. Widely used methods include the following.

- Rolling
- Dipping (immersion)
- Flow coating
- Continuous coating
- Dip-spin coating
- Tumble or barrel coating
- Curtain coating
- Roll coating/coil coating.

Rolling

Manual painting with a roller is faster than painting with a brush, but it too has drawbacks. A roller cannot get into hard-to-reach areas, and the application often requires touchup with a brush. In addition, rollers generally absorb—or "load up"—a substantial amount of paint, most of which cannot be saved when the job is done. Devices that pump paint through a hose to the roller are somewhat helpful, I'm told, in speeding up the rate at which paint can be applied. While it is not often used in manufacturing to apply paint, there is still some legitimate industrial use of paint rollers. An example is the black paint applied to plate glass for opaque glazing sections in high-rise buildings. Opaque black paints are also rolled onto the thin silver, aluminum, or chromium layer deposited on glass in mirror making. The black coating protects the reflective metal layer and blocks any light that might come from behind the mirror.

Dipping (Immersion)

Painting by dipping, as shown in **Figure 9-1**, is much faster than brushing or rolling and much less labor-intensive. With dipping, however, the paint thickness on parts is extremely dependent on the paint viscosity used. Dipping can be somewhat messy, especially with low viscosity paints, and is potentially hazardous if flammable solvents are present in the coating. The viscosity of the paint in a dip tank should remain practically constant if the deposited film quality is to remain fairly consistent over a period of time. Even with constant viscosity control, immersion-coating methods are not conducive to high quality paint appearance as a result of runs and thick paint deposits from variable drain-off from the parts. Drain lines are usually visible on parts, particularly if they are at all complex-shaped. The only way to avoid them is to rotate the part continuously until the paint dries.

The film thickness in dipping is controlled primarily by regulating paint viscosity. Higher viscosity causes more paint to cling to parts. To a lesser extent, thickness is dependent on the rate at which parts are withdrawn from the dip tank. Thinner films are produced by a slow withdrawal rate; and conversely, increased film thickness will result from a more rapid withdrawal of coated parts.

Dipping is not suitable for many hollow or dome-shaped items. Hollow items would, of course, tend to float in the paint bath and not get fully covered. Domed parts retain trapped air and do not get coated inside the dome. Difficult-to-sink parts must be carefully hung or they could easily become disengaged from their hangers or hooks by the buoyancy factor. Cupped shapes are not suitable since they would fill up and carry too much coating out of the paint tank.

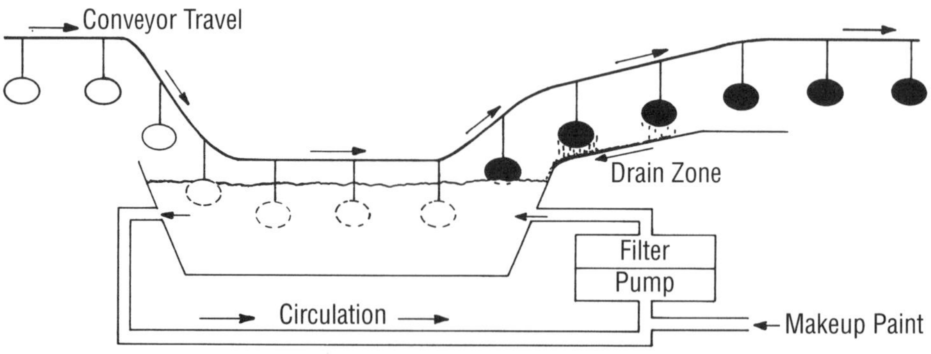

Figure 9-1. Side view of painting by dipping.

Color change with dipping is very slow because of the volume of paint to be replaced. Thus, color change is generally not feasible, although in some operations multiple dip tanks mounted on rollers are used. As needed, the various color tanks are rolled out of and into position under the conveyor line. While not many plants employ this technique, a number of high-volume paint users find this a convenient way to produce parts in multiple colors. A few companies also pump out the color they've just finished using into a holding tank, then after cleaning the dip tank, they refill it with the next color to be used.

An open tank of flammable paint would constitute a major fire hazard, so sometimes safety dump tanks beneath the dip tank are required by local fire codes. For all such tanks, an efficient fire-extinguishing system such as carbon dioxide fog must be installed as a safety measure. Accumulations of partially dried paint that has dripped from parts exiting the dip tank create an additional fire hazard, as well as a mess on the plant floor.

Flow Coating

In a flow-coat system, as shown in **Figures 9-2** and **9-3**, a number of individual streams of paint are directed at the parts. Paint is liberally applied to all surfaces of the parts. The number of streams needed varies according to the parts being painted, as one might expect, but commonly there may be 20–100 of them directed to cover all surfaces of the parts. An overflowing excess of paint is delivered onto the parts since the excess runs off and is recirculated. The flow-coat paint streams may be delivered through holes drilled

Figure 9-2. The coating chamber in flow coating.

in pipes or through the crimped ends of short pipe outlets. The paint headers and outlets are arranged so that all surfaces of parts carried through the flow coater on the conveyor are painted. A variation uses stainless steel trays with multiple holes in the bottom into which paint is pumped. The paint continuously "rains" down onto parts below the tray. With either method, the excess coating flowing off the parts drains into a sump, where it is pumped through filters and recirculated to the flow coat paint reservoir. Viscosity controls, often automatic, are often part of the circulation loop. These feed solvents as needed into the paint supply to replace evaporative losses that occur during flow coating operation.

Both dipping and flow coating tend to be used to coat items whose appearance is not class-A important, and also to apply primers that will receive some sanding. Topcoats having more than modest appearance requirements are not commonly applied by dipping or flow coating. Coatings applied by these two methods have only poor-to-fair appearance unless parts are rotated during the drippage period to avoid run lines. Currently only a limited number of manufacturers are using dip or flow coating, and then rotating the newly coated parts on special racks designed for this purpose. One such plant uses flow coat to apply clear protective coating to metal handles on furniture; another flow coats a basecoat for items to be vacuum metallized, and afterwards uses flow coat to apply clearcoat over the metal layer. The vacuum-deposited metal film is extremely thin and delicate, so the clearcoat is needed to resist abrasive rub off. In some cases, the clears are tinted to enhance the color of vacuum-deposited aluminum. Bluish clearcoats are used to make vacuum-applied aluminum appear more like chrome plating.

Both dipping and flow coating have the disadvantage that essentially the only way to control dry-film thickness is by the viscosity of the paint. The amount of paint that will drain from the parts depends heavily on viscosity. If the viscosity is allowed to go too low, insufficient paint will remain on the parts. This could cause a large number of parts to be

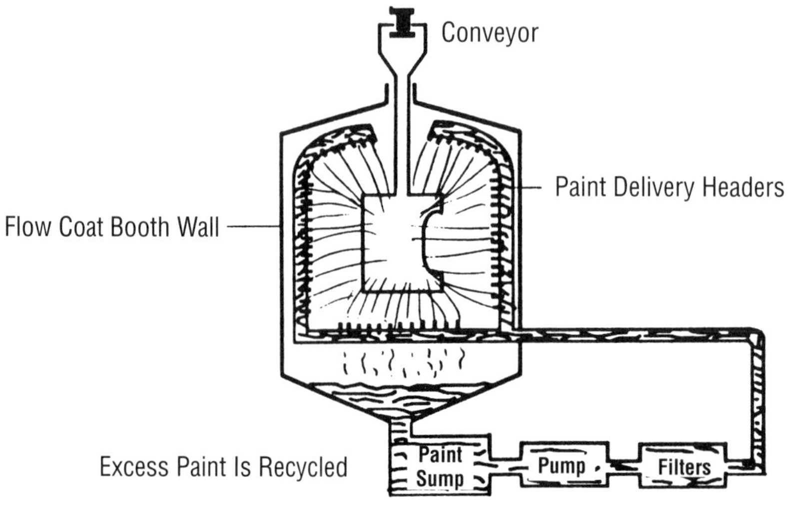

Figure 9-3. End view of flow coating.

rejected for low film build, and necessitate repainting. On the other hand, if the paint viscosity should rise too high, then extra paint will be applied. This not only increases paint costs, but extra thick paints drain poorly and so can plug small holes in the part. Automatic viscosity controllers to add solvent to the paint as necessary are therefore valuable, and can be cost effective for these reasons.

When items are dipped or flow coated, the drain-off pattern doesn't always contribute unsightly runs or heavy flow lines on the parts. A lot depends on the configuration of each part. Unattractive drain lines can usually be seen below horizontal edges of through-holes large enough to avoid being plugged by paint. These lines give a "droopy mustache" look to the holes. Sometimes extra paint draining down builds up as beads of paint along the bottom of parts; these are called "fatty edges." At times the excess paint will dry as it drips, forming elongated dried paint accumulations that resemble icicles. (Should these be called "paintcicles"?)

The most significant disadvantage of dipping and flow coating is that the finish is not as attractive as those produced by more sophisticated methods. Yet despite their limited flexibility, the cost effectiveness benefit of dip and flow coating makes them highly desirable methods for coating certain kinds of parts, particularly those that need complete paint coverage but for which appearance is not highly important. Nonappearance items such as metal flat and coiled springs inside upholstered furniture, and underbody car and truck parts, are just a few examples.

A strong plus is that both methods achieve very high paint transfer efficiencies, often 90% and higher. These techniques are well suited for automation with overhead conveyor paint lines. Manual flow coating is nonexistent, and hand dipping of parts is relatively rare. The latter is normally suitable only when a very limited number of parts needs to be painted. An exception is golf club shafts, a high-volume product frequently coated by manually inserting and then withdrawing each shaft through septum in a reservoir of paint. As the shafts are withdrawn, the septum opening wipes the paint to a uniform thickness. The club shaft is first pushed through an opening in an elastomeric squeegee (septum) on a small paint reservoir. The reservoir is tilted to let paint run to the squeegee end, so that as the shaft is withdrawn it passes through the paint supply and the septum opening squeegees off excess paint. The coated shaft is then hung and continues on through flash-off to the bake oven where the paint is cured. The club head and grip are attached later.

Painting by dipping or flow coating is fast and easy. Both involve a relatively low installation cost, require relatively little maintenance, and have low operational manpower requirements. In addition to a paint attendant to monitor the paint process, the only other labor requirement is to hang parts on the conveyor for pretreatment and painting, and later to remove them after they finish curing. Thus, in addition to a very low labor requirement, the worker skill required is also low compared to all forms of manual paint spraying.

Continuous Coating

A variation of the flow-coat process is the continuous coater, exemplified in **Figure 9-4**. An enclosed tunnel-like painting machine captures and reuses the excess paint that runs off coated parts. It uses directed atomized sprays rather than streams of paint. Tennis racquets, metal panels, and engines have been coated by this technique. Reuse of the paint

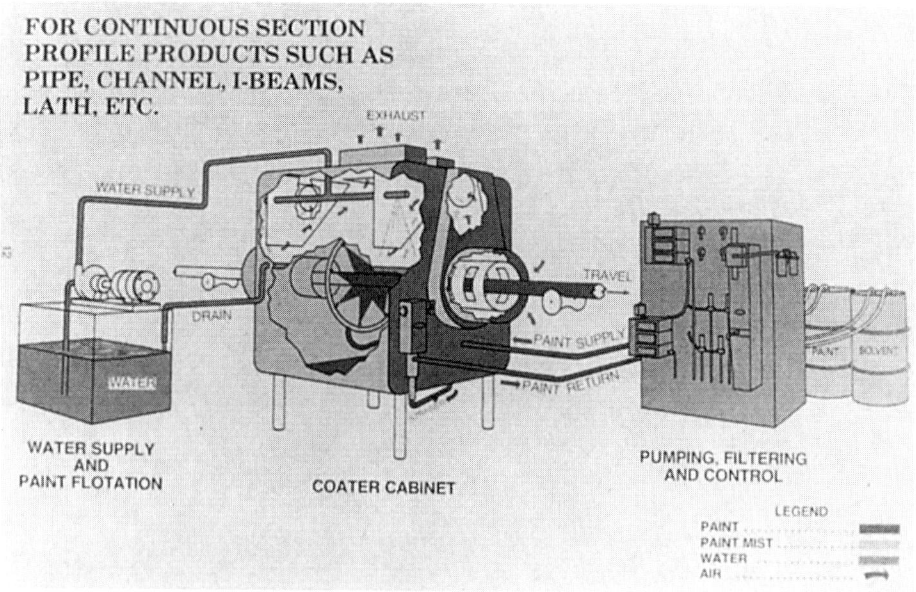

Figure 9.4. Nordson continuous coater.

within a confined space contributes to high paint transfer efficiency and reduced VOC emissions because of the solvent vapor capture capability of continuous coaters.

One drawback to the method is that all parts being coated must be roughly the same size and shape. This is because the gun-to-part distance (target distance) has to be kept uniform for best part appearance. Aluminum softball bats and long lengths of pipe are examples of items that are well suited for finishing using this type equipment. The continuous coater is in many ways analogous to a vacuum coater, yet each has enough of its own distinctive characteristics as well to warrant a separate consideration of the two processes.

Dip-Spin Coating

Dip-spin coaters, also known as centrifugal coaters, are designed to paint large-quantity batch loads of smallish parts that are not amenable to being hung on hooks for painting; for example, items such as hairpins, screws, bolts, various clips, or corrugated fasteners. There is often absolutely no practical way to hang them. In addition, the huge numbers of parts to be coated would make hanging them cost prohibitive anyway. Instead, a wire basket that might contain anywhere from half a pound to fifty pounds of parts is fully immersed in a reservoir of paint. The basket is then raised into an enclosed chamber with an open bottom. The parts are allowed to drain for just a brief period, then the basket is spun rapidly to remove excess paint. Next, the parts are shaken out onto a mesh conveyor belt and carried through a paint bake oven (see **Figure 9-5**). After a brief cool down section, the parts are allowed to drop off the end of the belt into containers. The entire process is done automatically.

The main advantage of dip spinning is the extremely high productivity rate. A disadvantage is that some of the painted parts shaken out onto the conveyor may stick together when baked in the oven, leaving paint void spots when they are separated. If this is not acceptable, parts can be run through the process a second time to eliminate the defects. The void points are small and this extra step is rarely taken.

Tumble or Barrel Coating

A variation on dip-spin is the coating of parts inside a rolling hopper. Some wood parts are coated (poorly and unevenly) by tumbling the paint and parts inside a sealed oak barrel. After brief tumbling, the barrel is opened and the paint air dries. Sometimes an open metal barrel is heated and the parts are sprayed with the desired coating as they tumble. The heat dries the paint as parts are being sprayed. Very small symmetrical parts get coated much more uniformly than larger and nonsymmetrical items. Most companies have abandoned these primitive coating techniques long ago, but a few still cling to them.

Curtain Coating

Curtain coating is designed for coating relatively flat and low profile stock only. Wall paneling, for example, would be perfect for coating by this method. In curtain coating,

Figure 9-5. After being coated in the dip-spin basket, fasteners are automatically shaken onto a belt that carries parts through the paint cure oven.

a paint reservoir overflows at a controlled rate through a slot whose opening size can be varied to control the amount of paint released. This miniature paint "waterfall" or curtain falls directly onto work moving horizontally on a belt conveyor through the flow of paint, as shown in **Figures 9-6** and **9-7**. The volume of paint released and the speed of the conveyed workpiece determine the thickness of the applied coating. Curtain coating overflow through the slot opening is highly viscosity-dependent. So viscosity control is important here also, just as it is for dip and flow coat.

Curtain coating is suitable only for items that are flat or at least relatively so; they should also be at least 6 inches (15.2 cm) long and wide. The maximum width of the curtain coater can be constructed to be any dimension, depending totally on the width of the material to be coated. A few curtain machines have been made that are as much as 60 inches (1.5 m) wide, but the width is usually between about 18–30 inches (45–76 cm).

Gaps between the sheets or pieces of material being fed through the curtain coater do not cause a problem—they do not need to butt up against each other as is necessary in vacuum coating. The paint that drops between the parts from the continuously falling curtain is captured in a collector and recirculated; thus, virtually no coating material is wasted. Extremely uniform coating thickness is possible, but only the topside of any item can be painted. Therefore, nearly everything coated this way requires just one-side painting; but if both sides require coating, that is not a problem. Items can be flipped over and the other side coated, should that be desired, once the first side is coated and cured.

People have told me they feel there are similarities between roll coating and curtain coating, but I'm not certain I agree. It's true they both work well on flat stock, which is the main factor I see they have in common. Because it is simpler and uses less complex

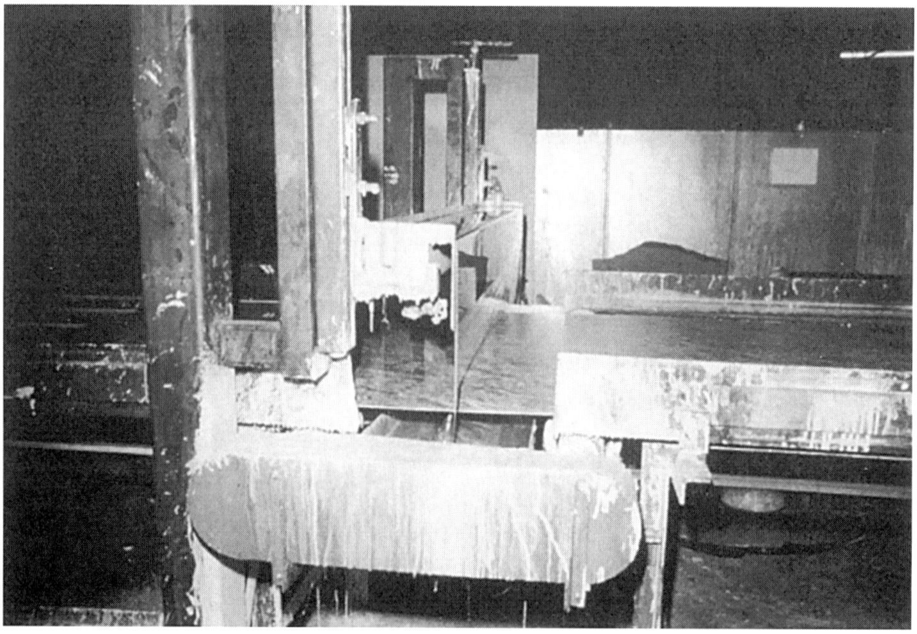

Figure 9-6. Curtain coating line.

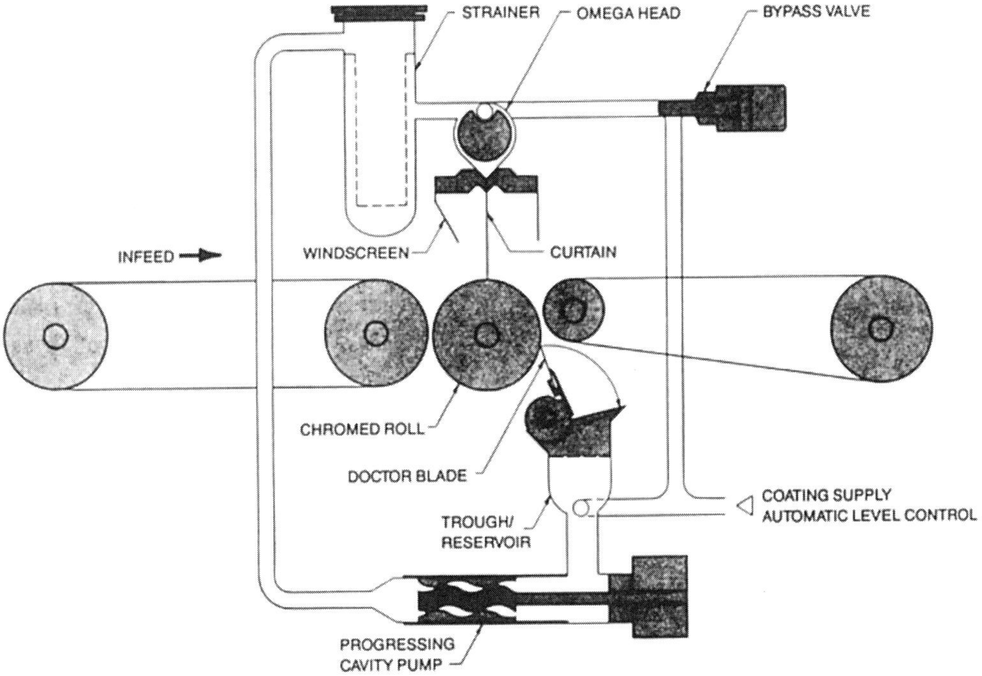

Figure 9-7. Curtain coater.

equipment, curtain coating may be preferred over roll coating for lower production runs. But it is limited to one side coating at a time, whereas many application methods can coat two sides of flat stock or the entire surface of complex shaped parts.

Roll Coating/Coil Coating

A roll coater is employed as the paint application device for coating large coiled substrates such as rolls of sheet steel, but roll coating is certainly not restricted to applying paint to coil stock only. Roll coating can be done on any predominantly flat items. The term *coil coating* refers to the painting of a coil strip, which can be anywhere from $^{3}/_{8}$ inch (9.5 mm) up to as much as 72 inches (1.8 m) wide. Coil coating was developed around 1930 to paint steel in a thin, narrow, continuous strip for Venetian blind manufacturing. Today nearly 200 coil-coating lines are operating in North America alone, with many more worldwide. It is in many ways similar to rotary printing and so lends itself to applying repeated designs and patterns. Food-can and bottle-cap designs are produced economically by this painting process that is performed prior to metal stamping. Multiple passes using different colors are used to create steel tape measures and folding wooden rulers. A coil coating line, as shown in **Figure 9-8**, unwinds a roll of metal, cleans it, applies a conversion coating, paints it, bakes it, and rewinds the painted metal into another coil, all automatically. It can be done rapidly—some coil-coating lines operate at speeds in excess of 600 ft/min (182 m/min). The large volume output possible with a fast running coating line is an important factor in the economics of precoated coil. Painted metal coil is purchased by fabricating plants who shear, punch, form, and assemble pieces of it into various products. For some manufacturers, the main advantage in fabricating from prepainted metal is that someone else does all the painting. The problems associated with meeting a variety of environmental regulations have made it attractive for them to stop painting in their own plants and to manufacture their products with parts fabricated from prepainted metal. Because the coiled metal is actually cleaned, conversion coated, roller coated, and cured while flat, products made from these materials are finished uniformly. Although no in-house paint line is required when prepainted coil is used, a small touch-up operation is generally still needed to repair the inevitable minor scrapes and mars. Prepainted metal is used on autos, appliances, cans and related containers, architectural panels, and on numerous other products.

When fabricating metal parts using pieces cut from precoated coil, precautions must be taken to hide the cut edges because they have no coating on them. Something must be done to avoid the visible corrosion that readily forms on unprotected metal. Protection or concealment is especially important for cut edges of coil steel since they can form unsightly red rust; the white corrosion that forms on zinc and aluminum is not nearly as obvious. Folded seams can often hide cut edges, but if bare edges cannot be hidden they need to be painted. Should painting of exposed edges be required, much of the advantage in using prepainted metal can be lost.

Because coil-coating lines may be over 1,000 feet (305 m) in length, threading a strip of metal through a coil-coating line is a fairly time-consuming process. If every new coil had to be threaded, the process would be slow and inefficient. To avoid repeated threading, the coaters use a scrap coil as the first and another scrap coil at the end of any given color run. The scrap coil is left in the coating machine when it is stopped to change colors. As the line is started, the leading edge of the new coil is fastened to the trailing edge of the scrap coil. When the scrap coil is wound, its trailing edge is sheared, and the leading edge of the production coil is threaded onto a new take-up roll.

Accumulators at the beginning and end of the coil line allow the new coils to be started, and the painted coils to be moved away without having to halt the line. The entry accumu-

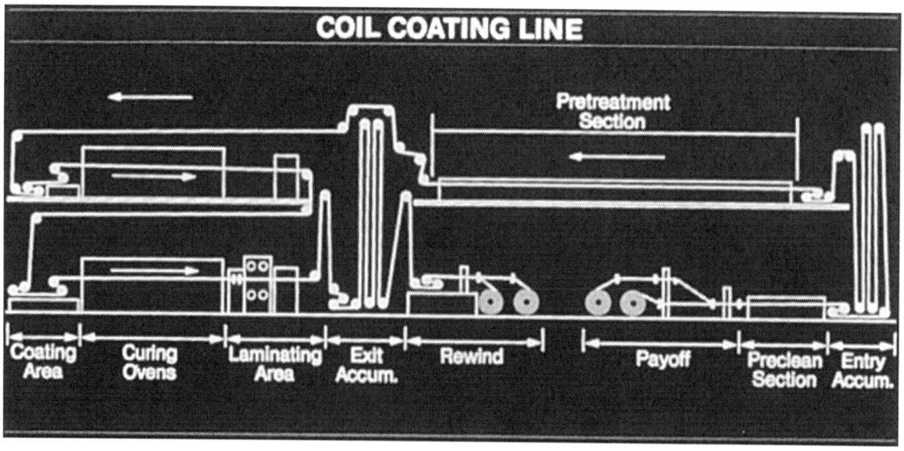

Figure 9-8. Coil coating line.

lator stores some 300 feet (90 m) of strip to feed the line while a new coil is being loaded and fastened to the end of the in-process coil. The exit accumulator runs empty until it is time to change coils at the painted end. That accumulator then collects the painted stock until the painted coil is moved out and a new take-up roll is started.

The process begins with a spray application of hot alkaline cleaner onto both sides of the coil strip to remove all oil, grease, and any other soils that may be present. Next is a rinse with warm water, followed by abrasive brushing (cleaning). This can be done on one or both sides of the strip if needed. Then, to ensure complete contaminant removal, the alkaline cleaning and rinsing is repeated. A final immersion rinse in hot water takes away any alkaline residues and completes the cleaning and rinsing operations.

The strip is then ready for dip, spray, or roll-on of aqueous chemical conversion solutions to improve corrosion resistance and adhesion of the paint to the metal. The type of pretreatment chosen depends largely on the metal substrate and the service requirements of the products to be manufactured from the prepainted stock. For cold-rolled steel, conversion coatings are microcrystalline iron phosphate or zinc phosphate. As we learned in a previous chapter, iron phosphate is less expensive, thinner, and more flexible, but zinc phosphate offers superior corrosion protection.

Two conversion coatings used on the zinc surface layer of either hot-dipped or electro-galvanized steel are zinc phosphate and chromate-sealed oxide. Zinc phosphatizing is done as explained in Chapter 8, but the oxide coating is formed differently using a three-stage process. First, the zinc surface is allowed to react with an alkaline solution, which results in the formation of an oxide layer. A cold-water rinse follows. Unless this rinse is thorough, blisters and paint peeling can result, for any salts left from the first step can react with the metal in high humidity. The third and last step is a dilute chromic acid rinse. This reacts with the initial oxide layer to yield a chemically complex substance that promotes paint adhesion and flexibility. Its corrosion resistance, however, will not be as good as zinc phosphate. Chromate conversion coatings like those detailed previously can be used as well on all types of galvanized steel. These are as good or better than zinc phosphate from both adhesion and corrosion resistance standpoints.

A special tri-metal combination with steel is prepared by hot-dipping steel in a molten aluminum-zinc alloy mixture. This alloy-coated steel requires chromate conversion treatment to achieve the best corrosion protection, although other processes may be used. With any conversion coating method, "react-in-place" and "no-rinse" processes are easier and less costly to operate. Not having to rinse avoids the problems and expense associated with wastewater treatment needed to remove environmentally objectionable substances. Unfortunately, pretreatments that eliminate rinsing have not equaled the corrosion protection of the phosphate or chromate coatings. (See Chapter 8 for more explicit information on conversion coatings.)

After exiting the selected pretreatment, the metal strip passes through an air knife section (high-intensity air blow-off) that completely dries the strip before it enters the coating section. Here a forward roll coater (more uniform film thickness) or a reverse roll coater (more level film) applies the paint. Either a single side or both sides of the strip can be painted in one pass, as dictated by specific product requirements. **Figure 9-9** shows a coil-coating line roll coater applying a single color; **Figure 9-10** shows a coil-coating line with multihead roll coater system capable of applying several colors. Epoxies, vinyl plastisols, fluorocarbons, polyesters, and silicone-modified polyesters are resins often used in roller coating. From being freshly coated by the rolls, the painted strip goes into a bake oven to drive out solvents and cross-link (cure) the paint. To shorten the oven dwell time and especially to reduce the length (and cost) of the oven, an unusually high bake temperature (about 750° F, or 400° C) is maintained. Oven length can be a problem because a 15-minute bake on coil strip traveling 10–12 ft/s (3.0–3.6 m/s) requires an oven two-thirds of a mile (1,060 m) long. This helps one to understand why bake times are normally held to about 5 minutes or less. Even so, at the above line speed, 5 minutes of baking requires an oven over 1,000 feet (305 m) long. The oven length could, of course, be halved for one-side coated work by reversing the coil direction at the mid-cure point. More than one direction reversal cannot easily be done because to do that the rollers would have to touch the uncured paint or spiral the coil somewhat. It is normally preferable (and wiser) just to build a longer oven so the line can handle either single or dual side coil coating.

After baking, the strip can be air-cooled or water-quenched and then coiled up again. When a second coat or multiple layers of coating are necessary, the coating and curing operations are repeated. The coating is inspected visually (and sometimes instrumentally) as the painted strip is wound onto the take-up roll. The finished coil is taken off the machine and moved from the area for shipment, storage, embossing, or slitting. Samples are cut from the completed coil at periodic intervals for quality control tests. If required, samples are sent to customers for preshipment approval. The coated coil can be slit to any width, cut into flat sheets of the desired size(s), or sold as an intact coil—whatever the customer desires. Coil lines may also be equipped to print designs or patterns; wood graining ef-

Figure 9-9. Coil coating line roll coater.

Figure 9-10. Multihead roll coater system.

fects are frequently used, for example. Painted metal coil can be embossed into various textures with appropriate rollers.

Many coil lines are able to laminate decorative or functional films onto the metal as well. Acrylic film laminated onto galvanized steel or onto aluminum is put to use for building applications; and various solid-color and decoratively printed vinyls are used on appliances, cabinets, lighting fixtures, and many related items. Even cork, paper, and rubber are being laminated onto specialty goods.

Two perennial problems in coil coating are the fastening methods for assemblies made from prepainted coil parts; and corrosion prevention on the bare metal edges where coil strip is cut. Bare cut edges can sometimes be located in nonvisible areas with proper product design. Hem folds, lock seams, plastic caps or plugs, and plastic beading are common ways to hide cut edges from view. Many potential users of prepainted stock depend on welding for joining parts. Although a few select paints (weld-through zinc-rich primers, for example) do not interfere with welding, nearly all other organic paints interfere with effective welding. The weld-through zinc-rich primers, although they allow welding, will emit zinc fumes that constitute a health hazard during welding. So with painted coil, joining methods that avoid welding are used. Adhesives, self-tapping screws, rivets, and spring clips are the most common of these.

Alternative Decorating Methods

A number of application techniques have been developed in more recent times to improve the effectiveness of paint application and the appearance of the cured coating. Some are hardly painting at all and may be thought of as decorating, but the aim is to both protect and enhance the appearance of an item.

Hydrographics

Any relatively nonporous substrate can now be coated with a prepared graphics film instead of paint. The environmentally friendly process is called *hydrographics.* This decorative method provides a limitless variety of single or multiple-color finishes in simple to highly esoteric designs. Application of the graphics film begins by floating the graphic in a water bath; its position in the bath is maintained using adjustable slides. The floating film is then sprayed with a chemical activator to dissolve the gelatinous support layer. Next the part to be coated is immersed and moved under the floating graphic. When the part is then raised up through the water, it captures the film on all its surfaces, even any curved or recessed portions. The freshly applied film is rinsed with water to remove any excess gelatin and then baked dry at a fairly low temperature. At this point it is normally necessary to apply a clearcoat to protect the graphics film. A precoating may also be applied prior to the decorative graphic application if additional part smoothness is needed. You may have seen some of the many interesting and striking designs (I have a cell phone cover with the American flag artfully rendered on it) that have been applied to baskets, vases, housewares, sporting goods, etc., using the hydrographics method. Thousands of designs are available, but unique designs from photos, drawings, or computer-generated artwork can also be custom made. The original pattern is digitized and printed onto the support film.

Adhesive Film

In another somewhat similar process, adhesive-backed decorative films can be applied to decorate products. This process has been used on various solids color interior and exterior auto parts for over 15 years. More recently, large "paint" panels have been applied to portions of military aircraft: C-130 Hercules Transport and S-3 Viking, and F-16 Fighting Falcon. The films have undergone testing to gauge the value of the process for military purposes. Use of decorative graphics on bus doors and along exteriors, on trains, delivery trucks, etc., is becoming more common. Intricate shapes are not easy to coat, however, with adhesive film technologies. Parts with complex configurations are better suited to coating using hydrographic films.

Vacuum Coating

Vacuum coating is used to coat items that tend to be long and have a relatively small and uniform cross-section from end to end. Simple examples of these type materials are L-shaped steel angles and wooden boards. Of course, complex cross-section stock can also be finished using vacuum coating as long as the cross-section is constant. A scalloped or carved board would not be suitable, therefore. The vacuum coater consists of a coating chamber through which the material must be fed without gaps; successive pieces must butt against the previous piece. What is necessary is that flow of material be continuous through the vacuum coater so that the coating inside the chamber doesn't dry out or emerge from the coating chamber. Precise silhouette openings in the chamber are sized just slightly larger than the cross-section of the material being coated, allowing entrance and exit of the workpieces. Each cross-section shape and size, naturally, will require its own entrance and exit silhouettes; these are frequently made of teflon or similar material.

Excess coating is applied within the coating chamber, but as the cross-section leaves the vacuum chamber through the silhouette, the vacuum prevents excess coating from escaping the chamber. So only a controlled amount of paint leaves the chamber on the workpiece. The gap between workpiece and the exit silhouette controls the thickness of the coating. Vacuum coating has been used for many years on items such as window frames and lintels, various decorative moldings, picture framing stock, and flat or shaped boards. Recently vacuum coating has been used for coating wood with UV-cure coatings, and this combination works extremely well. If desired, coating can be applied just to some sides or to all sides of the material.

What has been described is also done in a modified process called "rotary vacuum coating." When smaller numbers of shorter pieces are to be coated, they can be made to enter and exit through the same silhouette. The part enters the vacuum chamber where it receives paint, turns 180°, and then exits the chamber through the same silhouette. Except when parts are traveling through the silhouette, the opening is sealed off to contain the coating within the chamber. The reason for the term "rotary" in this process is the U-turn that parts take.

Safety Procedures in the Paint Shop

It is obvious to all that safety is critically important in any manufacturing operation, but when processes involve highly flammable materials it is especially true. Paint facilities need to be purchased, installed, and operated with personnel safety as a foremost consideration. All materials, interlocks, and moving equipment must meet or exceed local and national safety standards. Companies need to make sure they thoroughly know and understand proper safety procedures, and should manage employees in a manner that enforces strict observance of them.

Fire Safety Precautions

No electrical device, smoking material, or any other spark or flame producing device is permitted within 3 feet (1 m) of the spray booth, or within 10 feet (3 m) of where flammables are mixed or stored.

Nonessential combustible materials should not be brought into the booth. For example, if parts are wrapped in paper, they should be unwrapped before being brought into the spray booth.

Fire precautions are particularly important while spraying is in progress. Nearly all solventborne coatings and thinners are flammable, as are the dried overspray deposits adhering to booth surfaces and on filters. When atomized for spraying, solventborne materials burn very rapidly and should be considered "potentially explosive" according to National Fire Prevention Association Regulation 33. However, if applicable safety procedures are followed, the likelihood of injury is not much greater using flammable coatings than with operating most other types of manufacturing or shop equipment.

Cleanliness

In a spray-coating facility, safety depends on maintaining a neat, well-organized area. Nothing should be in the booth that is not required for spray application.

Overspray deposits on booth and filter surfaces must be monitored to avoid build-up.

Paint filters must be replaced when airflow falls below specification.

Collected overspray deposits and strippable coating may be hazardous waste and must be disposed of properly in approved containers.

Spills must be cleaned up immediately. Dried deposits should not be permitted to collect on anything. Paint containers, hoses, or measuring and mixing equipment need to be cleaned regularly.

No flammables are allowed to be stored within 3 feet (1 m) of a paintbooth.

Flammables are to be stored in fireproof cabinets.

Safe Paint Mixing, Handling, and Spraying Techniques

Mixing paint or cleaning with solvents must always be performed in well-ventilated areas irrespective of their flammability. All ignitable mixtures must be maintained in tightly covered safety containers (those which will not release any of their content if accidentally overturned) when not in use. Transporting open containers of any flammable liquid is a serious safety violation. Do not move paint that was mixed in the paint room to the line unless it is in tightly sealed containers.

When spraying, keep as much of the coating as possible on the parts and as little as possible on the walls or floor. This may sound simple enough, but in a typical spray-painting operation, only half of the paint or less ends up on the product.

Workers should operate near the center of the spray booth and away from the walls so that the cloud of atomized overspray is carried away from their breathing zone as rapidly as possible.

As much as possible workers should spray in the direction of the booth airflow. This direction is normally toward the water wall or particulate filters so that the paint overspray cloud moves away from the sprayer.

Workers should wear an approved, and *properly fitted*, respirator or fresh air mask. Facial hair may prevent correct mask fit and, worse, give the wearer a false sense of security against harmful vapors and particulate.

Waste Disposal

Any paper waste containing dried overspray deposits should be removed from the booth area on a daily basis for local disposal. When empty, containers for paint, solvent, and catalyst must be removed from the booth. They can be sent to local domestic disposal areas only if they are free of all traces of flammable material. Bake them in the paint oven if necessary to ensure the absence of flammable substances.

Combustible liquids such as unused paint or thinner should be removed from the booth area immediately after use. Liquid wastes need to be transferred to a properly identified waste container. The waste container should be stored in a fireproof cabinet located in a secure area; the cabinet must provide sufficient secondary containment. All flammable liquids temporarily maintained in the booth must be kept in sealed, spill-proof, electrically grounded containers.

Chapter 10

Paint Application II: Electrocoating and Autodeposition Coating

Coating by Immersion

As in ordinary dipping, both electrocoating and autodeposition require immersing the part to be painted into a quantity of liquid paint. Although dipping can be used with solventborne as well as waterborne finishes, electrocoat and autodeposition paints are always waterborne. However, the immersion of a metallic part into a waterborne coating is about the only thing these two techniques have in common. As may be deduced from its name, electrocoating involves the use of electricity to apply paint onto the parts. Autodeposition, on the other hand, utilizes a series of integrated chemical reactions to achieve coating deposition.

Electrocoating

Electrocoating is also called electrodeposition coating, electrophoretic coating, electropainting, the electrodeposition of polymers, or, most commonly, simply ecoat (preferred over the hyphenated spelling, e-coat). Ford Motor Company, under the guidance of the late Dr. George Brewer, pioneered much of the early work on ecoating. Possibly for this reason, at General Motors, the name ecoat was avoided, instead the "el" and the "po" from ELectrodeposition of POlymers was combined as a shorthand term. For over 35 years this process has continued to be known within GM as ELPO more often than as Ecoat, ecoat, or e-coat. But no matter, the shorthand term used—*ecoat*—involves the electrical deposition of a coating film from a waterborne organic solution onto an electrically conductive part. The process requires that the workpiece part be one of two electrodes that form an electrochemical cell. The ecoat electrical cell is comprised of a source of electricity, a positive electrode, a negative electrode, and the coating solution. Three of these components—the part to be painted (which is one electrode), the second electrode (which is charged electrically opposite to the part), and the ecoating solution—must all be electrically conductive in order to complete the requirements of a cell.

Ecoat application is very similar to electroplating of metal—a process that also involves a source of electricity, two electrodes, and a conductive solution (an inorganic electrolyte in plating instead of an organic coating solution). But plating deposits a film of metal, whereas ecoat deposits a layer of paint. The electroplated film of metal needs no baking, but the electrodeposited ecoat layer must be cured. Although there are a few air-dry ecoats and even some low-bake ones, most ecoats are baked at substantial temperatures. The reason is that high-bake ecoats are considerably more durable and corrosion resistant, and thus are much more likely to be used.

The ecoat process is extremely efficient, depositing a highly uniform paint film on all surfaces that can be reached by the electrical field strength. This includes all surfaces except those in confined and physically restricted areas of the parts, such as inside long, narrow recesses. These are termed Faraday cage areas. The Faraday cage effect can be illustrated with an ecoated pipe. With a pipe, the electrodeposition occurs preferentially at the high-current-density areas—that is, the outer pipe surfaces, especially those closest to the oppositely charged electrode. Little coating is deposited inside the pipe where current density is low; inside the pipe only the ends are coated somewhat. The rest of the inside is bare. Even where some coating is deposited, the film is thinner inside than outside the pipe. A rule-of-thumb is that ecoat will deposit inside the end of a pipe only to about a distance equal to the internal radius.

The basic components found in a typical ecoat system are illustrated in **Figure 10-4**. Parts to be painted are conveyed into and out of the ecoat bath using an overhead conveyor. This is shown in **Figure 10-1**. Sometimes, however, parts are lowered into an ecoat bath with an "elevator style" vertical hoist such as pictured in **Figure 10-2**. A few systems

Figure 10-1. Parts conveyed into ecoat bath by overhead conveyor.

even raise the entire ecoat tank up with a scissors lift to douse the parts hanging on the conveyor line in the manner shown by **Figure 10-3**. This method is a bit awkward and cumbersome but it does have an advantage in that it allows easy changing of paint colors by switching different ecoat tanks into position under the line. Very few such systems are used, however.

Electrical Considerations in Electrocoating

Ecoat power supply requirements vary in direct proportion to the total surface area of parts immersed in the solution at a given time. The power supply must be able to provide a current density of from 1 to 4 amperes (amps or A) per square foot of surface of the part being coated. This current density will normally produce a film build of approximately 1 mil (25 µm) thickness in about 90 seconds. A typical high-capacity power supply can provide 50–900 amperes of direct current at 75–500 volts.

The parts being coated as well as the conveyor carrying them comprise one of the cell electrodes. The conveyor together with the parts is always electrically grounded in all ecoat systems. This greatly simplifies safety and eliminates the need to isolate the con-

Figure 10-2. Parts lowered into ecoat bath by elevator-style conveyor.

Figure 10-3. Ecoat tank hydraulically raised to parts.

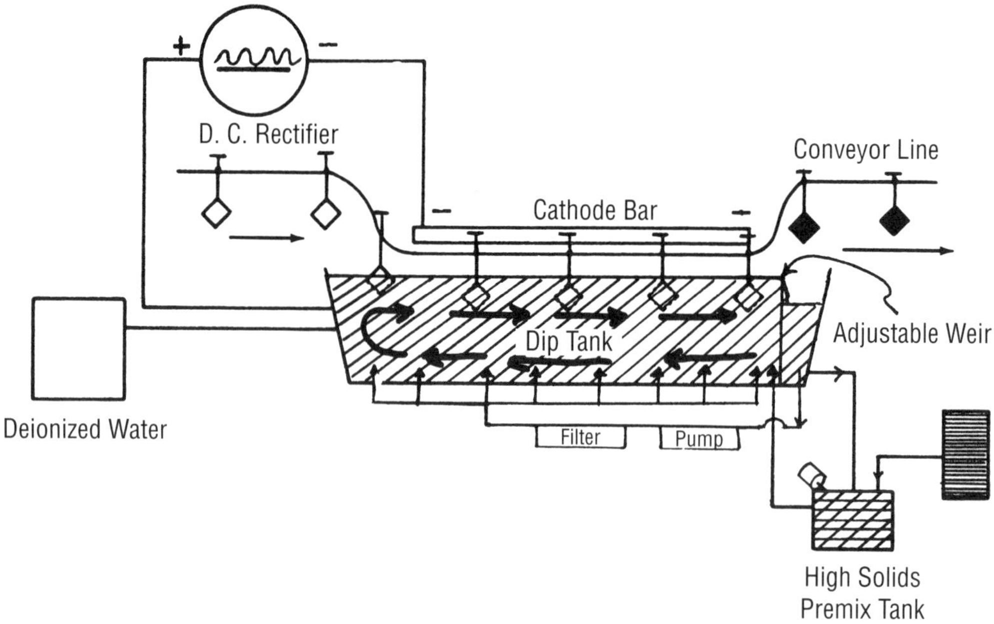

Figure 10-4. Basic components of an ecoat system.

veyor from ground. The opposite electrode of this cell, however, must be charged either *above* "ground" potential (making it the cathode), or *below* ground potential (making it the anode). In either case, this creates a difference in electrical potential that results in current flow and the deposition of the paint film. In a *cathodic*, or *cationic*, ecoat system, the part being coated is the grounded cathode. The opposite electrode, the anode, is charged positively relative to ground potential. In an *anodic* ecoat system, the part being coated is the grounded anode, and the cathode is charged negatively compared to ground potential.

Anodic ecoatings turned out to be simpler to formulate and so they were developed first. They were sold commercially for at least 10 years before any successful cathodic ecoats were marketed. The essential deficiency of anodic ecoats is the result of the fundamental nature of the chemical reaction, oxidation, which takes place at the anode in all electrochemical cells. The circumstances can be more clearly illustrated by studying a copper electroplating cell. During operation of the plating cell, the copper anode slowly dissolves, and dissolved copper ions go into solution. These copper ions are positively charged, and therefore are attracted to the negative electrode, the cathode. The copper ions move through the solution to the cathode; when they reach it they are neutralized, depositing on the cathode as a layer of copper atoms. In like fashion, with anodic ecoating the metallic part being coated (the anode) is dissolving to some degree while it is being coated. The dissolved ions of the anode end up in the paint film where they tend to discolor the ecoat. This is especially noticeable with light-colored ecoats on steel stained by red rust. The metal ions also lower the corrosion resistance of the film so it is not only an appearance problem.

These drawbacks encouraged experimental efforts that led to the eventual development of cathodic ecoats that proved superior to anodic ecoats. The basic defect of anodic ecoats was eliminated in cathodic ecoats; the cathode—the part being coated—does not tend to dissolve. Thus in addition to eliminating discoloration, the cathodic ecoats also improved corrosion resistance. Actually this higher corrosion performance is due mainly to the inherent chemical superiority of the cathodic resins that markedly improves paint film integrity, and is only to a lesser extent due to the absence of metal ions.

Further discussions in this chapter will focus more on cathodic ecoats than anodic because most systems used today are cathodic. Anodic systems are still in use, and some new anodic systems are being installed, but the main trend is toward cathodic ecoats because of their superiority. Yet cathodic deposition paints have a disadvantage in that they are more costly than anodic types. So if top performance is not required, an anodic ecoat may be the better choice in some instances. For a while certain color shades of anodic paints were still used because their pigments were stable only in alkaline (anodic) paint baths, but not in acidic (cathodic) baths. Some red colors and the bright whites were for a long time impossible to get in cathodic ecoat, but that's hardly true any longer. Now just about any shade or color tone can be formulated for cathodic paint.

Ecoat Paint Constituents

Paint to be used in ecoating must be specifically manufactured exclusively for that purpose; additionally, it must be designed as either an anodic ecoat or a cathodic ecoat, not both. Ecoat paints are always formulated as true solution paints. A typical cathodic ecoat resin is somewhat similar in nature to a conventional waterborne paint, but the polymer molecules are chemically modified to enhance their solubility in mildly acidic solutions. (Anodic resins are made to be soluble in alkaline formulations.) For this reason, ecoat paints cannot be used in ordinary dip coating; the paints must be electrically deposited or their film appearance and durability would be unacceptable.

Like nearly all conventional paints, an ecoat bath contains resin, pigments (unless it's a clearcoat), solvent (mostly water plus an organic cosolvent), and additives. The relatively high water content (60% and higher) puts ecoat in the family of waterborne coatings. A typical ecoat bath will also contain from 2–6% of a high-boiling water-soluble solvent, such as butyl cellosolve or hexyl cellosolve (ethylene glycol monobutyl ether or ethylene glycol monohexyl ether) to give smoother films. Controls on ecoat operation are designed to keep the cosolvent in this range. Moderately higher cosolvent levels do not greatly affect the coating process, but if excess cosolvent is present the film can be soft and plant VOC emissions may become unacceptably high.

We have seen that not just any waterborne coating can be used as an ecoat bath; the resins must first be modified to contain certain solubility-enhancing chemical groups. This makes ecoat paint distinctly different from other waterborne paints. To make resins suitable for the ecoat process, they need to be treated with appropriate solubilizers to form positively or negatively charged molecular species called polymer ions. Acid solubilizers are needed for cathodic ecoats, and alkaline solubilizers for anodic ecoats. In their ionic form these resins are polar, and thus soluble in water since water is a polar solvent. Anodic and cathodic resins, however, are never interchangeable.

For the anodic resins, carboxylic groups are chemically attached at intervals along the resin molecules. The modified resin will be mixed with alkaline substances such as potassium hydroxide, sodium hydroxide, triethyl amine, or other amine compounds to solubilize the resin molecules. Organic salts are formed by reaction of the alkaline solubilizer with the resin. These resin salts are polar and appreciably water-soluble.

Cathodic ecoat resin molecules are chemically modified whereby tertiary amine groups are added along the backbone carbon chains of the molecules. Without ionizable groups such as amines, the resin could not be used in an ecoat formulation. Tertiary amine groups, when treated with dilute acid solubilizers such as acetic, lactic, or formic acid, will form the necessary positively charged resin ions needed for cathodic ecoat.

In theory any chemical resin type can be modified for use with ecoat. Yet, in fact, epoxies and acrylics are used almost exclusively. Epoxy ecoat paint films, like any other epoxy, tend to chalk when exposed to ultraviolet (UV) light. This is generally not a problem if the epoxy ecoat film is to be a primer. If the epoxy is to function as a topcoat, then it shouldn't be exposed to UV light unless chalking does not present a problem. Ecoat paint films that require outdoor exposure and relatively high appearance standards use acrylic resins. These are resistant to attack by UV light and have excellent weatherability.

In formulating ecoat resins, chemists select cross-linking agents, such as aromatic and aliphatic isocyanates, that are appropriate to the end use of the paint film. The aromatic cross-linkers are the less costly of the two. But the aromatic cross-linkers tend to cause yellowing of the paint film if given outdoor exposure, whereas the aliphatic do not. To avoid darkening problems, acrylic ecoat resins are cured with aliphatic cross-linkers.

Since both aromatic and aliphatic isocyanates react with water, and ecoat baths contain water, chemists "block" the isocyanates to prevent water reaction. The blocking is done chemically by "capping" the water-sensitive ends of the isocyanate molecule with a blocking agent. The blocked isocyanates deposit within the ecoat film, but later the high bake temperatures drive off the blocking agent "cap" in the cure oven. At this point the isocyanate groups are to cross-link the ecoat resin molecules and cure the ecoat film.

Chemical Reaction in the Ecoat Tank

For the initial fill of the ecoat tank, insoluble neutral resin is premixed with a solubilizer (most often acetic or lactic acid) to produce soluble positive resin ions. Formed simultaneously are negative solubilizer (acetate) ions according to the reaction:

$$\text{neutral resin} + \text{acetic acid} \rightarrow \underset{(\text{positive})}{\text{polymer ions}} + \underset{(\text{negative})}{\text{acetate ions}}$$

During deposition in both anodic and cathodic ecoat systems, the bath reactions are essentially similar except for different chemical groups and the opposite electrical charge on the polymer ions. When the part to be ecoated is immersed in the paint, direct current is passed through the solution. The passage of current causes an electrochemical reaction that converts the water-soluble polar ionic resin polymer into a neutral nonpolar (and therefore insoluble) molecule that can no longer remain dissolved in the paint bath. As a result, resin molecules are deposited directly onto the surface of the parts to be painted.

Such large amounts of direct electrical current pass through the bath that some of the water simultaneously decomposes into oxygen gas and hydrogen ions ($H+$) at the anode, and hydrogen gas and hydroxide ions ($OH-$) at the cathode. Normal plant air circulation sweeps away the flammable hydrogen gas to prevent the possibility of an explosion or fire. The amount of hydrogen generated is generally of low volume.

The soluble positively charged (cationic) resin ions in the paint bath interact with the negatively charged hydroxide ions generated at the cathode (the part being painted). Combination of these two oppositely charged species forms water and neutral insoluble paint molecules, and in this way items are quickly coated with paint. For a while, as direct current continues to pass through the ecoat, increasing amounts of resin become insolubilized. **Figure 10-5** depicts what happens at the electrodes in a cathodic ecoat system. As more and more neutral paint molecules deposit on the metal surface, the part becomes increasingly electrically insulated. This gradually reduces the current flow, slowing the rate at which the paint film is deposited, until it finally stops almost completely. In most cases, about 0.5–1.8 mils (13–45 μm) are deposited in 1–4 minutes. After about 4 minutes, virtually no additional coating will be deposited. Because of the self-limiting effect, it is not possible to produce single ecoat films over 2 mils (50 μm) thick.

Only substrates that have relatively good electrical conductivity can be ecoated, so this process tends to be done only on metal parts. Just a single coat of paint can be applied by ecoat unless the existing layer of paint is strongly conductive. If an object has a layer of any ordinary paint film, the surface probably will be too limited in conductivity for coating to deposit. Double ecoating is possible; a highly conductive paint is used for the first

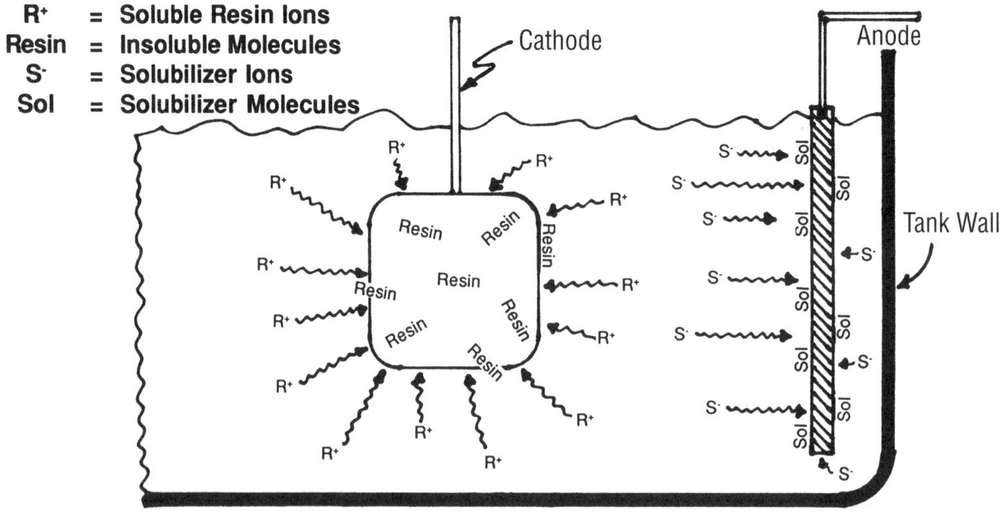

Figure 10-5. Cathodic ecoat electrode reactions.

coat and completely baked before applying the second ecoat. Double ecoating can provide excellent salt spray resistance on phosphated steel, often in the vicinity of 1,500–2,000 hours.

The voltage setting regulates the current flow, typically 1–4 A/ft^2 (11–43 A/m^2); greater voltage is used to obtain higher film thickness. If the voltage is set too high, however, excess current will flow between electrodes, and gases formed by electrolysis of water can be generated under the newly formed paint film. This causes the paint to lift off in spots, sometimes as large as an inch in diameter. The voltage at which this film lifting occurs is called the *rupture voltage*. Rupture tends to occur on projections or other areas of parts that are closest to the electrodes because current density will be greater in those areas.

At the other electrode—the anode—a process similar to resin neutralization occurs during ecoating with the acetate ions from the solubilizer. Oxygen gas and hydrogen ions are generated at the anode, and the attraction of opposite charges causes negative solubilizer ions to be drawn to the positive anode. Positively charged hydrogen ions are generated by electrolysis of water in the vicinity of the anode. The negative solubilizer ions (acetate ions) combine with positive hydrogen ions to regenerate neutral (in the sense of electrical charge) acetic acid (solubilizer) molecules. The reaction is:

$$CH_3COO^- + H+ \rightarrow CH_3COOH \text{ (acetic acid).}$$

Although neutral, the acetic acid molecules are small and mildly polar so they are water soluble. As coating continues, resin, solubilizer, and pigment must be added periodically to replenish the paint bath. But when the parts enter and then leave the bath, they carry out with them only resin and pigment, not solubilizer. If nothing were done, the solubilizer concentration would continually rise and as a result the electrical efficiency would drop. Three methods can be used to prevent excess solubilizer from accumulating in the bath.

Any single one (or all) of the following may be used in a given system to keep solubilizer concentration at an acceptable level:

- Anode semipermeable membrane
- Solubilizer-deficient makeup resin
- Ultrafiltration/reverse osmosis.

Anode Semipermeable Membrane

If the anode is encased in a semipermeable membrane, small molecules and ions (including acetate ions) will be able to pass through the membrane. Small negative ions can travel through the membrane pores rapidly because they are electrically attracted to the positive anode. Paint resin ions and pigment particles will be excluded because they are too large to pass through the membrane pores. When acetate solubilizer ions (CH_3COO^-) are neutralized by combining with H^+ ions, they become concentrated to a large extent within the anode enclosure. An aqueous flush called the *anolyte solution* is introduced into the bottom of the anode compartment. The anolyte is essentially water that slowly flushes excess acetic acid out of the cell directly to drain or to a holding tank for recirculating the anolyte, as shown in **Figure 10-6**.

As this flush water is recirculated, its acetic acid content slowly increases. Normally, the pH of the anolyte flush ranges from 2.5 to 5.0, and the electrical conductivity ranges from approximately 250 to 950 micromhos. Periodically, some of the dilute acetic acid solution in the tank must be released to drain. (Acetic acid is so mild in dilute form that it is used in a large number of foods; vinegar, for example, is a 5% water solution of acetic acid.) Deionized water is used to replace the amount of anolyte solution released to drain. If the dilute acetic acid is neutralized with caustic soda in waste-treatment holding tanks, it forms sodium acetate, a harmless soluble substance that in dilute form can be drained directly into the sewer.

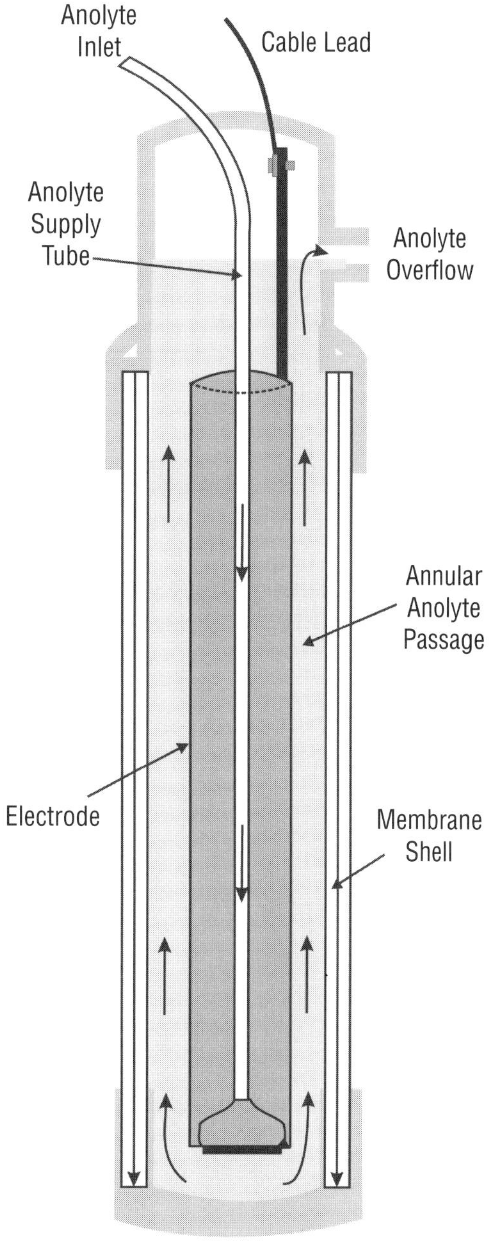

Figure 10-6. An anode cell.

Solubilizer-deficient Makeup Resin

Another way to reduce the solubilizer content is by using solubilizer-deficient resin in the paint makeup additions. If replenishment resin is added that is not fully neutralized with solubilizer, it is possible to force the excess solubilizer in the bath to become the source of the needed solubilizing acetic acid. If added slowly and with very thorough mixing, the solubilizer-deficient makeup resin can be stirred and blended into a large quantity of the tank's contents. There it reacts with the excess solubilizer in the bath to form fully solubilized resin.

Ultrafiltration/Reverse Osmosis

Ultrafiltration and *reverse osmosis* are terms that describe forcing molecules of a liquid through a very fine filter membrane. Both techniques are essentially the same, differing only in the pressures and membranes used. Reverse osmosis uses higher pressure than ultrafiltration. Ultrafiltration is used to separate resin and pigment from water (and solubilizer) for rinsing parts. This water contains some solubilizer, however, and if part of the ultrafiltrate is discarded it lowers the solubilizer concentration of the paint bath. A few systems use reverse osmosis to further purify the ultrafiltration-generated rinse water by removing solubilizer and ionic contaminants. These removed materials are released to the wastewater drain. Any discarded ultrafiltrate or reverse osmosis liquid must be replaced, of course, by an equivalent amount of pure water.

Part Rinsing After Ecoating

A part emerging from an ecoat bath has a deposited paint film that is covered by a wet solution consisting primarily of water and bath paint solids (drag-out). An elaborate rinse system is employed to remove the drag-out before the coated parts are baked. Extremely large quantities of rinse water are required to remove all drag-out materials that may cling to the part. If this drag-out were not thoroughly rinsed away, the ecoat film quality would suffer. For example, poor rinsing causes pronounced ecoat roughness that can require sanding of the film after curing. Sanding ecoat films is possible, but it is not easy to do because they are extremely hard. Any slow, tedious sanding would increase labor costs and slow production so thorough ecoat rinsing is, therefore, absolutely essential.

This need for thorough rinsing, however, does cause a major dilemma. The rinse water volume must be large, and yet the total amount of rinsed drag-out is relatively low. To discharge all the rinsed-off bath solids to drain would be cost-prohibitive and cause an excessively high biological oxygen demand on the sewer system. The problem is complicated because the small amount of paint solids and the large quantity of rinse water cannot economically be separated. Filtration, distillation, and related methods are all far too costly for separating paint solids from the rinse water. If this problem could not be resolved, nearly all ecoat processes would be stymied; most plants simply could not afford to discard that volume of solids along with the rinse water.

A possible answer might be to use fresh rinse water and allow the rinse water plus the small amount of drag-out to run directly back into the ecoat tank. But if continuous fresh water were used for rinsing, the ecoat tank would rapidly be diluted and accumulate such a large amount of extra water that the tank would quickly overflow. But, continually adding fresh deionized water to rinses to remove the drag-out and then discarding all of the ultrafiltration would also be cost-prohibitive.

The remedy to the problem is found by using ultrafiltration to withdraw (temporarily) a small portion of the mostly-water permeate from the ecoat bath. The *permeate* consists of water plus a small amount of solubilizer and bath salts that is "squeezed" out of the ecoat bath by ultrafiltration. It is called permeate because it penetrates through the ultrafilter pores. This borrowed permeate is then counterflowed through the rinse system. All of the rinse permeate is collected and reused in several (usually three) separate counterflowing rinse stages. The rinse permeate, having been used and collected for reuse several times, is finally returned to the bath with an initial rinse located directly over the exit end of the ecoat tank. **Figure 10-7** shows how ultrafiltration is tied into an ecoat rinse system.

Thus, the ultrafilter system temporarily appropriates a small portion of the water in the bath, uses it to rinse off the parts several times, and finally allows the permeate plus all the rinsed-off drag-out paint to flow back into the ecoat tank. This closed-loop filtration system continually provides enough permeate for rinsing without diluting the bath. The permeate material extracted from the bath by ultrafiltration is also referred to as *ultrafiltrate* and *flux*. Factors that can affect the amount of permeate that is produced for ecoat rinsing include:

- Concentration of paint resin and pigment
- Pressure of the bath flowing through the ultrafilters

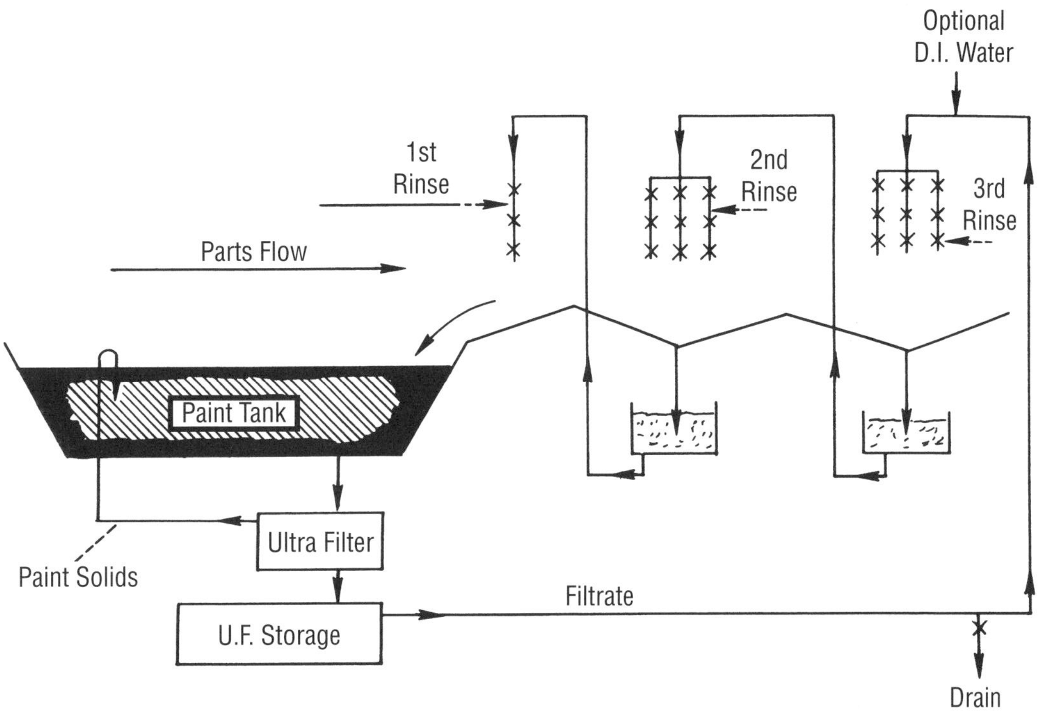

Figure 10-7. Closed-loop ultrafiltration.

- Flow rates of bath through the ultrafilters
- Bath temperature
- pH
- Co-solvent concentration
- Pigment-to-binder ratio of the bath solids
- Presence of foulants, such as chromates or phosphates, from pretreatment carryover and dissolved solids from the use of tap water rather than deionized water.

Under optimum conditions, the permeate production rate is about 1% of the total bath flow through the ultrafilters. As the filters gradually become fouled, this percentage slowly decreases until cleaning of the filter membranes becomes necessary. **Figure 10-8** shows a bank of ultrafilter cartridges in a fairly large ecoat installation.

Types of ultrafilter construction used in ecoat systems include spiral-wound, shell-and-tube, plate-and-frame, and hollow-fiber multiple-tube. Each has advantages and disadvantages that should be examined by those considering ecoat systems. Individual preferences and preventive maintenance patterns seem to be as significant as operational and performance factors.

Ultrafilters must be properly maintained because good ultrafiltration requires a sufficient flow rate through the filter elements. In many instances, manufacturers suggest flow rates of at least 35 gal/min (132 l/min) to prevent fouling of the thin filter membranes. Thorough rinsing of parts after pretreatment will help avoid bringing contaminants into the ecoat bath, because these contaminants will plug ultrafilter pores and reduce permeate generation. It is inevitable that some insolubilized resin molecules will not deposit but remain in the bath. These will also gradually clog the pores of ultrafilter membranes.

Figure 10-8. Bank of ultrafilter cartridges in a large ecoat installation.

In normal production, the ultrafilter permeate output can be expected to slowly decrease. Depending on throughput rates, normal cleaning of cathodic ecoat ultrafilters may be required at intervals of 6-20 weeks. When permeate generation falls to 80% (approximately) of initial output levels, the ultrafilters must be cleaned with a mixture of solvents and concentrated solubilizer solution. Cleaning the ultrafilters frequently will minimize rinsing problems. Extending the interval between cleaning is an invitation for ecoat trouble. If cleaning is postponed too long, permanent impairment of ultrafilter function can result, especially with hollow-fiber and multiple-tube ultrafilters. Flat-plate and spiral-wound types may survive this neglect. However, even these ultrafilters should have regular cleaning as soon as their output diminishes to the designated level specified by the cleaning guidelines of the ultrafilter manufacturer.

After single or multiple rinsing with ultrafiltrate, the parts may undergo still further rinsing with a small volume of reverse osmosis or deionized water to ensure that all contaminants are removed prior to baking. The RO or DI water rinsing is usually accomplished in two stages: a recirculating rinse and a final misting rinse with virgin RO or DI water. The final fresh water rinse of about 1–2 gal/min (3.8–7.6 l/min) should flow into the recirculating deionized rinse tank, from which it can then overflow either to drain or into a permeate rinse tank.

A large amount of DI or RO water is needed to prepare the ecoat bath and provide water for the ecoat rinses. Many companies insist that the drain-off from the final DI or RO rinse in the pretreatment section have maximum conductance of 30 micromhos. This is done to ensure that there is little possibility of contaminating the ecoat bath or fouling the ecoat ultrafilters.

Prior to the solubilizer-deficient makeup entering the bath, it must already have been nearly totally solubilized. This is essential because even tiny amounts of insoluble resin will quickly plug the ultrafilters and sharply curtail the production of permeate rinse. Solubilizer-deficient makeup resin is often added to the ecoat tank through injection nozzles, or ports, used to return the recirculating ecoat bath material. The entry ports, called *eductors* (see **Figure 10-9**) are located along the bottom and in the sloped exit end of the tank. **Figure 10-10** shows how replenishment materials are added to an ecoat system, and **Figure 10-11** shows banks of eductors at the bottom and exit end of an ecoat tank. Adding the makeup resin in this way helps lessen the chances of fouling the ultrafilters, compared to systems where makeup paint is introduced into the intake of the circulation pump. The eductor entry method helps to ensure thorough mixing and complete solubilization of the makeup material before it reaches the ultrafilter section.

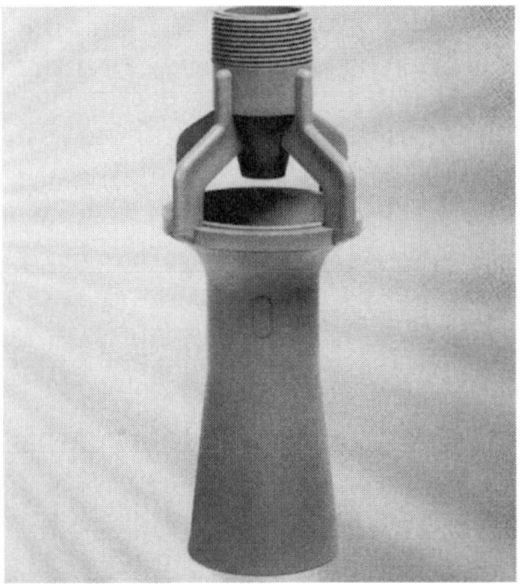

Figure 10-9. Numerous eductor nozzles, such as the one above, are used to agitate the ecoat tank and prevent pigment settling.

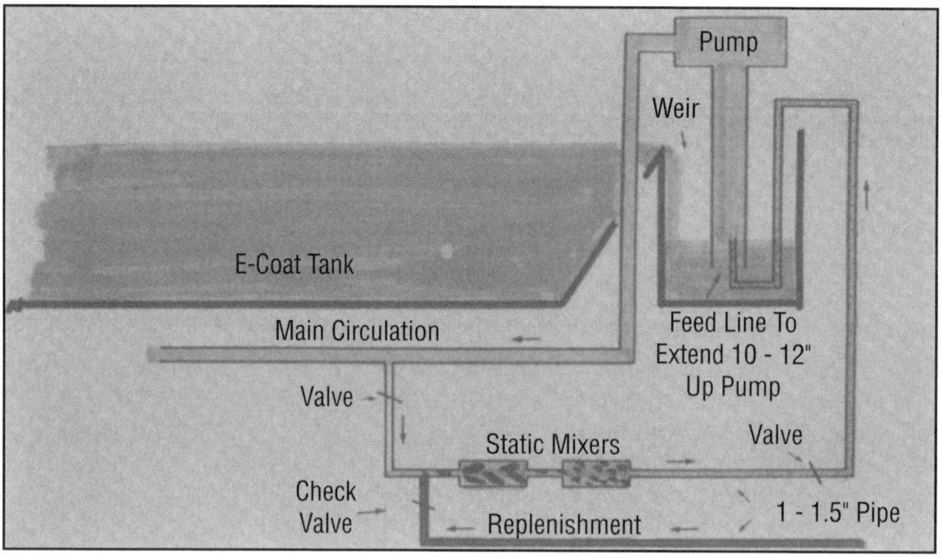

Figure 10-10. Ecoat single component feed.

The bath content must be monitored closely for the buildup of excessive chemicals that can interfere with efficient ecoat operation. For example, an excess concentration of ionic substances in the bath can raise conductivity and interfere with electrical efficiency. To prevent such buildups, a small amount of permeate in some systems is continuously released to drain and replaced with an equal amount of RO or DI water. As a consequence, water hardness salts and ionic contaminants from pretreatment baths, both of which are present in small amounts in the permeate, are released to drain along with the discharged permeate. Dumping permeate to drain also eliminates some excess solubilizer from the ecoat bath. Solubilizer buildup in the permeate rinse must be avoided because the solubilizer will dissolve the freshly deposited paint film. Too much solubilizer in the permeate rinse has the effect of stripping off the paint immediately after it is applied.

Pigments. Pigment particles are mixed uniformly throughout the bath by vigorous circulation. During the ecoat process the pigment particles become entrapped among resin molecules as the resin is attracted toward and being electrodeposited onto the parts. Ecoat pigments never undergo any change in electrical charge; they remain uncharged throughout the process. The same is true for any additives. Only the ecoat resins undergo change; they are electrochemically converted during ecoating from their soluble ionic state to the neutral (and hence insoluble) form.

Bath Parameters. An ecoat bath needs to be monitored closely to ensure satisfactory coating deposition and good quality film formation. Meticulous control and records should be kept on film build levels and on the following ecoat bath parameters:

- Percent total solids
- pH
- Solubilizer concentration
- Solution conductivity
- Temperature

- Relative amounts of pigment and binder
- Voltage and amperage.

The pH for many cathodic systems with flushable anodes is held at a slightly acidic level of around 6.0 to 6.5. A number of other systems, especially those that eliminate excess solubilizer primarily by flushing ultrafiltrate to drain, may operate at pH levels as low as 3.0 to 3.5.

A bath sample must be titrated periodically to check the concentration of acetic acid solubilizer. The concentration is usually expressed in units of milliequivalents (meq) of solubilizer per liter of bath material. A typical operating solubilizer concentration is 90 meq.

Overall ecoat bath conductivity affects the electrical efficiency. If it is too high, electrical efficiency of the ecoat deposition drops. Normally, a maximum bath conductivity of roughly 900 micromhos is appropriate. Above this point, more permeate and/or more anolyte flush should be sent to drain to lower conductivity.

A substantial amount of direct electrical current is used in ecoating. This tends to raise the temperature of the paint bath because an amount of conversion of electrical energy to thermal energy always occurs. Chillers and heat exchangers are used in the main bath circulation system to keep the bath temperature in a normal range of 70–95°F (21–35°C). Excessively high temperatures can cause difficulty with the ultrafiltration process and deteriorate the bath.

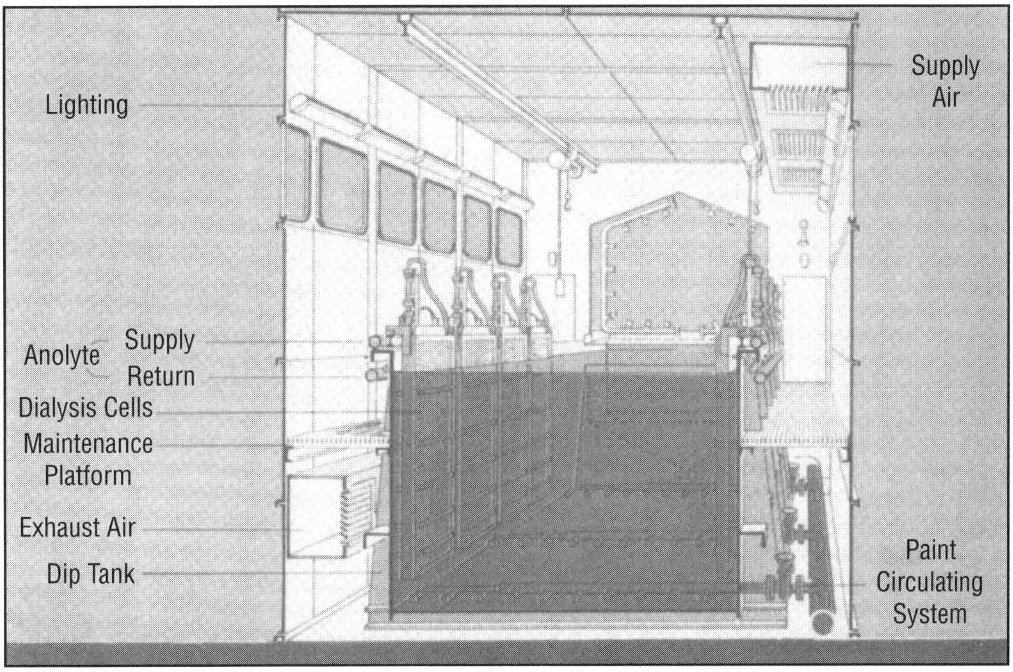

Cross Section Through Dip Zone

Figure 10-11. Eductor entry method using eductors at bottom of tank to ensure thorough mixing of makeup material.

As resin and pigment are removed from the bath by being deposited as a paint film onto the parts being coated, additional resin (presolubilized with acetic acid) and pigment has to be added periodically. Sometimes the resin and pigment are replenished in a single concentrate. However, since the resin and pigment may be used up at different rates as they are deposited on parts, it may be necessary to add varying amounts of each. This requires adding each separately rather than as a constant ratio makeup blend of both. At times a system may use up resin faster than pigment, while at other times the reverse may be true. Paint supplier assistance in monitoring the resin and pigment levels and in establishing the correct pigment-to-binder ratio is usually available. Improper ratios can quickly and easily be readjusted by the appropriate makeup composition.

Continuous circulation and filtering of the bath are necessary to prevent pigment settling and to keep the bath clean. This ceaseless filtration is vital to avoid deposition of foreign contaminants along with the resin and pigment. Pumps should circulate the bath at a rate of four to six turnovers per hour. The bath is taken from the area behind the *weir* (an adjustable dam to control the level of the paint in the tank) and is then filtered. The paint is next fed back into the tank along the bottom through a series of pipes fitted with eductor nozzles. The jet force out of the nozzle orifices and the venturi action of the eductors on the surrounding bath help prevent pigments from settling and maintain good pigment dispersal in the bath. Eductors produce a mild scouring action that also reduces the amount of residue that can accumulate on the bottom of the tank. If adequate circulation is not maintained, extremely hard resin/pigment aggregates can build up on the tank bottom. As much as 14 inches (35 cm) of deposit have been found in ecoat tanks due to grossly neglected circulation problems. Sizeable parts that fall off the conveyor should be removed immediately from the bottom of the tank so that no dead circulation spots are formed that allow pigments to settle out of the paint. These would contribute to residue deposits and small bits of resin/pigment aggregates floating in the paint tank. These small pieces will deposit into the ecoat film and contribute unwanted visible roughness to the paint film.

Tank Configuration

The ecoat tank must be large enough to hold the parts to be coated and the necessary quantity of paint (see **Figure 10-12**). Some tanks are only 8 gallons (30 l), and some have a capacity as high as 120,000 gal (454,000 l). To paint parts moving continuously along a conveyor line, the tank must be considerably longer than for parts held stationary. For example, if parts need two minutes submersion time for ecoating and the conveyor is line traveling at 14 ft/min (4.25 m/min), the total tank length must be roughly 50 feet (15.25 m) long including the portions used for gradual part immersion and withdrawal.

A number of anodes are usually located vertically along both sides of the ecoat tank, but in some tanks anodes are also located horizontally near the bottom as well for better film build on the lower portions of parts. In most cathodic systems, each anode is enclosed within a flushable cell. These cells enable continual flushing with water (anolyte) to remove excess solubilizer produced by electrochemical reaction at the anodes. Tubular anodes (5 inches, or 13 centimeters, in diameter) or rectangular anodes (6" × 24", or 15 × 61 cm) membrane cells of the needed length can be used, but the ease of handling has made the tubular type cells increasingly more popular.

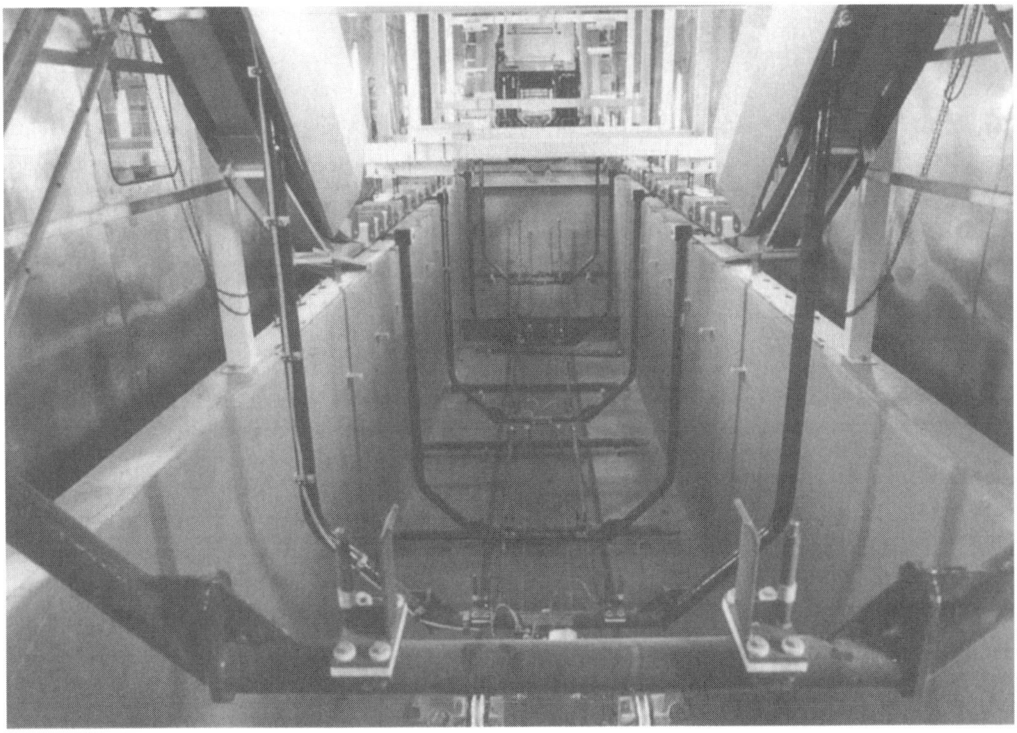

Figure 10-12. Empty ecoat tank.

Voltage at Tank Entry. Several options are available concerning the method of applying DC ecoat voltage while parts are entering the tank and after they are fully immersed. The two options are as follows.

1. The parts may receive no voltage during immersion into the ecoat. This is called *cold entry* or *dead entry*. After the parts are fully immersed, the voltage can then either be raised gradually to the specified level or be applied to that level all at once. Raising the voltage gradually after immersion will bring a slow, even, and optimum deposition of paint film. Applying full voltage suddenly after full immersion may cause current spiking and the formation of excessive hydrogen gas at the part surface, which can interfere with even paint film buildup, causing film rupture.
2. In the second procedure, the voltage is fully energized before parts begin to be immersed, during immersion, and after immersion. This is termed *hot entry* or *live entry*. As a part is being immersed during hot entry, a paint film buildup begins on the wetted portion of the part's surfaces. Part surfaces that are in the paint the longest will have the greatest film buildup. Sometimes the progressive film buildup can leave "hashmarks" or noticeable lines of distinct film buildup around the part, which can become pronounced if the conveyor movement is jerky during immersion. If anodes are left out at the tank entry section, the current density as parts enter the tank will be low and hashmark problems can be averted. If a separate control permits using low voltage in the front tank section, that will also prevent hashmarking.

Two different kinds of systems may be used to lower parts into the ecoat bath: a vertical hoist and an ordinary overhead conveyor. In the hoist type, a single part or rack of parts is lowered into the ecoat and the voltage is applied. With a hoist the tank need not be much bigger in length and width than the dimensions of the part or rack. After a specified bath deposition time, the part or rack is raised and is advanced sequentially to the rinse stages. The cycle repeats with each subsequent part or rack. With this type of system, it is relatively easy to have cold entry; the voltage can be raised gradually for optimum paint film buildup with no voltage spiking.

Using an overhead conveyor system, the tank will have a narrow width but an extended length. This often permits a relatively large number of parts to be immersed and ecoated at any given moment. To eliminate mechanical complexity, this type of system usually has hot entry, and the bath electrodes are generally spaced equidistant along each side of the tank. Some overhead conveyor systems wire the power supply so that the first few electrodes near the tank entry have reduced voltages. In this way, the parts do not come near an electrode with full specified voltage until a predetermined time after full immersion. Other overhead conveyor systems omit electrodes near the bath entry to permit a gradual paint film deposition both during and immediately after parts immersion. All these methods reduce the current spiking that otherwise would occur with hot (or live) entry of parts.

Because ecoat direct current electrical power may exceed several hundred volts and hundreds of amperes (amps), an ecoat tank is extremely hazardous during production operation. A person falling into an energized ecoat bath could be instantly electrocuted. A locked protective safety enclosure around the tank is necessary to prevent entrance by unauthorized personnel.

Ecoat Curing Cycle

The final ecoat step is in most cases oven curing of the deposited coating. Although low-temperature curing and even air curing are possible, the most durable films require a moderately high-temperature bake to cross-link the resins in the binder. This temperature is often in the order of 300–350° F (150–177° C) for anywhere from 12 to 35 minutes, depending on the design, configuration, and heat capacity of the part being coated. Convection ovens are used almost exclusively for curing ecoat paints, although infrared ovens can be used as well if the part configuration is simple.

Ecoat Advantages

Some of the advantages of ecoat include:

- Coating thickness uniformity and exact film build control
- High production rates and high transfer efficiency
- Primer/one coat/two coat options
- Zero to low environmental pollution potential.

Coating Thickness Uniformity

In electrical deposition systems such as electroplating and ecoat, the consistency of film thickness varies with the separation distance of the part being coated and the other electrode (or electrodes). Part surfaces closest to the charged electrode will receive the greatest film buildup because these are high-current-density areas, where film thickness will always be the greatest. Properly positioned charged electrodes will minimize differences in deposited film thicknesses. The deposition power (throwing power) is usually sufficient to deposit at least some coating thickness into many hidden areas. Auxiliary electrodes can be positioned to improve throwing power into these areas. By setting the ecoat voltage and regulating deposition time, film thickness can be precisely controlled. Edge coverage is excellent because in many instances edges become high current-density areas and receive slight extra film deposition.

High Production Rates

Ecoat is adaptable to very high production rates because of the fast film deposition rate and the rather dense line racking that is possible. Ecoat lines coating automotive wheels have operated at a conveyor speed of 48 ft/min (14.6 m/min), for example. Few paint lines run this fast except for some coil coating operations.

Primer/One Coat/Two Coat Options

Ecoat can be deposited to act as a primer to provide corrosion resistance and to serve as a durable base for topcoats or other primers. Ecoat primers tend to be mainly epoxies; others that have seen very limited use include alkyds and polyesters. Ecoat formulations—generally acrylics—also can be deposited as singlecoat systems. Many items used outdoors are coated this way; the acrylic resists chalking to which epoxies are susceptible.

Since the mid-1990s it has been possible to double ecoat parts. The first coat must be a special conductive coat, which is fully cured before the second ecoat is applied. This allows manufacturers to use both epoxy primer and acrylic topcoat and apply both by the electrocoat process. Currently a number of U.S. truck plants are applying two-coat ecoat priming on truck bodies using nonblack ecoats.

Zero to Low Environmental Pollution Potential

Ecoat paints are waterborne types with little cosolvent (in only a few cases with zero cosolvent) so VOC and HAP emissions are nil to low. New lead-free epoxies have eliminated the heavy metal problems of earlier epoxy ecoats. Heavy-metal-free ecoats have become commonplace today.

Ecoat Disadvantages

Ecoat disadvantages include the following:

- Substrate limitations
- Color change difficulty
- High costs
- Masking problems
- Bulk small-part coating difficulty
- Corrosion-resistant equipment requirement
- Conveyor stoppage problems
- High deionized/reverse osmosis water purity requirements

- Sophisticated maintenance requirements
- Sanding/stripping difficulty
- Second coat restrictions
- Air-entrapment pockets
- High deionized/reverse osmosis water volume requirements
- High energy demand
- Suitable only for high parts volume
- Coating thickness limitations.

Substrate Limitations

Ecoat cannot be used for plastics, wood, and other materials that are either nonconductors or poor electrical conductors. Even some high-silicon-containing aluminum alloys cannot be ecoated due to their lack of sufficient electrical conductivity.

Color Change Difficulty

Color change for most ecoat installations would be so slow that few even attempt it. More than 98% of ecoat installations paint only a single color or use a separate tank for each color. Quick color change would require a separate tank for each different color used. This is not generally installed for economic reasons, although a couple of plants have three different color tanks. Using more colors than three gets to be rather awkward and expensive. Most plants that change ecoat colors usually pump out the ecoat into a continuously agitated holding tank, and after cleaning the tank and all the ancillary equipment, fill the tank with the new color material.

High Costs

The initial capital cost and the continuing operating costs are extremely high when compared with simple systems such as dipping and spraying. This is not a problem as long as parts volume is fairly high; ecoat economics are poor if production numbers are low.

Masking Problems

Since ecoat is an immersion process and requires a subsequent high-temperature bake, the masking requirements are unique. Masking tape, plugs, and caps must be resistant to the ecoat bath as well as the oven temperatures. In many instances, these masking items also need to function as watertight seals.

Sophisticated Maintenance Requirements

Because of the complexity of an ecoat system, sophisticated maintenance is required. Careful maintenance records must be kept; testing on an array of parameters must be done on a regular schedule. Additionally, a paint sample of the ecoat bath is usually sent to the paint supplier for detailed testing, including total paint solids and pigment-to-binder (resin) ratio. Samples are submitted biweekly at first, then when the bath is stabilized, testing by the supplier can be extended to monthly intervals. Color corrections, however, may need to be performed on a daily basis.

Sanding/Stripping Difficulty

Because of the hardness of the deposited coating, it is extremely difficult to sand ecoat rework. It also can be difficult to strip ecoat paint from hooks, hangers, and parts baskets. This is not to say it cannot be done, rather that ecoat sanding is to be avoided as much as possible due to its difficulty.

Second Coat Restrictions

No previously painted parts can be run through an ecoat tank to apply a second coat of paint. It doesn't work. The paint in the prior coat will act as an electrical insulator on the parts and prevent deposition of ecoat paint over it. In order to apply a second coat by using ecoating, the previous coat must be highly conductive. The only process in which this is done is called "double ecoating." The first layer of ecoat is designed so that after a full bake it will be conductive enough to permit a second ecoat application.

Air-entrapment Pockets

Since ecoat is a dip process, it can be subject to coating voids due to air entrapment in pockets on parts. Any part that tends to form air pockets when immersed may need to be redesigned if it must be ecoated. Sometimes a conveyor oscillation system is employed to eliminate the air, or else it will move the air bubble back and forth enough to permit some coating to deposit. These systems work quite well but are too slow and expensive; not surprisingly, then, only a couple of these setups worldwide have actually been built and used. Auxiliary jets of paint under the surface of the paint can often be used to minimize voids from trapped air pockets.

Bulk Small-part Coating Difficulty

Bulk ecoating of small parts can be done, but it tends to require a complex system. Only a few bulk ecoat systems have been built due to the inherent problems of operation. Recently, however, improved equipment has made it possible to bulk ecoat small fasteners in huge volumes.

Corrosion-resistant Materials Requirement

Because of potential rust problems associated with any water solution, it is frequently necessary to use corrosion-resistant materials for ecoat tanks, piping, valves, heat exchangers, and other components that contact the paint. The acidity of all cathodic ecoat paints will slowly attack and dissolve metal parts; steel tanks for ecoat are lined with $^1/_8$" to $^1/_4$" (3 to 6 mm) special epoxy coating for protection.

Conveyor Stoppage Problems

The susceptibility of freshly deposited ecoat paint to the solubilizer and dissolved salts in ultrafiltrate rinse requires special precautions if the parts stop in or between post-rinse stages. If the conveyor stops, ecoated parts remaining in the ultrafiltrate rinse may be adversely affected by solubilizer stripping some paint off parts. On parts hanging between rinse stages, any unrinsed bath solids will dry on the parts within 5 minutes and then cannot be rinsed off. Manual rinsing or special misting rinses between stages must be initiated if the conveyor stops for more than a minute or two. Similarly, during a line stoppage any parts that remain submerged in the ecoat tank or one of the rinses for longer than 5 minutes will start to undergo paint film degradation.

High Deionized/Reverse Osmosis Water Purity Requirements

Premium DI or RO water quality must be maintained. Its conductivity should not exceed 25 micromhos; the resistivity minimum should be about 40,000 ohms/centimeter, and the dissolved solids content must be no greater than 10 ppm.

High Deionized/Reverse Osmosis Water Volume Requirements

The amount of deionized water used can be significant. It is normally necessary to produce reverse osmosis or deionized water in the plant. Equipment for this adds to the capital cost when installing ecoat systems.

High Energy Demand

The ecoat process requires a significant amount of electrical energy, and most systems need a high temperature bake. An elevated cure temperature is especially common for high corrosion resistance ecoats that are likely to be applied above 1 mil dry film thickness.

Suitable Only for High Parts Volume

Because the elaborate equipment that is required is expensive, ecoat is generally profitable only when finishing a large volume of parts. Amortization over a low number of parts would make alternate coating methods more competitive.

Coating Thickness Limitations

Ecoat film thickness is limited to about 0.5–1.8 mils (12–45 μm). As more and more paint is deposited, the coating acts as a stronger and stronger electrical insulator. This slows further coating deposition and finally halts it totally. It should be noted that maximum obtainable thickness varies widely among ecoat paints, and the value is dependent also on parts configurations. Many ecoats have maximum builds well below 1.8 mils.

Autodeposition

Organic paint films can be deposited onto iron, steel, and zinc parts by an oxidation-reduction precipitation process known as *autodeposition*, or chemiphoretic or Autophoretic® coating. The autodeposition process, which unlike ecoat uses no external source of electricity, is available primarily in black; however, several colors such as brown, orange, and light blue have been mentioned by the supplier as potentially available if high enough demand warrants it. The autodeposition film has a dull or low-gloss appearance and is primarily protective but not particularly decorative. The largest application areas for autodeposition coatings have been nonappearance and under-hood parts for cars and trucks. Excellent anticorrosion properties and black color make it highly appropriate for this application. It is also used on drawer slides for office furniture, replacing zinc-plating at much reduced cost.

An important advantage of autodeposition is its 100% coverage on all part surfaces wetted by the coating bath. Faraday cage areas, which hinder ecoat deposition, are nonexistent with autodeposition. "Where it wets, it coats" is a slogan used by the Autophoretic supplier. Applying coating to the inside of a pipe is virtually impossible with ecoat; no current density can be generated inside the pipe. (Oh, it could in theory be done utilizing an electrode inside the total pipe length, but that is so intricate a process it cannot be done except at prohibitive expense.) Both outside ***and inside*** pipe of any length is easily coated by Autophoretic. It only needs to have the Autophoretic bath in contact with the surfaces to be coated in order to deposit the coating.

Figure 10-13 shows the main process stages in an autodeposition system. The process begins with a heated aqueous alkaline spray cleaning for about 1 minute at 160° F (71° C). Then a dip cleaner follows for 1–3 minutes at 185° F (85° C). These cleaners are the only stages in the entire autophoretic process that are heated; all other stages operate at room temperature.

Thorough cleaning is extremely important because the process tends to be highly intolerant of contaminants. After multiple rinses, ending with a DI or RO water rinse, the parts go into the autodeposition bath for about 2 minutes. Two autodeposition paint resins are available: vinyl and acrylic. The acrylic is less common since a chromic sealer rinse must be used on coated parts before oven curing. The acrylic resin has a much higher thermal stability than the vinyl so it can be used as a primer for powder coat paints, whereas the vinyl cannot. The bath, typically held at 68–72° F (20–22° C), is a waterborne material containing about 10% of a resin emulsion, plus hydrofluoric acid and hydrogen peroxide. The coating deposition reaction is nonexothermic (does not give off heat), and the rate of deposition slows as increased coating covers the metal surface.

Coating deposition begins immediately upon immersion of the parts into the bath, which is maintained at a pH of 2.5–3.5. After about 2 minutes, 0.75–1.0 mil (19–25 μm)

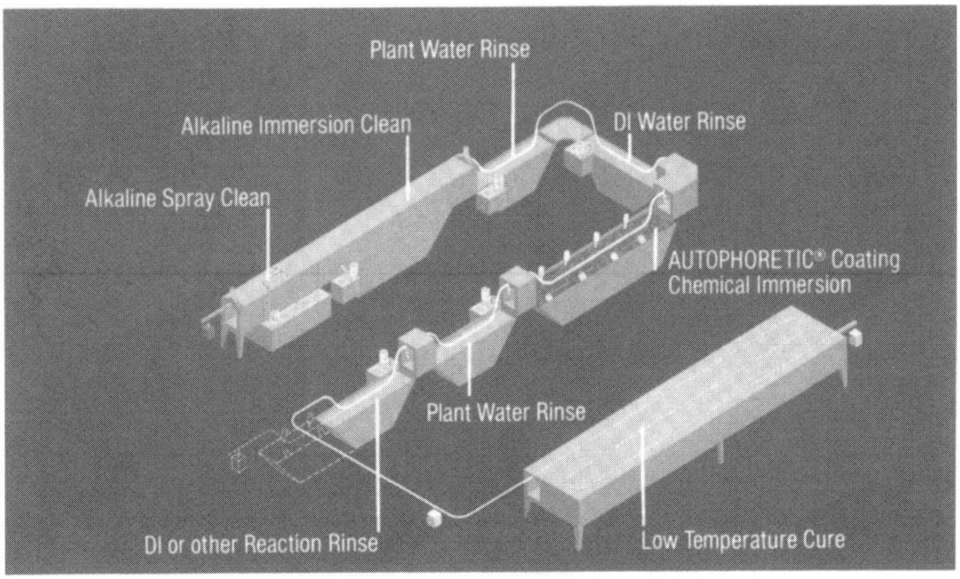

Figure 10-13. The main process stages in an autodeposition system.

of coating is deposited. Hydrofluoric acid etches and removes metal ions from the part by chemically attacking its surface. Hydrogen peroxide converts the solubilized iron ions from the +2 to the +3 (ferrous ion to ferric ion) oxidation state. The Fe^{+3} metal ions combine chemically with the emulsion polymer to form an insoluble resin that precipitates onto the surface of the metal. Even after parts are removed from the bath, unreacted resin is soon insolubilized by the continuing chemical action. Thus, no unreacted paint needs to be rinsed from the parts. The inorganic chemicals remaining on the parts are removed from the paint film by a 30 second water immersion rinse. **Figure 10-14** shows automotive brake parts emerging from one of the baths in an autodeposition process. When zinc is coated, a recent possibility with autodeposition paint, close control of the metal ion concentration becomes necessary.

After the coating is deposited, several options exist. One option is the use of a dilute chromic acid rinse, followed by an oven bake at about 250° F (121° C) for 15 minutes. Another option is to cure the coating in water at 180° F (82° C) for about 8 minutes. The water cure tends to yield a coating with less corrosion resistance than a coating cured in the hotter bake oven temperatures. The hot water cure method may include the use of chromic acid in the water for greater corrosion resistance if desired, but it is not required.

A second autodeposition process cannot recoat an autodeposition coated part because no exposed metal is left for the acid to attack. A major advantage of autodeposition is that no organic solvents are needed and thus no VOC is emitted. Autodeposition is used both by various OEM product manufacturers and also by a few custom-coating shops. It works well on thinner gauge metal but some problems have been experienced when autodeposition has been applied on heavy gauge steel. There are current process versions applicable to zinc and ferrous metals also, but no versions are available yet for coating aluminum.

Figure 10-14. Automotive brake parts emerging from autodeposition bath.

Chapter 11

Paint Application III: Spray Guns

Components of Air-Atomizing Spray Guns

The methods of paint and decorative film application described in the two previous chapters rely on various forms of spreading or splashing paint over the parts and immersing parts in various types of paint and film coating baths. As was detailed there, each method has its own unique advantages and disadvantages, thus any one of these techniques might become the most appropriate choice for coating a particular product. But there are many more paint application methods yet to be examined.

A different (and also very commonly used) group of processes for applying paint involves atomizing the paint into tiny mist-like particles and directing them to travel onto the surface to be coated. If a sufficient number of the particles is applied, they will create a continuous coating film. Paint can be atomized to various degrees and by several different techniques. The most common of these is with an air-atomizing spray gun, often referred to as *conventional air spray*.

The essential components of an air-atomizing spray gun are:

- Gun body
- Fluid inlet
- Fluid nozzle or fluid tip
- Fluid needle and seat assembly
- Air inlet
- Air cap or air nozzle
- Air valve
- Fan or spray pattern control
- Trigger.

Gun Body

The gun body consists of the handle for the operator to grip and the spray gun barrel. The main thing that almost all of the countless models of air-atomizing spray guns have in common is a grip and trigger configuration designed for operator hand comfort. The handle and barrel housing serve as a platform onto which are attached the gun's various air and liquid paint delivery components.

Fluid Inlet

The fluid inlet is an opening, usually below the tip of the barrel, that allows the paint to flow into the gun. The fluid paint inlet opening is threaded to allow attaching either a siphon cup or a paint hose to it. In some instances, the inlet is atop the gun. This allows a cup to be mounted on the top of the barrel so paint can flow by gravity into the gun just ahead of the fluid tip.

Fluid Nozzle or Fluid Tip

The fluid nozzle is a small device with a precision opening to permit the paint to flow out of the gun at a determined rate. One end of the nozzle is externally threaded and screws into the internally threaded barrel tip. An assortment of nozzles is available with different diameter fluid orifices to accommodate paints of different viscosities and delivery rates. The fluid needle seat is at the back of this part.

Fluid Needle and Seat Assembly

This assembly inside the gun serves as a valve to control the flow of paint through the fluid nozzle. The needle is attached to the external trigger. By using this trigger, the sprayer can start and stop the paint flow and also to a degree control the quantity of paint released from the gun. By turning a knurled knob at the back of the gun, the sprayer can alter the extent to which the trigger is retracted with a full trigger pull.

Air Inlet

This is a threaded opening at the bottom of the gun handle to allow the attachment of a hose connected to a source of compressed air.

Air Cap or Air Nozzle

The air nozzle is a small device that allows compressed air to be directed at the paint stream exiting the gun to produce optimum atomization. The nozzle, also called the air cap, is internally threaded to attach to the externally threaded gun barrel tip. Air nozzles

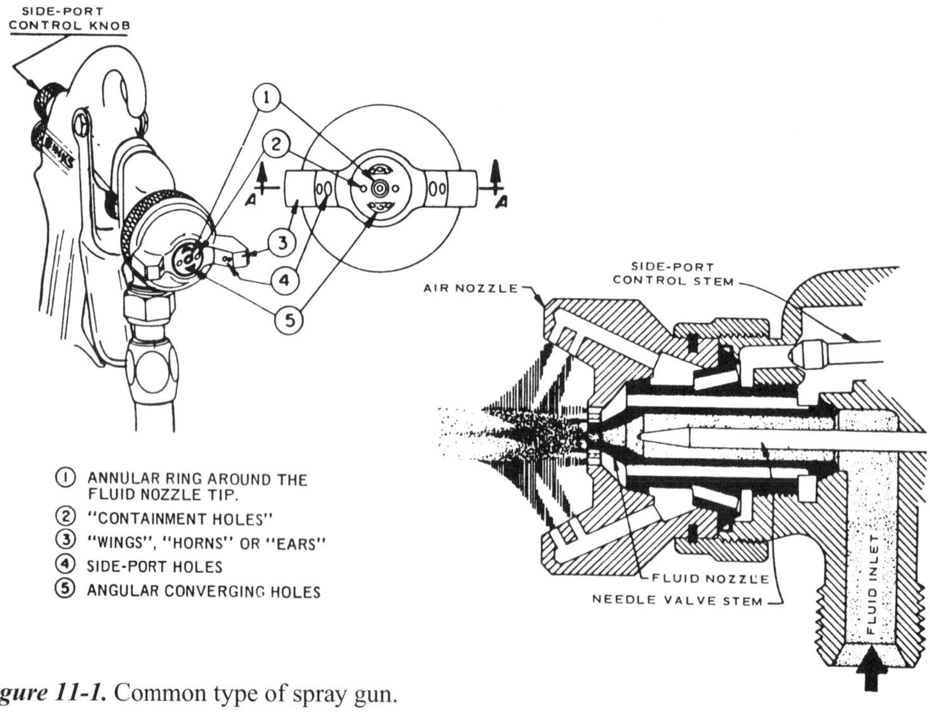

Figure 11-1. Common type of spray gun.

are available with many different configurations of openings to allow various paint atomization patterns. Air nozzles typically have two opposed "wings" or "horns" with precision air openings that can be utilized to vary the shape of the spray pattern into variably elongated configurations. **Figure 11-1** shows the details of an air nozzle on a common type of spray gun.

Air Valve

A simple valve gives the operator a means of controlling the flow of compressed air used for atomization and spray pattern control through the gun. Variable airflow cannot be accomplished by triggering with this valve; it is an on-off valve only.

Fan or Spray Pattern Control

This control permits the operator to regulate airflow through the air nozzle horn openings to vary the paint droplet spray pattern. **Figure 11-2** shows how turning a knurled knob at the back of the gun, which varies the airflow to the gun's air cap "horn" or "wing" openings, can alter a paint spray pattern.

Trigger

The trigger is connected to both the compressed air and fluid flow controls in the most commonly used type of spray guns called "nonbleeder guns." When triggered, the air valve opens first, and then the fluid valve; when the trigger is released, the paint valve closes just before the air valve closes. This assures that all paint exiting the gun will be atomized by the airflow. Partial triggering activates just the air valve and allows the sprayer to blow dust off parts before painting if desired.

In "bleeder guns," the trigger activates only the fluid controls; compressed air will escape the gun continuously, even when the trigger is released. Bleeder spray guns are far less prevalent than nonbleeder guns, except for heated-air high volume low-pressure (HVLP) spray guns. These are discussed later in this chapter.

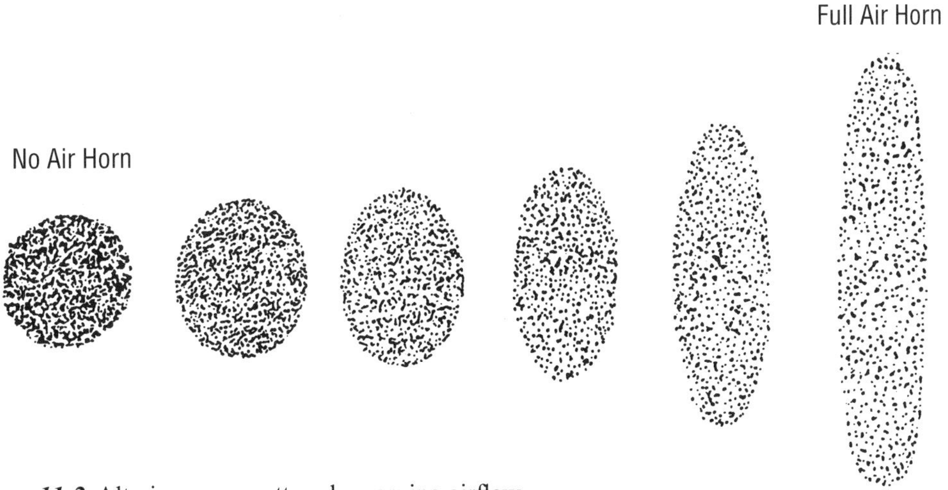

Figure 11-2. Altering spray pattern by varying airflow.

Compressed Air Supply

To prevent paint contamination, the air supplied to a spray gun must meet three requirements.

1. The air must be dirt free. The air must be filtered to remove dust, lint, and other foreign contaminants. The air is generally filtered at the inlet port of the air compressor and again just downstream of the compressor after oil and moisture removal.
2. The air must be oil free. Air compressors usually send out oil vapor with the compressed air because oil is used to lubricate compressor rotors. Such oil vapor is generally removed with an oil-absorbing or oil mist coalescing filter.
3. The air must be moisture free. The solubility of water vapor in air is such that warm air is able to hold far more moisture than cool air. When air is cooled, moisture tends to be wrung out of the air in the form of condensation or water. Since air increases sharply in temperature when it is compressed, it normally holds a considerable amount of water vapor. En route to the spray gun, the compressed air temperature drops, which produces condensed moisture. This water must be removed from the air lines or it will cause defects in the applied paint film. Water removal can be accomplished using refrigerated air-cooling systems and water cold traps, or by passing the wet air through beds of chemicals that will capture airborne moisture by absorption and adsorption. **Figure 11-3** shows the capability of deliquescent, refrigerated, and desiccant air dryers to remove moisture from compressed air streams.

Air can be supplied to an air-atomizing spray gun either by an air compressor or an air turbine. Air compressors of various design types are utilized; they may include diaphragm, rotary, or reciprocating pumps. They generally operate in a range of 75–160 PSI (5–11 bar). Both single and double compression is done. The compressed air in one design of air compressor is additionally compressed using a second compressor. In contrast to the above types of compressors, an air turbine is a fan-like device that operates at pressures usually much lower than those generated by air compressors. Turbines are used to supply relatively large volumes of air to paint application equipment at pressures rarely exceeding 50 PSI (3.5 bar).

Drying Medium	Dew Point	Moisture Downstream
Deliquescent	40° F to 70° F	30 to 80 gal/day
Refrigerant	35° F	20 gal/day
Desiccant	-40° F to -100° F	0.04 gal/day

Compressor inlet air 90° F, 3,000 standard ft^3/min generated at 100 PSI

Drying Medium	Dew Point	Moisture Downstream
Deliquescent	4.5° C to 21° C	115 to 300 l/day
Refrigerant	1.7° C	75 l/day
Desiccant	-40° C to -73° C	1.5 l/day

Compressor inlet air 32° C, 85 standard m^3/min generated at 6.9 bar

Figure 11-3. Dryer performance chart.

The pressure of compressed air generated by any style compressor is generally controlled through an air regulator to maintain a steady pressure for the gun(s). For example, the line air pressure in a manufacturing plant may be 100 PSI (7 bar), but perhaps only 50 PSI is desired at the spray gun. The air regulator is set to drop the pressure to 50 PSI and consistently maintain it at this level. Even though the plant line air pressure might fluctuate between 85 and 112 PSI (5.8 and 7.7 bar), the paint gun air will remain steady at 50 PSI to give consistent paint atomization performance.

Paint Supply

Gravity, siphon, or pressure systems can supply paint to an air-atomized spray gun. To avoid excessive strain on the sprayer's wrist and arm, gravity and siphon cups are small. Some hold just a few ounces, and larger ones rarely hold more than about a quart of paint. Weight becomes a negative factor for the sprayer's arm and hand if too much paint is in the gravity or siphon cup. Thus they are acceptable only for applying relatively small amounts of paint at a time; frequent stopping for refilling these small containers would waste time. In gravity feed guns, as shown in **Figure 11-4**, the paint supply is above the gun and flows down by gravity. The paint container must, of course, be covered to keep out dirt and dust, and it needs an air inlet vent to allow air in to replace the paint that flows out of the cup. Paint in a *siphon cup* is pushed upward into the gun by atmospheric pressure. This happens because atmospheric pressure is reduced at the exit of the fluid tip by the flow of compressed air through the specially designed air cap. Thus, paint is pushed up to the gun only when compressed air flows through the gun. Siphon cups also need to

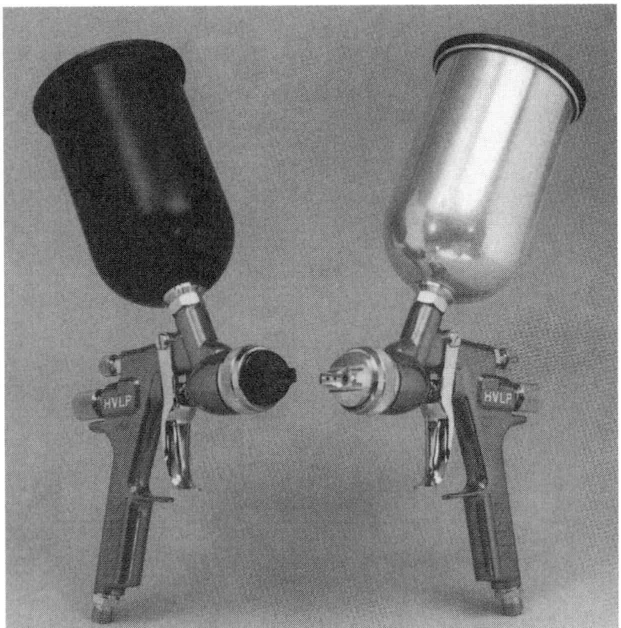

Figure 11-4. Two gravity-feed liquid spray cup guns. Powder guns in this feed style are also available.

be vented; like gravity cups, they come in various sizes ranging up to about a quart (.95 l) or just slightly larger. **Figure 11-5** is a drawing of a spray gun using a siphon cup for the paint supply.

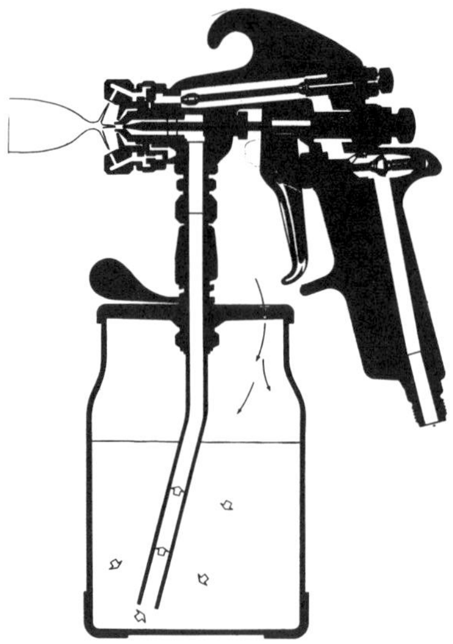

Figure 11-5. Siphon-feed spray gun.

Because of the paint quantity limitations of both gravity feed and siphon feed devices, the paint supply systems in predominant use are pressure-fed types. *Pressure-feed* systems can be either of two basic types: 1) compressed air is applied to the surface of the paint inside a pressurized container, forcing paint out through an outlet hose to the gun; or 2) paint is physically compression-pumped to the gun by one of the various styles of liquid pumps such as diaphragm, gear, piston, or rotary pumps.

The pressurized paint container may be a cup that is mounted at the bottom of the gun, but it is usually a *pressure pot* (tank) connected by a 5–10 ft (152–305 cm) flexible hose to the spray gun. Pressure pot sizes range from about 2 quarts (1.9 l) to 50 gallons (190 l). **Figure 11-6** shows a drawing of a pressure tank with a double regulator so that the atomizing air and the air inside the paint tank can be controlled independently. **Figure 11-7** illustrates the cross-sectional difference in an air nozzle for a gun using a siphon cup and one using pressure feed.

Paint can also be pumped to the spray gun out of any unpressurized container such as a pail, drum, or tote tank. The paint flow can be either a simple dead-end system, or it may loop back to the supply tank in a recirculating system. Paint pumps that supply air-atomizing spray guns can be grouped into four basic categories:

1. Reciprocating piston (single- or double-acting)
2. Diaphragm
3. Rotary (cam or gear)
4. Centrifugal.

The pumps may deliver paint to just a single gun, or to a larger system from which paint can be piped to two or more guns. With any of these systems, the paint delivery can be dead-ended at the gun(s) or recirculating with a return to the paint supply reservoir. Paint re-

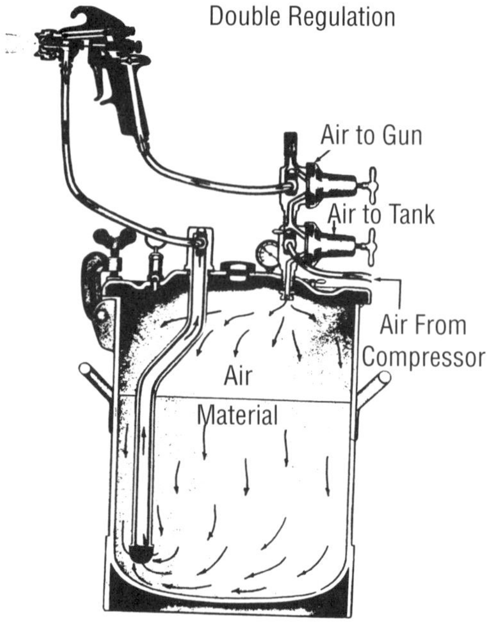

Figure 11-6. Pressure-feed paint tank.

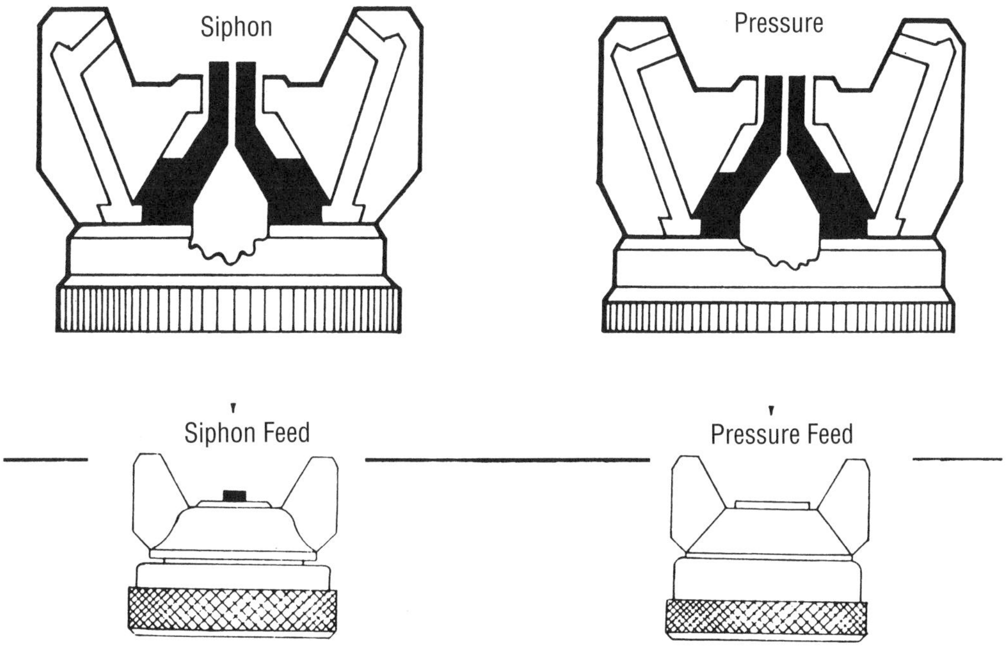

Figure 11-7. Cross-sections of siphon feed versus pressure feed.

circulation tends to be used with paints that have pigments that tend to settle out rapidly. Recirculation of paint is also done to prevent localized hot spots when paint heating systems are used.

The paint delivered to a gun is generally filtered several times to remove contaminants that might cause rejects if sprayed onto a product. It should be filtered before it is poured into the container, as it exits the container, and again just as it enters the atomization device. The micron size of the filters must be small enough to remove foreign materials, but at the same time large enough to permit passage of pigments and metallic particles and not filter out these essential coating components.

Gun Operation

The rate of paint flow through a spray gun is a function of several interrelated factors: the fluid pressure (driving force), the paint viscosity, the size of the fluid nozzle orifice, and the distance to which the gun's fluid needle is retracted from the seat in the fluid tip. The fluid pressure is set to deliver the required amount of paint through an appropriately sized fluid nozzle orifice. High production requirements need high rates of paint flow and large orifice paint nozzles to be able to apply sufficient coating volume for the production parts moving at conveyor speeds past the spray gun(s). For lower parts volumes, paint-flow requirements are not as large, so lower paint pressures and smaller fluid tips would be used.

The degree of atomization produced by an air-atomized gun depends on how efficiently the atomizing air breaks up the paint particles. Increasing air pressure and increasing air volume both increase the degree to which paint is atomized. But it isn't the air pressure alone that determines the degree of atomization of the paint. Instead it is related to several factors including the air pressure (PSI, or bar), atomizing air volume, and the precise merging configuration of the air and fluid streams built into the spray gun design.

Types of Guns

Air-atomizing spray guns can be categorized into three types according to the volume and pressure of the atomizing air:

- Low-volume high-pressure air spray (usually called *conventional air spray*)
- High-volume low-pressure air spray (HVLP)
- Low-volume low-pressure (LVLP) and high-efficiency low-pressure (HELP).

Low-volume High-pressure (Conventional Air Spray) Guns

This category of air-atomized paint spraying has been called "conventional" to distinguish it from the modified gun versions that first appeared roughly 20 years ago. **Figure 11-8** is an exploded drawing of a conventional air spray gun; and **Figure 11-9** shows a conventional gun being used. This style spray gun (which has on rare occasions also been called an LVHP gun) was developed early in the 1900s. It has remained essentially the same over all these years except for some refinement in gun construction materials and in the design of the air and fluid nozzles.

These spray guns use compressed air supplied by an air compressor. The air pressure will typically range from about 25–30 PSI (1.7–2.0 bar) to about 90–100 PSI (6.2–6.9

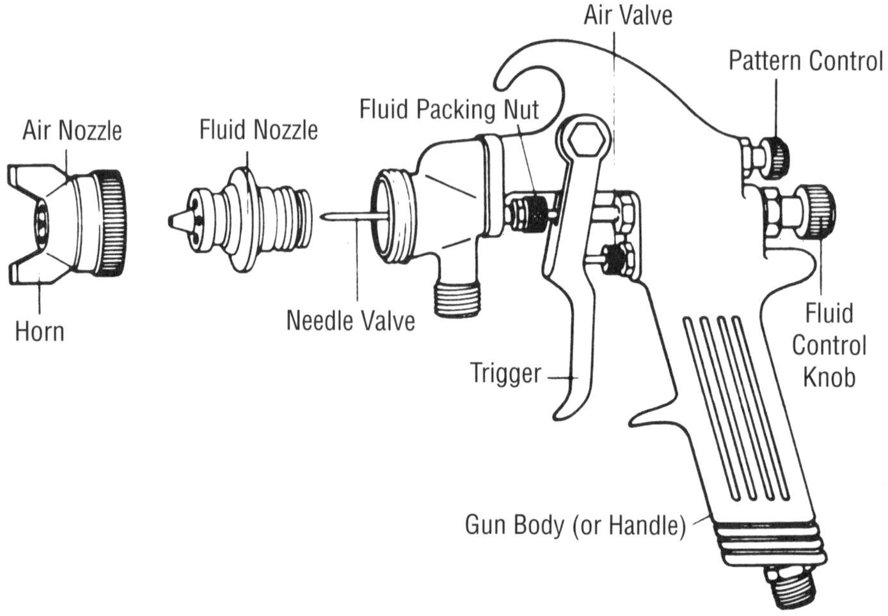

Figure 11-8. Conventional air spray gun.

Figure 11-9. Conventional air spray gun in use.

bar), and the air volume from about 5 cubic feet per minute (ft^3/min) to 20 ft^3/min (142 to 566 cubic decimeters per minute, dm^3/min). The compressed air volume tends to be low when spraying with low air pressures, but the total air volume rises somewhat proportionately as the pressure is increased. The air pressure selected is tied in closely with the air nozzle that is used. The size of the openings in the air nozzle orifices must not deplete the air compressor capacity. The relatively high air pressure typically used with these guns gives exceptionally fine atomization up to a maximum paint flow of around 1 quart/min (0.95 l/min). The flexibility of conventional air spray will allow use of high rates of paint flow to meet large volume production requirements, but it can also be used for production situations in which only a few drops of paint per minute are needed. Both high and low flow rates can be accommodated with conventional spray guns (although the guns will be sized differently in each case) and both will produce outstanding finish quality.

At all times when spraying, the paint flow and the air pressure must be balanced and proportional to each other. Should insufficient air pressure be used, the finish will be rough and tend toward runs and sags. If the atomizing air pressure is too high, droplets as small as five microns in diameter can be formed and this may create a paint fog that will decrease overall application efficiency. The greater surface area of tiny paint droplets will also cause increased solvent evaporation, and produce "dry" spray, giving the parts being painted a dusty appearance.

High-volume Low-pressure (HVLP) Air Spray Guns

The HVLP category of air spray gun uses appreciably lower atomizing air pressures than conventional air spray guns, most often 10 PSI (69 centibar) or less. The figure "10 PSI" has become important as a maximum pressure because of air quality regulations. In many states in order to be considered as part of valid compliance efforts, the air-atomized guns may not exceed 10 PSI air pressure measured at the gun tip. Actual air pressure values for industrial painting are typically in the 2–5 PSI (13.8–34.5 centibar) range. Excellent atomization can be obtained at these pressures. Air volumes range from about 15–65 ft^3/min (425–1,845 dm^3/min) or even higher.

HVLP guns are further classified according to whether the air is supplied by an air compressor or by a turbine. The type of gun that uses an ordinary air compressor typically works in conjunction with an air-regulating device located in the airline or inside the spray gun itself to ensure that no more than 10 PSI of air pressure reaches the gun tip. The tendency is to use specially chambered guns that will not exceed 10 PSI air pressure at the gun tip rather than regulators on the compressed air lines. In this case, a nonbleeder style HVLP gun is used.

The other type of HVLP gun uses air (usually warm) supplied by a turbine. An air turbine can typically put out about 200 ft^3/min (5.7 m^3/min) of air at 10 PSI. This means that up to eight HVLP guns can be operated using just a single turbine. An advantage of a turbine is that its air output can be heated to as high as 180° F (82° C), and this helps provide easier atomization by heating the paint in the end of the gun to lower its viscosity. Bleeder style spray guns are used with heated turbine air so that the tip of the gun remains warm. An air heater would need to be used with the air compressor type of HVLP gun to provide the same heated air—an option that was offered by an equipment supplier but did not find acceptance in the market. When turbine air is too hot, however, it can make the handle of a manual gun uncomfortably hot. Introducing the hot air into the gun at the tip instead of the handle (as seen in **Figures 11-10** and **11-11**) circumvents this problem.

Figure 11-10 compares the air source for a conventional gun with an HVLP gun using a turbine. As mentioned, an air compressor can also supply air to the HVLP gun.

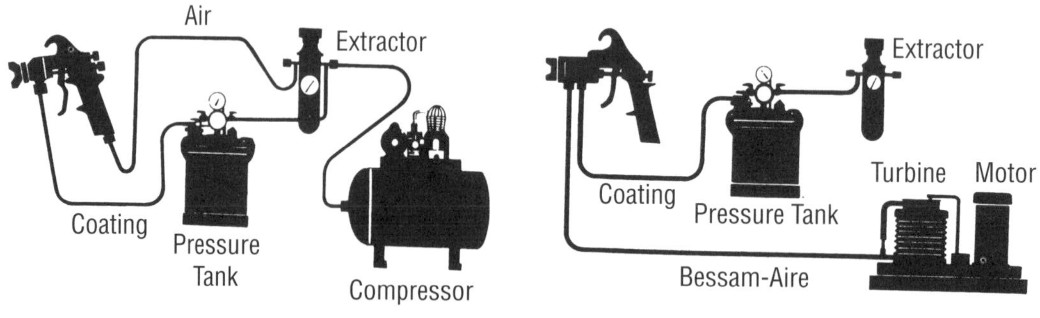

Figure 11-10. Air source for a conventional gun versus a gun using turbine air. Note the supply tank in the turbine system still requires pressurized air from a compressor (not shown).

Both the air compressor and turbine types of HVLP guns are characterized by an air nozzle with a relatively large-diameter opening for atomizing air. At 10 PSI the air compressor typically supplies 15–30 ft^3/min (425–850 dm^3/min). Turbine-type HVLP guns are recognizable by the large-diameter air hose connecting to the gun. This is evident in the guns and hoses shown in **Figure 11-11**. Compressor-type HVLP guns are usually not easily distinguishable from conventional guns by their outer appearance or by the hoses used to supply atomizing air.

The low-atomizing air pressures characteristic of an HVLP gun tend to minimize the amount of "bounce-off" paint fog and reduce the amount of atomized paint that is blown past a part as overspray. The improved HVLP transfer efficiency helps hold down operating costs by reducing paint waste. As the solids percent of paint is increased, the need to minimize overspray increases accordingly to hold down costs. However, such reduced atomizing air pressure also tends to decrease the fineness of atomization, which in turn reduces the finish smoothness capability. The low atomizing air pressure of HVLP guns also tends to require reduced paint flow to the gun, which limits production speeds until the sprayers become accustomed to using it. **Figure 11-12** summarizes the differences between high- and low-pressure air-atomized spray guns.

Low-volume Low-pressure (LVLP), and High-efficiency Low-pressure (HELP) Guns

Some conventional high-pressure guns modified to operate below 10 ft^3/min (285 dm^3/min) with an internal mix air cap are labeled by their manufacturer as LVLP and HELP guns. These guns may be suitable for some lower viscosity paints applied at rates of less than 10 oz/min (285 cm^3/min). The coarser atomization of such guns compared to the conventional and HVLP guns is acceptable only for some applications.

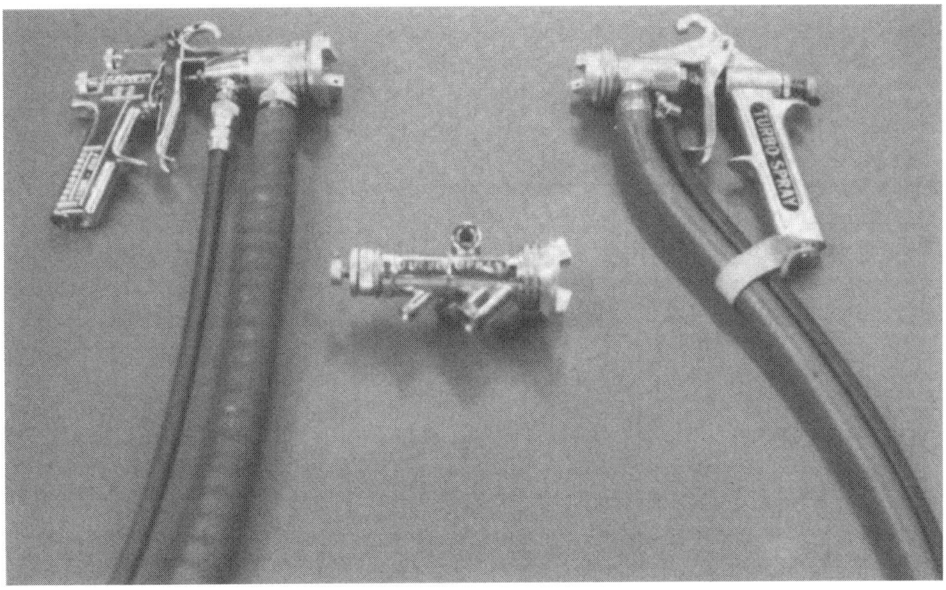

Figure 11-11. Comparison of air hoses used in conventional versus turbine type guns.

High-Pressure Air Spray

Advantages	Disadvantages
Excellent atomization permits fine finishes.	Extensive overspray wastes paint and increases VOC emissions.
The overspray increases booth cleanup costs.	The overspray increases exhaust filter replacement costs.
High production rates can be accommodated.	
The overspray increases water-wash reservoir treatment costs.	

Low-Pressure Air Spray

Advantages	Disadvantages
Minimal overspray reduces paint waste and increases transfer efficiency.	Atomization may not be sufficient for fine finishes.
Reduced paint usage lowers VOC emissions.	High production rates may not be possible.
Minimal overspray lowers booth cleanup costs.	
Minimal overspray cuts exhaust filter replacement costs.	
Minimal overspray decreases water-wash reservoir treatment costs.	

Figure 11-12. Comparison of high-pressure and low-pressure advantages and disadvantages.

Spraying Techniques for Air-Atomized Guns

Proper spraying techniques with air-atomized guns are extremely important because of the cost of the paint being applied and the expense of having to rework reject parts. Good spraying techniques will minimize overspray, ensure uniform film builds, and result in the correct paint application for optimum appearance.

Before a person begins to spray paint onto a product, a number of operational details should be reviewed. The spray gun should be clean and in perfect working order. It should yield a symmetrical spray pattern with reasonably defined boundaries. The optimal spray pattern is achieved using 1) the lowest air pressure that still gives good atomization, and 2) the minimum amount of fluid pressure to provide sufficient paint to meet production rate requirements.

Good spraying technique requires adhering to the following basic principles.

- The gun should be put in motion before the trigger is squeezed.
- The gun should be kept a constant uniform distance from the surface being coated.
- The gun should be moved at a uniform speed across the surface to be coated.
- The gun should be triggered on at the beginning and off at the end of each stroke.
- Spraying for each part should begin at the top, and at the same point every time for each identical part.
- Each previous stroke should be overlapped by the same amount (50%).
- The same number of strokes should be used on identical product surfaces.
- The final stroke on identical products should be ended at the same location.
- For an optimum coating, the gun distance from the product surface being coated should be 6–8 inches (15–20 cm). Moving the gun closer will increase the wetness and film build. Backing the gun away will decrease wetness and minimize film build.

Basic gun movement principles should be followed as much as possible when manually spraying products with complex shapes. For example, the ends of vertical flat panels should be sprayed first, followed by back-and-forth horizontal spraying, beginning at the top. Large panels should be sprayed this way in strokes up to about 5 feet (1.5 m) long. When edges are being sprayed, the gun should be aimed so that as much overspray as possible lands on uncoated surfaces. Exterior edges should be sprayed first. Spray should not be directed directly into internal corners; each of the three flat surfaces that make up the corner should be sprayed separately instead with the gun perpendicular to the surface in each case. Some variance from these rules will of course be necessary when complex shapes are involved.

Gun variables should be monitored closely while spraying. These variables are the paint flow rate, fluid pressure, paint viscosity, air pressure, fan pattern, and distance of the gun from the work.

The solvent evaporation rate from atomized paint particles as they are sprayed from the gun toward the product needs to be considered. Excessively slow solvent evaporation will yield an applied coating that might be excessively wet and cause paint runs or sags, especially on vertical surfaces. Excessively fast solvent evaporation, on the other hand, can produce coating that is too dry. Such films may have a dusty or a pebble-grain appearance.

Air Spray Characteristics

Air spray can be totally suitable for a single-gun system applying only 2–3 oz (55–85 cm^3) of paint per hour. Yet air spray can be equally suitable if scaled up for use by a system involving a dozen painters working out of a common paint supply line, with each of them spraying a liter of paint per minute. Air spray is readily adaptable to any size coating operation and a wide range of coating application volume.

Quick-disconnect fittings on paint hoses and the spray gun provide fast color change capability. The color hoses can be switched quickly even in manual operations. The old color spraying is stopped, that paint hose is disconnected, and the new color hose is attached. The new color paint is used to flush the little bit of remaining old color out of the gun, and spraying with the new color can begin. A solvent hose could also be attached briefly for the color flush if desired. Automatic color changers are also available, principally for use with multiple- or automatic-gun systems. These use manifolds with valving to control color changes and solvent purges.

The readily interchangeable fluid nozzles and air caps of conventional air spray guns permit the application of paints having a fairly wide range of viscosities. Paint heaters can be added to the supply tank or paint line to reduce the viscosity of viscous coatings. Regulators, fluid valves, and variable orifice size tips make it easy to regulate the air and fluid pressures at numerous points in the system.

The spray guns can be mounted onto various mechanical motion devices for automatic coating application. These devices include reciprocators to move the guns back and forth, or up and down, as parts are conveyed in front of them, or robots that will move the guns through programmed paths. Also, various rotational or indexing systems can be incorporated to position the part with respect to the paint application device. This is done to improve the application ease and efficiency.

Chapter 12

Paint Application IV: Airless Spray and Air-Assisted Airless Spray

Airless Spray Guns

In addition to using compressed air as the paint-atomizing force as in air spraying, another often used method of atomizing paint sharply increases the paint's fluid pressure. The pressurized paint is then atomized when it expands as it is released from the airless spray gun. This fluid pressure may range into the thousands of PSI. It therefore requires the gun fluid flow control to be a heavy-duty type such as a ball-and-seat style control valve. This is not a variable flow device but an on-off valve only. A redesign of the fluid tip nozzle to a much finer orifice than for air atomized spray is also necessary. With these changes, the paint is then atomized without introducing a pressurized airflow. This type of spray gun is termed an *airless spray gun*.

The overall appearance and parts of the construction of the airless spray gun (**Figure 12-1**) are much the same as that of the air-atomized spray gun. The main differences are the fluid control and the complete elimination of the air inlet, air nozzle, air valve, and fan control. All air-related features are unnecessary because of the absence of any air supply to the gun. Like the air-atomized spray gun, the airless paint gun also has the following:

- Handle and barrel
- Fluid inlet
- Fluid nozzle
- Fluid control (on or off only) assembly
- Trigger.

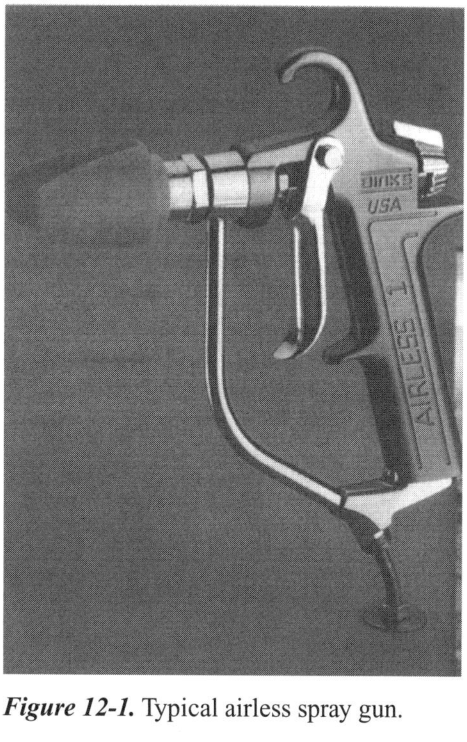

Figure 12-1. Typical airless spray gun.

Handle and Barrel

These parts perform the same function as in the air-atomized gun: the handle allows the operator to grip the gun, and the barrel provides a support for the fluid nozzle and trigger.

Fluid Inlet

On some guns the fluid inlet is located under the end of the barrel, and on other guns it is located at the bottom of the handle. The former design is much more common.

Fluid Nozzle

The fluid nozzle on an airless spray gun differs substantially from the fluid nozzle on an air-atomized spray gun. The airless fluid nozzle orifice is elliptical in shape but is rated in equivalent circular diameter, typically from 0.007–0.072 inch (0.18–1.83 mm). The orifice is beveled or fanned out at various angles, typically in increments from 10–80 degrees. This determines the somewhat flattened paint fan spray pattern width.

Fluid Control (On/Off) Assembly

This part starts and stops the flow of paint through the fluid nozzle. A rather powerful spring is used with a ball-and-seat valve rather than a needle-and-seat style valve. This type valve is necessary because the fluid pressure is so high—often several hundred times higher than air spray pressures. Some airless guns use a tungsten carbide ball and seat, but steel and aluminum construction are more common.

Trigger

This part gives the operator a convenient means of operating the fluid-control assembly. While triggering will control the amount of paint flow through an air spray gun, the airless gun has only an on/off capability. A trigger lock is a normal part of the airless gun construction. The trigger lock should be used to prevent accidental release of paint from the gun. Paint pressures can be high enough to inject paint through skin.

Other Features

An external duckbill device (see **Figure 12-2**), shaped like its name suggests, or a similar device is often mounted at the end of the barrel as an additional safety device. It is designed to prevent the operator from accidentally touching the high-pressure paint stream emerging from the fluid nozzle. The paint stream exits the nozzle with great force

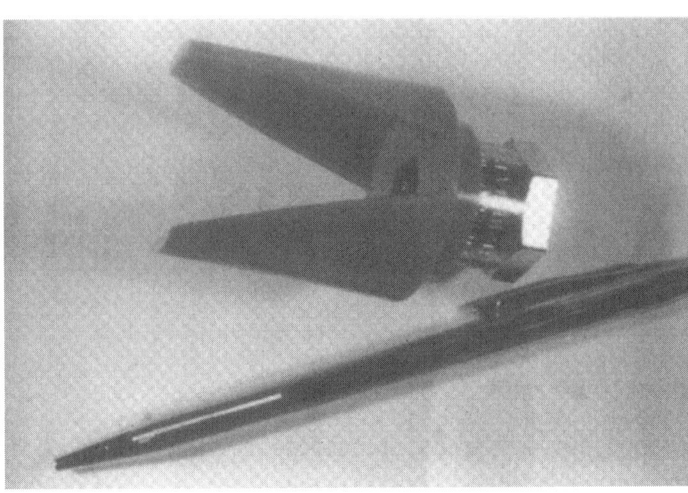

Figure 12-2. "Duckbill" safety guard for airless spray gun.

and can penetrate the skin and cause serious injury so this protection is vital. Manual airless guns should never be used without these devices in place at the tip.

An air-driven multiplier reciprocating piston pump typically supplies paint to an airless spray gun. The pressure exerted by the pump is theoretically directly proportional to the ratio of the area of the pump piston and the area of the air piston. For example, if the pump piston is 20 times as large as the air piston, and if 100 PSI (6.9 bar) is applied to the air piston, then 100 × 20, or 2,000 PSI (138 bar) will be applied to the pump piston or to the paint. In actual practice, frictional losses through the paint lines will lower the actual pressure at the gun tip. Pressure loss is greater with long paint lines and narrow diameter paint hoses.

Airless spray paint delivery systems can be of two types: either dead end from pump to gun(s) or recirculating. In the dead end type, paint is pumped directly from a container to the gun without any return line. In a circulating system, paint is circulated through a paint loop continually. Some systems circulate paint all the way from the supply tank to the gun; others circulate it only through a main line loop, and the spray guns are connected to the circulating loop by short dead end paint hoses. A fluid filter may be located just ahead of the gun or elsewhere in the paint line to minimize plugging of the small orifice tips used for airless spraying.

Airless Spray Gun Operation

Paint is pumped to airless guns typically at about 900–1,200 PSI (62–83 bar), although this may also range anywhere from 500–6,500 PSI (35–450 bar). When the paint exits the fluid nozzle at these high pressures, it expands slightly and encounters air resistance. Thus, it is atomized into tiny droplets without the impingement breakup from atomizing air pressure. The high velocity of the exiting paint propels the droplets toward the work being painted.

The width of the spray fan from the fluid nozzle is determined by the nozzle's fan angle. With the gun tip positioned about 12 inches (30.5 cm) from the part being sprayed, the spray width at the part may vary from about 5–17 inches (13–43 cm), depending on the fan angle of the fluid nozzle being used.

Airless fluid delivery rates are high and not amenable to low volume flow. It ranges from about 25–75 oz/min (0.71–2.14 l/min) as compared to one drop to nearly one qt/min (0.95 l/min) for air spray. It is the paint pressure plus the size of the tip orifice that determines the quantity of fluid sprayed. The tip orifice size and shape determine the paint fan angle. Two nozzle tips having different spray angles—but the same orifice size—will deposit the same amount of paint, but over different sized areas. While spray technique is the major way to control paint thickness, all of these factors help determine applied coating thickness.

Spraying Techniques for Airless Guns

Recommended spraying techniques for airless spray guns are nearly the same as for air-atomized spray guns. The basic differences relate to the high paint volume and flow rate of airless guns, to the larger paint droplet size with airless atomization, and to the absence of compressed air that propels paint droplets toward their target.

The high paint flow requires a consistent spray procedure to prevent an uneven film build, and the associated problems of runs and sagging. It is extremely important in spraying to point the gun directly perpendicular to the target and to move the gun laterally or vertically in strokes that as closely as possible keep the gun in a plane parallel to the surface. Overlapping of the spray pattern from one spray stroke to the next must be consistent to prevent fluctuations in film thickness.

The absence of pressurized airflow in the vicinity of the target allows airless guns to spray more readily into corners and hard-to-reach areas. When air-atomized spray is directed into these restricted areas, the airflow builds a cushion of air turbulence that noticeably repels the movement of atomized paint particles toward the target. **Figure 12-3** compares the efficiency of an airless spray gun with an air-atomized spray gun in painting a product interior, such as a box-like part. The paint sprayed from an airless gun nicely penetrates the cavity, while the air-atomizing spray gun has considerable atomized paint blown back out of the cavity.

Figure 12-4 shows the identical phenomenon happening to a lesser degree when comparing airless and air-atomized spray on a flat surface. The high air pressure associated

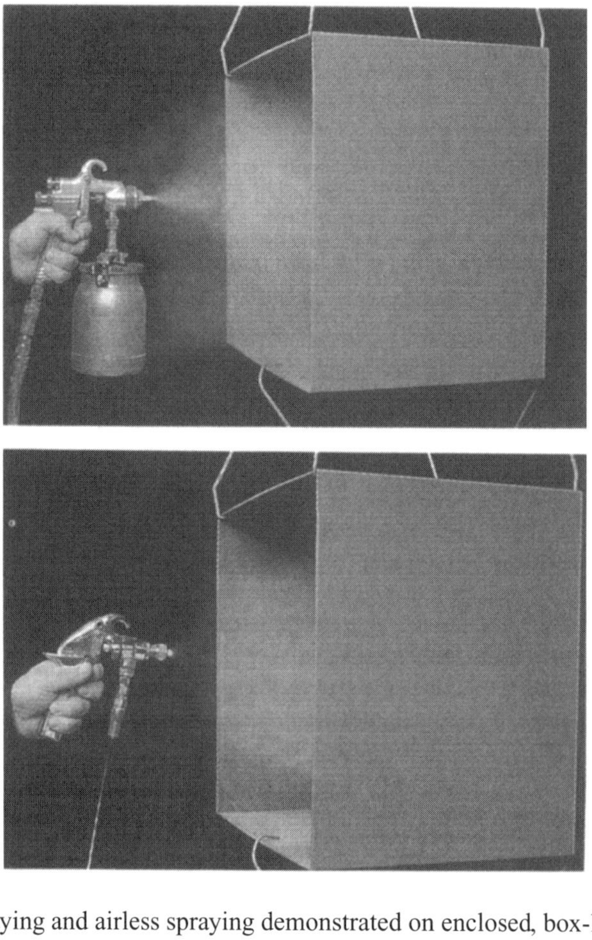

Figure 12-3. Air spraying and airless spraying demonstrated on enclosed, box-like products.

with an air-atomized spray (conventional) gun creates air turbulence and atomized paint *bounceback* (**Figure 12-4A**). The absence of air pressure along with the larger droplet size in airless spray eliminates to a substantial degree much of the turbulence and bounceback (**Figures 12-4B** and **12-4C**).

Advantages of Airless Spray Guns

The advantages of airless spray guns include:

- High rates of paint flow
- Relatively high paint transfer efficiency
- Gun-handling versatility
- Ability to spray highly viscous coatings.

A.

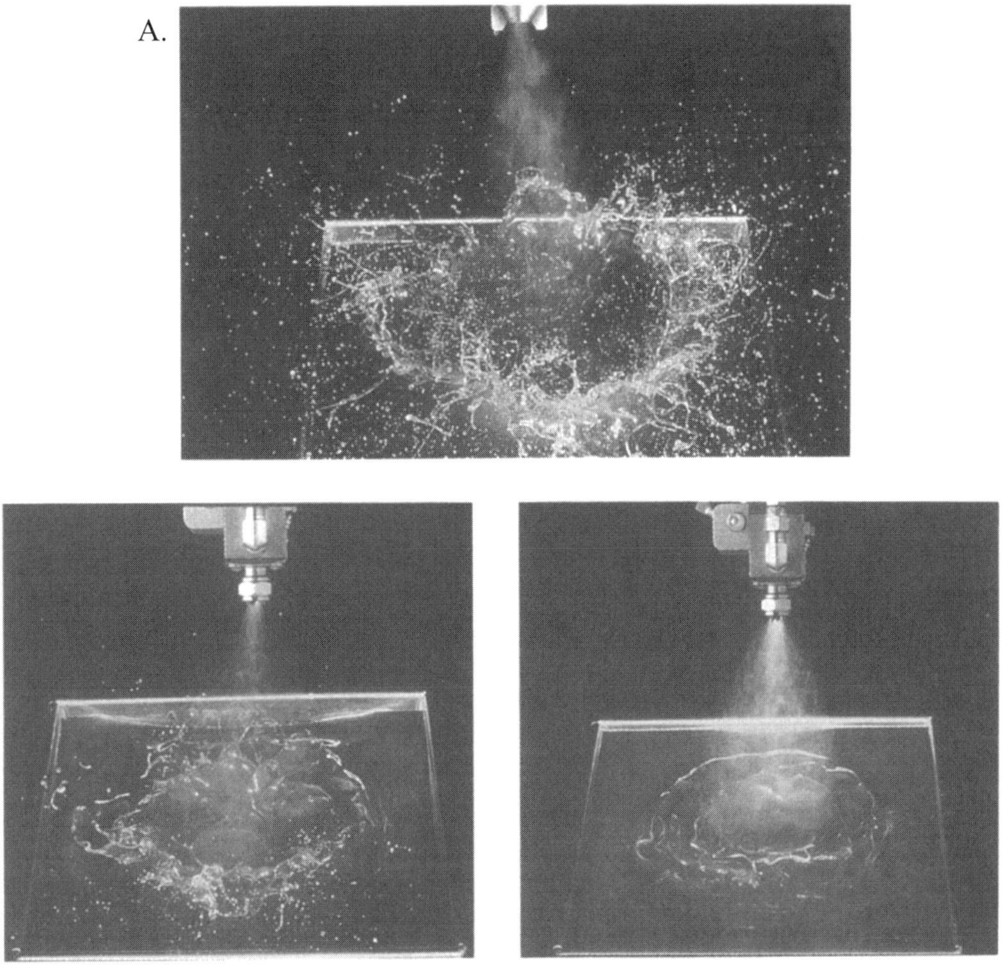

B. C.

Figure 12-4. Spraying velocity differences in droplet size and spray bounceback between air spray (A) and airless spray at 1500 PSI/103 bar (B) and 600 PSI/41 bar (C).

High Rates of Paint Flow

The high fluid pressure and the availability of nozzle tips with various-size openings permit high rates of paint flow. This allows airless guns to be used advantageously on very high-speed production lines and where large square footage of surface is to be coated.

Relatively High Paint Transfer Efficiency

The availability of nozzles with various fan openings helps optimize application efficiency. The absence of blowing compressed air associated with the gun operation simplifies application. Most important in the high transfer efficiency achieved by airless spray, however, is the relatively coarse paint droplet atomization, resulting in the near absence of fine spray droplet "fog" from atomizing air usage and droplet "bounce-off" from the target.

Gun-handling Versatility

The absence of an air hose improves gun-handling versatility. One fewer hose connected to the gun lightens the gun and facilitates improved gun maneuverability, both of which reduce sprayer fatigue.

Ability to Apply Highly Viscous Coatings

Fluids that are simply too viscous to be applied with air spray guns are often readily sprayed with airless guns. Obviously, there are limits beyond which even airless cannot spray if the material is too thick to flow well in the lines. But even thick, syrupy materials, such as urethane roof sealing mastics, are frequently put on as thick as 50–65 mils (1.27–1.65 mm) using airless spray.

Disadvantages of Airless Spray Guns

The disadvantages of airless spray guns include:

- Relatively poor atomization
- Nozzle wear
- Limited fan pattern control
- Coatings with particulate limitation
- Tendency for tip plugging
- Skin injection danger.

Relatively Poor Atomization

The atomization with an airless spray gun is distinctly less fine compared to that obtained with an air-atomized spray gun. The appearance quality of the final film is noticeably less smooth, which restricts the use of airless spray application to surfaces or parts that do not require fine finishes.

Nozzle Wear

The hundreds of pounds of fluid pressure (measured in PSI, or bar) used with an airless spray gun delivers a high rate of paint flow through the very tiny nozzle opening, which tends to cause wear that enlarges the orifice, increase the flow rates, and change spray pattern characteristics. This is especially true at very high pressures and particularly with paints that contain large concentrations of pigments or paints with highly abrasive-type pigments.

Limited Fan Pattern Control

The only real control of the fan pattern is in the design tip put on the airless spray gun. To change fan patterns with airless spray guns requires the comparatively slow process of changing out one fluid nozzle for another.

Coatings with Particulate Limitation

To maintain the fluid pressure at the needed high pressure for atomization, the fluid tip orifice must be much finer than with air spray guns. The generally small fluid nozzle openings limit the materials that can be sprayed to those in which all components are fluid or else quite finely ground. This rules out airless spray for fiber-filled coatings, coarsely pigmented paints, and similar finishing materials.

Tendency for Tip Plugging

The small orifice of the nozzle is easily plugged during spraying. Carefully filtering the paint at the supply tank and again just before it enters the gun will help to reduce this problem.

Skin Injection Danger

Probably the biggest disadvantage of airless spray is the considerable danger of injecting paint through the gun operator's skin. The high fluid pressure creates a paint stream that can possibly penetrate the skin if the gun is triggered directly against or close to the skin. Sprayers have accidentally injected paint into their fingers, hands, arms, and other parts of their bodies. As a consequence, some have required amputation of fingers, hands, and arms. Some have even died after their bodies went into shock from the physical insult of paint injection. When paint injection occurs, only a tiny opening may be noticeable in the skin. This can be deceptive, for the injury can be severe, despite the minor-appearing wound.

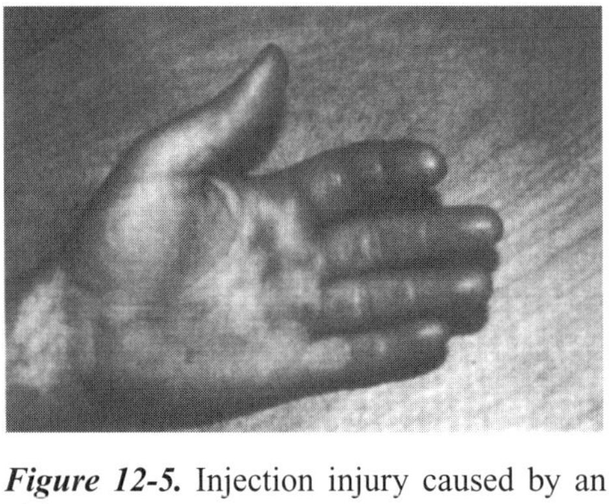

Figure 12-5. Injection injury caused by an airless spray gun.

Total flush out and surgical removal of such injected paint is extremely difficult. The body reacts to the injection by the formation of considerable amounts of fluid that may cause further tissue damage if the pressure is not relieved. **Figure 12-5** shows a tiny airless injection injury at the end of a forefinger. Note the small entrance wound and that the entire finger is swollen and discolored. To remove as much of the material as medically indicated, the finger will need to be surgically opened throughout its length, possibly causing further nerve and tissue damage.

Physicians knowledgeable in airless spray paint injection recommend immediate surgical examination. Keeping the affected area immobile after an injection will minimize the spread of the injected paint deeper into the body. Physicians report that paint injected into a finger can work its way past the wrist surprisingly rapidly if the accident victim continues to manipulate the affected hand and fingers. This obviously complicates the surgery and extends the patient's recuperation period. Unfortunately, many victims try to move their injured hand and clench their fist repeatedly in attempts to assess the extent of damage and physical motion impairment.

As aids to preventing injection accidents, devices such as the duckbill have been helpful. The extremely high pressure at the gun tip decreases rapidly with distance from the fluid nozzle. A pressure of 3,000 PSI (207 bar) at the tip will decrease to 25–200 PSI (1.7–13.8 bar) several inches or centimeters away. Safety tip guards should always be used. The safety on the gun trigger should always be activated when the spray operator moves the gun to a new location and at all similarly appropriate times.

Some manual airless guns are designed to be inoperable without the duckbill. However, sometimes operators will cut off the wings of the plastic safety device because the tips of the duckbill have a tendency to collect paint that then drips onto newly painted parts. This reckless behavior is rarely tolerated in larger plants, but this author has observed it—with horror—in more than a dozen locations.

Using the duckbill is a small inconvenience for the greatly added safety that it brings. Even with the duckbill in place, operators of airless spray guns still need to use extreme caution. Sprayers should be reminded frequently that fluids under great pressure can be dangerous, as evidenced by the use of high-pressure water jets to cut thick steel plates. Nor is injection danger limited to the gun. Airless hoses that burst when being moved during normal painting operations and forced paint into the victims' hands have caused permanent physical damage. It is redundant but certainly not at all superfluous to state again with emphasis: You must always be extremely careful with airless spray equipment!

Air-Assisted Airless Spray Guns

The air-assisted airless spray gun looks almost exactly like an air-atomizing spray gun. The handle, barrel, trigger, and tip look the same. An air hose attaches to the handle, and a paint hose connects to the bottom of the end of the barrel.

But beyond the similarity in looks, the air-assisted airless spray gun is much different in operation than the air-atomizing spray gun. The difference lies in the amount of fluid and air pressures that are used. An air-assisted airless gun uses from about 150–800 PSI (10–55 bar) of fluid pressure and only 5–30 PSI (35–208 centibar) of air pressure. The fluid pressure of an air-assisted airless gun is far greater than an air-atomized gun but considerably less than in an airless gun. The air pressure is far less than a high-pressure low-volume (conventional) air-atomized gun, and, of course, higher than in the airless gun, which uses no atomizing air pressure at all. When the assisting air is 10 PSI (69 centibar) or lower, the guns fall into a special compliance category in Federal rules and also in most states, just as the HVLP-type spray guns. One manufacturer typifies its guns of this style as high-efficiency low-pressure (HELP) air-assisted guns.

The major difference in gun construction between an air-assisted airless gun and an air-atomized gun is in the atomizing tip. The air-atomizing gun incorporates a juxtaposed fluid nozzle and an air cap. A concentric ring of atomizing air surrounds the round fluid orifice in the center of the tip. Contrast that with an air-assisted fluid tip that delivers a flat fan spray of partially atomized paint. Exiting from ports in small projections on each side of the tip (similar to the wings of air-atomizing guns), jets of atomizing air impact at a 90° angle into the spray. The large coarse droplets formed from reduced paint spray pressures are further broken apart by these air jets, completing the atomization. Thus the air jets are "assisting" the airless spray in this process.

The tips are available with various size fluid orifices and fan angles. These may range from about 0.009–0.036 inch (0.23–0.91 mm) orifice and from 15 to 90° angles. With no air and only paint exiting the tip, an air-assisted airless spray gun has a spray pattern with heavy paint "tails" on each side of the pattern. The paint-heavy tail portions of the spray pattern are eliminated by gradually increasing the assisting air atomizing pressure.

The gun tip also has two additional air vent holes, called "shaping air holes," located outboard of the slotted fluid opening. By varying the airflow from these shaping air holes, the operator can narrow the width of the overall pattern. The choice of tips and shaping air pressure variation enables the gun to achieve a spray width from about 4–19 inches (10-48 cm) at a typical target distance of 12 inches (30.5 cm).

The paint flow rate can be varied from about 5–20 oz/min (0.14–0.56 l/min) by changing the fluid pressure. The selection of tip and fluid pressure would be determined by production requirements. Paint is typically delivered to the gun from an air-driven reciprocating pump with about an 8:1 ratio of air piston to fluid piston size. Triggering is the same as with other guns. Fluid on/off control normally is regulated with a ball-and-seat valve inside near the end of the gun.

Advantages of Air-Assisted Spray Guns

The advantages of the air-assisted spray gun include:

- Low equipment maintenance
- Good paint atomization
- Varied fluid delivery
- Low paint bounceback
- Reduced paint injection danger
- High paint transfer efficiency.

Low Equipment Maintenance

The reduced assisted airless paint pressure in comparison with straight airless paint pressure cuts down on pump and fluid nozzle wear. While this advantage touted by air-assisted gun makers is real, it is a somewhat minor benefit.

Good Paint Atomization

Some gun manufacturers rate the atomization quality of air-assisted airless spray as superior to airless spray, but not as good as air-atomizing spray. In actual practice, however, air-assisted airless spraying is mostly used just like airless spraying would be—not to improve atomization and thus obtain a better paint appearance, but for rapid (and more controllable) film build plus reduced paint waste.

Varied Fluid Delivery

The paint flow rates can be varied considerably from about 5–50 oz/min (0.14–1.47 l/min). This greater fluid control (airless delivers at least five times this minimum amount of paint) results in the ability to paint with more film uniformity, in reduced chances for runs and sags on small and complex-shaped parts, and with less mess and less booth cleaning, and greater paint transfer efficiency.

Low Paint Bounceback

The extremely low assisting air pressure allows air-assisted airless guns to spray into corners and hard-to-reach areas. The reasonable air pressures minimize the rebound effect as coating droplets hit the target.

Reduced Paint Injection Danger

The danger of accidental paint injection into the skin with air-assisted spray is not eliminated, but it is considerably less than for the very high paint pressures used with regular airless spray. However, since the fluid pressure is relatively high, it still requires great caution on the part of the operator in order to avoid any chance of an accident.

High Paint Transfer Efficiency

Air-assisted airless spray has a low-end delivery rate of just 5 oz/min (0.14 l/min) compared to 25 ounces (0.73 l) minimum for airless spray. The added flow control allows air-assisted spray to accomplish a transfer efficiency that is even higher than airless spray. This is significant because airless transfer efficiency is already significantly higher than conventional air spray.

Chapter 13

Paint Application V: Electrostatic Painting

The Basics of Electrostatics

In spray application, the propelling force that pushes the atomized paint to the part to be coated includes various combinations of fluid pressure and air pressure. In the electrostatic application of coatings, the small coating particles are provided with an additional driving force—an electrostatic attraction that is made possible by electrically charging the coating droplets.

All types of coatings application can be designed to include electrostatically charging the particles of powder or atomized paint droplets. The spray coatings that were applied by the previously discussed spray methods—air atomizing with high or low air pressures, airless, and air-assisted airless spray—can be applied either with or without electrostatic charging. In addition to those methods, electrostatically charged rotary disc or rotary bell applicators discussed in the next chapter can also apply powder and paint. While the principles of electrostatic charging apply equally to liquid or to powder coatings, this chapter will deal mainly with the electrostatic charging of liquid coatings. Electrostatic powder charging was covered in Chapter 6 on powder coatings.

Sooner or later everyone has a personal encounter with static electricity and electrostatic charging. Who hasn't experienced a small, yet startling, electrical shock in low humidity after walking on a certain type of carpet and touching another person or a light switch? Many a child has observed that after combing his or her hair, the comb can attract small pieces of paper. Certainly everyone has seen flashes of lightning numerous times. These are all individual experiences and observations of electrostatics in action. To help us understand these phenomena, scientists have postulated the following basic electrostatic principles.

- All matter is electrostatically chargeable to various degrees.
- Matter is composed of positive protons, neutral neutrons, and negative electrons.
- Matter having an excess of electrons will be negatively charged.
- Matter that has a deficiency of electrons is positively charged.
- Objects with the same charge (either both positive or both negative) will repel each other.
- Objects with opposite charges attract one another.
- A grounded (neutral) object and a charged object will also attract each other.

Static electricity is the name given to the sudden transmission of surplus electrons from one object to another. This frequently occurs as a sudden spark that jumps the air gap between two objects. Static charges (electrons) can accumulate on all conductive materials of any size. The larger and more conductive the object, the greater the total number of electrons that can be retained on it. But of course, the surplus electrons will seek any opportunity to hop onto grounded materials and objects, such as machinery, cabinets, people, liquids, and even the air itself. Liquids can easily build up surplus electrons when flowing through ungrounded pipes and by being poured or mixed in ungrounded containers. Static electricity can then jump between any nearby grounded object. This will generate a spark that can easily ignite any flammable solvent vapors. Safety lids and electrical grounding of containers holding flammable liquids are always important, but especially so around electrostatic equipment of any sort.

Keeping in mind the basic electrostatic principles, it becomes easy to understand what happens in the three examples of static electricity mentioned previously. In the first example, the friction of a person's shoes moving across a carpet transfers an excess of electrons either to the person or to the carpet. The carpet and the person become oppositely charged. If the charged person fairly quickly thereafter (before the charge drains away) touches another person or a metal object, each of which has a neutral charge, a small "shock" is felt. The person's excess charge becomes neutralized by the spark of electrons, and thus grounded by the other person or object. Electrons flow from one object to another to neutralize the charges, many times with a visible spark or by a shock that is felt or even heard. This shock or spark seldom happens in humid weather because static electrical charges are drained away rapidly by the water vapor in moist air.

The forces of electrostatic attraction and repulsion are noticed more strongly, of course, on objects that are low in weight. In the second example, running a plastic comb through hair results in the plastic becoming charged. When the charged comb is brought near the neutrally charged pieces of paper, the difference in the charge on the comb and paper creates an attraction. The paper is drawn to and held by the comb. Usually soon after making contact, the charges tend to become neutralized, and the paper will then fall away from the comb. In the third example, clouds moving across the face of the earth similarly can become so highly charged that finally the excess electrons jump between the clouds and the ground in a dramatic lightning strike. Lightning (electrons) has been proven to jump up from the ground to the clouds as well as the reverse direction. It flows from whichever has the surplus of electrons. Thus lightning also will jump from one cloud to another.

Electrostatics in Painting

In electrostatic painting, each of the droplets in the atomized paint cloud is given an excess of electrons, thus, every droplet becomes negatively charged. The part to be painted is grounded through the hook and conveyor, and so is electrically neutral. As a result there are electrostatic attraction forces between the neutral (grounded) part and the negative paint droplets. Although the electrostatic forces attract the part to be painted and the droplet equally, the lighter object is the one that will move toward the other. So the droplet moves to the grounded part.

With electrostatic spray application, the droplets pick up charges generated by an electrically charged electrode built into the tip of the spray gun. The charged droplets are given their initial momentum from the fluid pressure/air pressure combination. As the charged droplets approach the vicinity of the electrically neutral part, the charges attract the droplets toward the grounded part. This attraction toward the part reduces the number of droplets that would otherwise travel past the part, increasing transfer efficiency. The attraction is so strong that many charged paint droplets hurtling past the part will actually be pulled out of their flight path to the part. Some droplets will curve and head toward the part; others will even turn around 180° and travel back to the part (wraparound), as shown in **Figure 13-1**. This "wrap-back" tendency allows electrostatic painting to fully coat the face and edges of a flat part, and even paint a portion of the part facing away from the spray gun.

As long as the spray gun trigger is pulled, a DC electrical power unit provides a continuous flow of electrons (electrostatic charges) to the electrode in the gun tip. The electrons are pushed by the power supply to this needle-like electrode at the gun tip, as shown in **Figure 13-2**. With the power supply energized, the excess of electrons leaks off the electrode end into the air in the immediate vicinity of the tip. This creates an invisible negatively charged cloud of air molecules called the *ionized air cloud.* (In a darkened room with the paint supply turned off and the gun trigger pulled, the ionized air cloud may be seen as a bluish glow around the electrode.) When paint begins to flow and atomized paint droplets are ejected from the gun tip, they must pass through the charged air cloud. In doing so, the previously neutral paint droplets pick up an excess of electrons. Except for the metal handle, electrostatic guns are constructed in large measure of nonmetallic materials, normally plastic. A long nonconductive plastic barrel physically separates the needle electrode that protrudes from the gun tip from the grounded metal gun handle. Nonconductive plastic is used as a separator so that charges from the ionized air cloud and the needle electrode at the gun tip cannot travel back to the grounded gun handle. Plastic

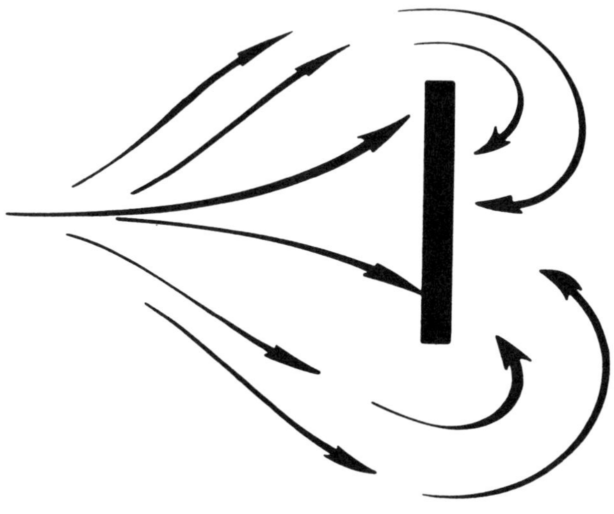

Figure 13-1. "Wraparound" effect in which electrostatically charged paint droplets are attracted to the front and back of a grounded target.

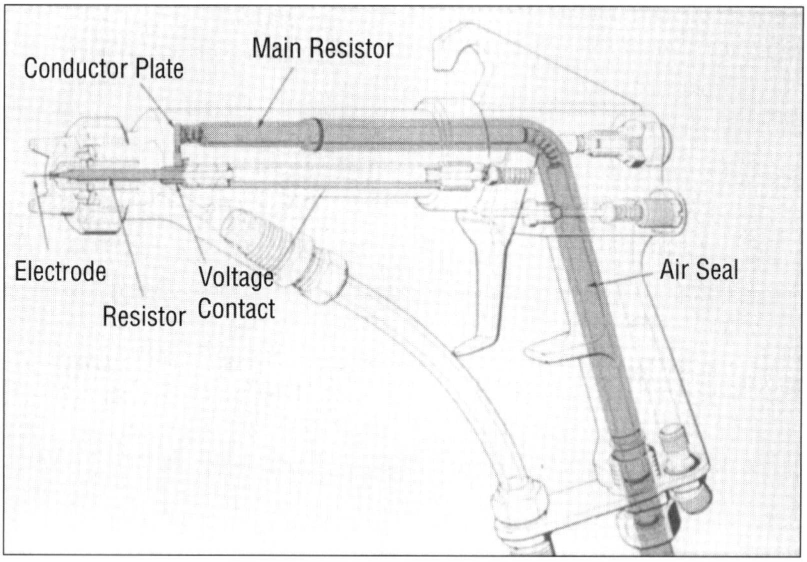

Figure 13-2. Electrostatic gun.

gun parts are also used because plastic does not hold electrostatic charges. This avoids both shorting to ground and sparks from electrical arcing discharges.

The electrical pathway is typically completed as follows. The part to be painted is suspended from a metal hook or part hanger attached to an overhead metal conveyor. The conveyor is electrically grounded since it is attached to the steel supports that connect to the building's foundation in the earth under the building. The negative side of the power supply is connected to the gun electrode, and the positive side is connected to ground via the building's steel beams and foundation. Thus, the electrons drawn from the ground by the power supply travel to the gun electrode, to the ionized air cloud, to the paint droplets, to the part being painted, to the conveyor, to the building's steel beams and foundation, and back into the ground from where they started their journey. An electrical power supply can be viewed as a generator or pump of excess electrons. All electrons that flow through the power supply eventually make the external electrical circuit complete and return back to electrical ground, that is, the earth itself.

The gun's electrode is shaped as a blunt needle to help the excess of electrons from the power supply to transfer to air molecules around the tip. These molecules then comprise the ionized air cloud. It is the nature of sharp points or edges to easily allow electrons to be attracted to them or drain away from them. (Therefore, grounded metal used to protect structures from lightning are pointed rods, not rounded or wide, flat metal pieces.)

The amount of electron flow through a gun electrode during electrostatic painting is intentionally kept extremely small, frequently in the order of only 5–100 millionths of one ampere (5–100 μamp). An *ampere* is a unit of current flow and is a measure of the strength of electron flow. A single lightning bolt may have a flow of millions of amps. A part being electrocoated (Chapter 10) may draw 100–300 amps. A lit 100-watt light bulb has about 1 amp flowing through it—that's 20,000 times more amperage than used by electrostatic painting.

The force in an electrical power supply that provides the "push" for the electron flow is called *voltage*. A residential power outlet has about 110 V. A bolt of lightning may have millions of volts. A typical electrostatic power supply for painting operates in the range of 30,000–120,000 V (30–120 kV).

Grounding and Safety Precautions

In electrostatic terminology, a building's steel frame that is in strong contact with the soil is said to be *grounded*. This is because the frame is either embedded into the ground or is physically connected to something that is embedded. The adjective "grounded" indicates that an object is electrically neutral with respect to the ground (earth). The terms *earth* and *ground* are synonymous in electrical usage.

One of the most important requirements for safe and successful electrostatic painting is to be certain that the positive side of the power supply is completely grounded, and also that the part to be painted is well grounded. The positive electrode of the power supply is hardly ever a problem because a tight mechanical connection such as a bolt or a soldered welded connection connects it to ground. The part to be painted may not be properly grounded, however, due to a weak electrical contact in the system. This could occur due to grease or dried paint between the part and the hanger, between the hanger and the conveyor, or between the conveyor wheels and the conveyor I-beam. In rare instances, the building's steel beams are not in full contact with the earth under the structure. For efficient electrostatic painting, all of these connections must be electrically sound. If they are not, electrical current flow will be restricted, and poor electrostatic attraction of paint droplets will result.

For top efficiency, the part to be coated should be the closest grounded object to the charged paint spray droplets coming from the spray gun. The charged paint particles are attracted to the nearest electrically grounded item; the larger the item, the greater the attraction. A painted hand or face can result when someone accidentally gets too close to the spray pattern.

We have seen that paint buildup on hooks and hangers can act as an insulator and block some or all of the electrostatic charges from completing the circuit to the conveyor and then back into the ground. The greater the paint buildup on the hook, the more severe the problem. If the grounding loss is only slight, wraparound may be only partially reduced. When paint buildup on hangers is heavy, the paint droplets that hit the part cannot lose any of their negative charges. From the repulsion forces between like charges, the incoming negatively charged particles are actually pushed away by the negative charges already present on the part. The result is thin paint on parts and lowered transfer efficiency from wasted paint. When this happens, parts may fail to pass inspection because their film builds are below specification.

Any ungrounded objects, especially large conductive objects (including human beings) in the vicinity of the charged gun electrode and the invisible ionized air cloud, can accumulate electrons and pick up a considerable electrical charge. This is extremely dangerous. The charges can build up until they suddenly discharge in a spark arcing to any grounded object near it. The intense heat of the arc is more than sufficient to ignite any solvent vapors coming from the freshly painted parts, and fire can then spread to other items in the spray booth. For fire safety reasons, grounding of all persons and equipment—of everything in the booth—is critically important!

Should complete loss of grounding of parts be experienced, the danger of arcing becomes great, and fires can result. Numerous instances can be cited where fires have occurred this way. To ensure good electrical connection between parts and hangers, some plants use square rather than round rods for hangers. The square rod is formed into hangers with the edge up so that the part hung for painting sits on the sharp edge. The part weight resting on the sharp edge tends to cut through any paint that might have accumulated on the hanger rods. Hangers and hooks should be regularly stripped or otherwise cleaned of paint buildup to maintain good grounding contact to both the parts and the conveyor.

All persons in the vicinity of the charged electrode (or any part of the spray gun) also need to be grounded. Spray paint operators should wear conductive footgear, or at least shoes with leather soles rather than rubber, to drain charges to ground. In some plants, operators are required to wear conductive safety shoes or to attach a conductive, carbon-impregnated rubber strap around one ankle above their sock. The strap trails on a metal grate to make certain the sprayer is continuously grounded. Sprayers need to have a gloveless (bare) hand on the grounded spray gun handle, a special conductive electrostatic sprayer's glove, or a glove with its palm removed. The metallic handle of the gun is always by design grounded through the power cable, and this will ground the sprayer if skin is in direct contact with the gun handle. These precautions will ensure that the operator does not accumulate electrostatic charges. Metal rings, watches, and other jewelry worn next to the operator's skin will not accumulate electrons on a grounded sprayer. But sprayers should watch out for metallic items such as coins, knives, keys, pens, etc., in pockets, and belt buckles since they do not touch skin and can accumulate static charges.

To complete the electrical circuit satisfactorily, the part to be coated must be an electrical conductor. The part need not be a highly efficient electrical conductor since the amount of charge is small, but at least some conductivity is essential. Part conductivity is required to allow charges from the paint droplets to flow to ground, and to prevent a charge buildup on the part since this would repel additional charged paint droplets. It is acceptable for only the part surface to be conductive; it need not be conductive all the way through. Some nonconductive plastic parts are molded with a conductive plastic skin to enable being coated electrostatically.

The conductivity of the paint itself can affect the path of the electron flow from the electrical power supply, to the gun's electrode, to the part, and back to ground. Solventborne paints tend to be rather poor conductors; but waterbornes and some metallic paints are excellent conductors. This increased paint conductivity, however, introduces a new and shorter pathway by which the electrons can reach ground. The conductive paint can carry the complete supply of electrons reaching the electrode back to the fluid in the gun tip and then all the way back through the paint line to the grounded paint tank. This shorts out the system internally and effectively "wastes" the electrical energy. So the spray droplets never have an ionized air cloud to pass through; all the charges instead follow the short circuit through the conductive paint in the hoses to ground. If this occurs, electrostatic painting is not taking place. Sometimes plants are blissfully unaware that this is happening with their system. The power pack is turned on but the charges are invisible so no one sees that the charges are draining right back to ground through the paint lines.

To prevent highly conductive paint from shorting (grounding out) the system through the paint hose and the paint pot, it is necessary to isolate the hoses and paint supply from ground. This is most commonly done using nonconducting plastic supports and isolation arms. Once the isolated paint supply gains enough electrons, it cannot accept any more, so all additional electrostatic charges then flow to the gun tip. This isolation of the paint supply from ground is called an *isolated system*, and is one of the best ways to enable electrostatic application of conductive paints. (Isolated systems should never be used with solventborne paints because the fire danger from current arcing to ground would be extreme with such materials.)

Notice, here, that an isolated system still permits electrons to travel back to the pressure pot, turning it into a very dangerous source of a potential high voltage. It would give a powerful electric shock to persons who might unknowingly touch the paint pot. As a safety precaution with isolated electrostatic application for conductive paints, the paint line and pressure pot are confined in a caged area. Paints that are not conductive, typically most solventbornes, do not transfer the electrostatic charges back to the pressure pot, and thus protective safety caging is not needed with those coatings.

To ensure that a solventborne paint remains a poor conductor, paint formulators recommend using low polar or nonpolar solvents. Nonpolar solvents tend to be poor conductors. Examples of nonpolar solvents are mineral spirits, VM&P naphtha, xylol (xylene), toluol (toluene), and N-butyl acetate. In contrast, polar solvents tend to be conductive because of the nature of their covalent and ionic atomic bonds. Examples of high-polarity solvents not recommended for paints when using electrostatic application are acetone, methyl ethyl ketone, isopropyl alcohol, methyl cellosolve, diacetone alcohol, ethyl alcohol, methyl alcohol, and methyl acetate.

A manual electrostatic gun used to spray conductive paint must, of course, not permit current to reach the person spraying paint. While the isolated fluid becomes charged and carries the full supplied voltage, the gun handle held by the sprayer is grounded and insulated from the charged paint. Although with the proper equipment it is possible to safely spray conductive coatings manually, for sprayer protection the manual systems operate at lower voltages and amperage draws than the automatic systems. Manual electrostatic spray typically uses around 80,000 volts maximum, while automatic electrostatic systems might use up to nearly 130,000 volts. The potential electrical hazard associated with electrostatic spraying of conductive paints tends to cause some plants to limit themselves to automatic application so that the possible shock hazards of manual spray operation are not involved.

The electrical shock and fire dangers inherent with high voltages used in electrostatic paint spraying have led to the development of numerous methods to lower the voltage at the gun's electrode when a person approaches, or as the electrode approaches an electrical ground. This prevents electrical shock and eliminates almost all chance of a spark that might ignite a solvent fire.

A number of patented protective systems are used by various equipment manufacturers in their power supplies as safety devices to restrict the total amount of current that can flow. Some of these current-limiting devices proportion voltage and current. If the current begins to rise, the voltage drops correspondingly. This is known as a *resistive system*. **Figure 13-3** depicts how the voltage at the tip of a current-protected electrostatic gun declines to zero as electrical current flow increases. This occurs because the increasing current flow causes the device to decrease the voltage at the gun tip. Another formerly often-used method is the so-called *stiff system* that maintains the voltage until a preset current flow limit is reached. At this point, the circuit is mechanically tripped off. Generally, stiff systems are not approved for use with manual spray guns but are appropriate for some automatic electrostatic spray operations. Other sophisticated protective devices measure how fast the change in current draw occurs and can shut off all current almost instantly, even though current draw has barely begun to rise.

The Effects of Humidity

Recall that two of the examples of static electricity—the charging of the comb (after combing one's hair) and a person's body (after walking across a carpet)—tend to occur only in low humidity or at least are more pronounced in low humidity. This is because the highly polar water molecules in humid air readily accept electrons. Humid air will tend to leak away charges quickly from the comb and from a person's body; while drier air has fewer water molecules and so is less able to do that. This also happens on a much larger scale in the case of lightning. The air in the vicinity of a thunderstorm tends to be humid,

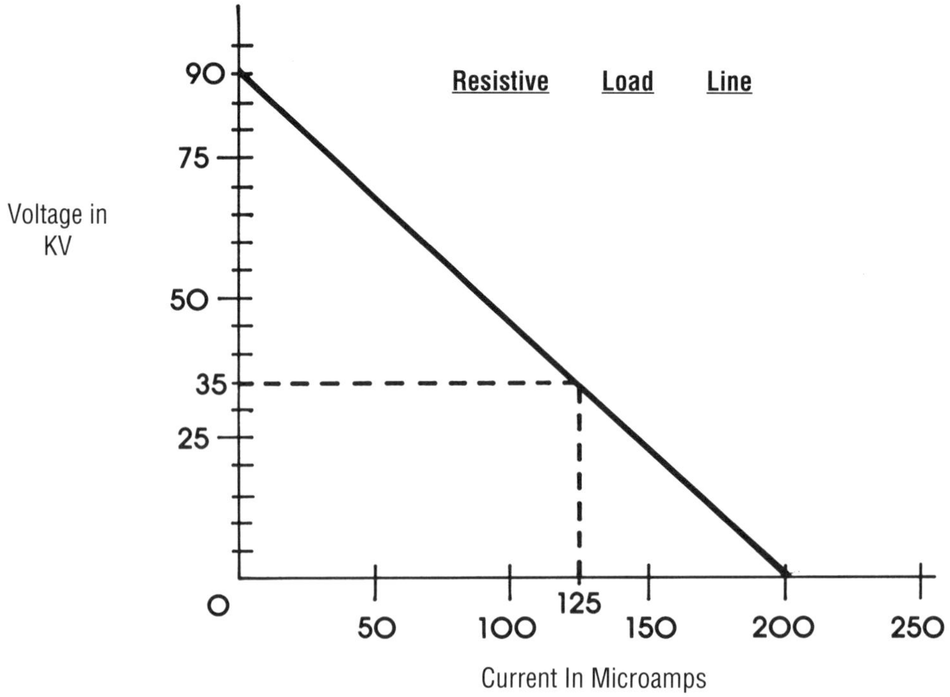

Figure 13-3. A safety feature in the power supply reduces voltages as current draw increases.

and this increases the tendency of cloud charges to discharge to earth or to another cloud in the form of the long visible sparks we call lightning.

Humid air in the vicinity of a gun's electrode will tend to increase electron flow away from the electrode. This is why electrostatic painting is sensitive to humidity fluctuations. You will notice that amperage draw from the electrostatic gun is often appreciably higher in humid weather.

Advantages of Electrostatic Spray Painting

The advantages of electrostatic spray include:

- Higher transfer efficiency
- Good edge coverage
- Paint wraparound
- Uniform film thickness.

Higher Transfer Efficiency

It is a rule of thumb in paint spraying that electrostatics will increase transfer efficiency by roughly 10–20%. Much of this is due to the wraparound effect.

Good Edge Coverage

Excellent edge coverage is achieved with electrostatic painting because electrons (and droplets with electrons) are attracted more strongly to edges than to flat surfaces, as shown in **Figure 13-4**. Sometimes this can be a mixed blessing. Occasionally, so much paint is drawn to the edges that they will undergo solvent popping in the oven due to excess paint buildup. At times, edge buildup can be so severe that paint runs are caused. As would be expected, this is most likely to cause problems on parts with vertical edges. Turning off the electrostatic voltage prior to the spray reaching the edges and altering the spraying stroke patterns can avoid such solvent popping and paint runs.

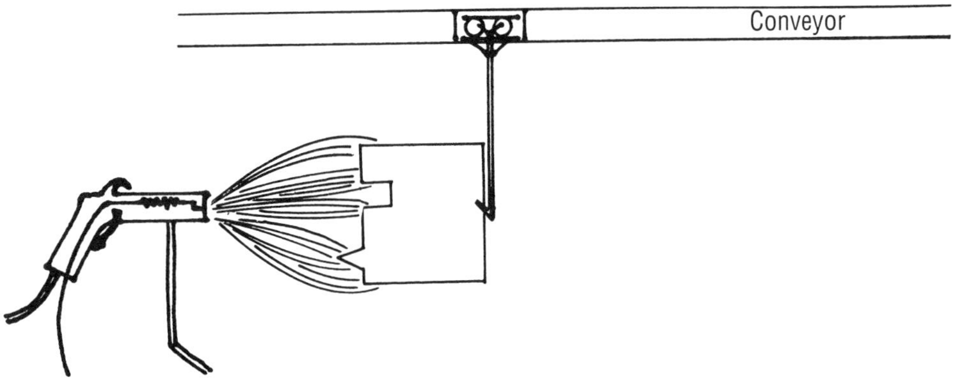

Figure 13-4. Drawing of electrostatic force lines showing attraction to product edges.

Paint Wraparound

The attractive force of electrostatics will, to some extent, draw paint droplets around to the backside of the parts, a phenomenon called *wraparound* or simply *wrap*. This reverse side coating will extend around on parts about 1–3 inches (2.5–7.6 cm) away from the edge, with diminishing thickness. Because of the wraparound effect, complete coverage of small C-shaped parts is often possible with just a single gun pass.

Uniform Film Thickness

A uniform coat is somewhat self-produced because the incoming paint droplets experience the best grounding where the paint is the thinnest. As the paint wet film thickness builds, it forms an electrically insulating layer. Therefore, newly arriving paint droplets are attracted most strongly to areas on the part where grounding is best, that is, the spots where the paint is the thinnest, as shown in **Figure 13-5**. In this way, an increased overall film thickness uniformity can result; but spray techniques are a more significant factor in dry film thickness uniformity.

Disadvantages of Electrostatic Spray Painting

Although the advantages of electrostatic spray are impressive, the disadvantages also tend to be pronounced, and relate to:

- Safety/fire hazards
- Gun construction
- Equipment cleanliness
- Faraday cage effect
- Conductivity
- Metallics
- Solvent selection.

Safety/Fire Hazards

These two factors are lumped together as a disadvantage because they tend to go

Self-Produced Film Thickness Uniformity With Electrostatic Application

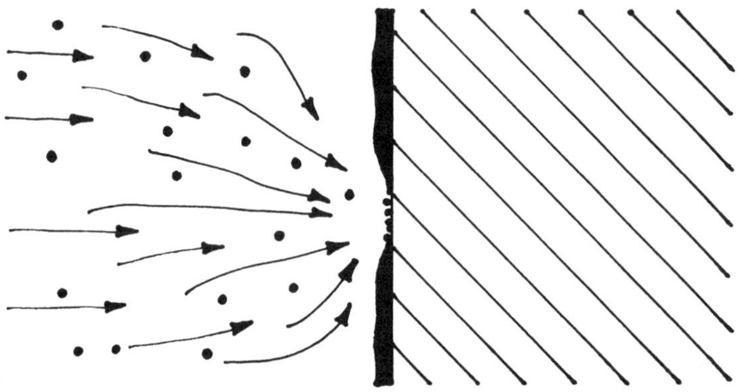

Figure 13-5. Grounding and electrostatic attraction are greatest in thinnest film areas.

together. A high voltage is hazardous and can cause a spark that could ignite a fire in a solvent-laden atmosphere. Everything near the tip of the gun must be grounded to prevent accumulating electrostatic charges. This includes machinery, personnel, and anything that can build up enough charge so that arcing of electrons to ground can occur.

The phenomenon of *capacitance* enables any ungrounded object that can conduct electricity to store extra electrons (at least temporarily). The better the conductor and the larger the mass, the greater the quantity of electrons that the object can store. Thus, the human body and metallic objects can build up large amounts of static charges unless they are continuously grounded. The connection to ground provides a path for the electrons to flow from the objects. If objects near the gun electrode are not grounded, charge accumulation can continue until suddenly the electrons jump to a grounded object in the form of a spark. The energy of the spark, as the electrons drain off in a miniature form of lightning, can be more than enough to ignite solvent fumes. For this reason, spray gun operators should be cautious about ungrounded metal items on their person. It was mentioned earlier that coins and keys in pants pockets, metal buttons, large decorative metallic belt buckles, and similar items can cause this danger, although in these cases, the charge usually discharges through the fabric of clothing to the wearer's body in a very noticeable shock.

Ignition of paint caused by improper operations during electrostatic application is not at all uncommon. A number of people, almost always painters, have died from the resultant fires. When the proper safety procedures are followed, fires will not occur. But the danger is always there and must be recognized. Unfortunately, doing things properly hundreds of times does not afford protection should there be just one instance when a mistake is made.

Gun Construction

Electrostatic guns tend to be bulky and delicate. Their bulkiness reduces maneuverability, and their susceptibility to physical damage requires they be handled more carefully. The bulky feature is simply inherent with the long insulating plastic gun barrel.

Gun bulkiness is increased by the necessity for many electrostatic guns to have an attached high-voltage power cable. The higher the voltage carried in the cable, the heavier and thicker the cable has to be. Some guns require the full high voltage to be brought to the gun. Others utilize just a low-voltage cable and then rely on electronic transformers within the gun to step up the voltage to application levels. The advantage is that low voltage cable is quite thin and more easily maneuvered than a high voltage cable. Other electrostatic guns, termed *cartridge guns,* do not require an electrical cable at all because they have an air turbine inside the gun that generates the high voltage, as illustrated by **Figure 13-6**.

The needle electrode protruding from the gun tip can be damaged easily. Therefore, the gun requires careful handling. A bent electrode will reduce the efficiency of the electrostatic charging. In addition, the relatively soft plastic parts, such as the fluid tip and air cap, tend to wear and require frequent replacement.

Figure 13-6. Cartridge electrostatic spray gun.

Equipment Cleanliness

Extra cleanliness is essential for efficient electrostatic painting operation. This is important for hooks and hangers, but also for the outer surfaces of spray guns. The latter item is frequently overlooked or its importance not recognized. Dirt or oversprayed paint can form a conductive track on the surface of the plastic gun tip and barrel back to the grounded metal handle, thereby shorting out the system. Sometimes people mistakenly think they are painting electrostatically; but actually, because of accumulated overspray on the application device, the system is not applying paint with any appreciable electrical potential. Sometimes spray operators, not recognizing the problem, will simply turn off the power supply and use the electrostatic gun as a nonelectrostatic gun. They do this because no electrostatic charging appears to be taking place. To a trained observer, the reason for the lack of charging is obvious almost every time: it is caused by dirty guns (charge leaking to ground) and/or dirty paint hooks (charge not getting to ground). These are the usual suspects.

Faraday Cage Effect

Electrostatic application is not able to coat recessed areas as well as nonelectrostatic application. This is due to a phenomenon termed the *Faraday cage effect*, as shown in **Figure 13-7**. Charged paint particles always seek the nearest grounded surface. Thus, when coating a recessed area, the charged droplets tend to be attracted toward the sides of the recess, and especially toward sharp edges that may be present, instead of traveling to the bottom of the recess. Nonelectrostatic manual touch-up is often used to get coating into these problem areas. A common procedure in conveyorized painting is first to spray using

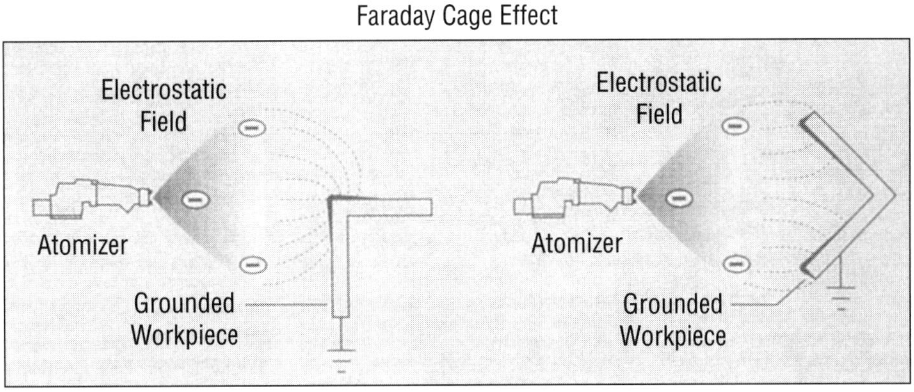

Figure 13-7. Faraday cage effect.

automatic electrostatic application, then touch-up Faraday cage areas by manual nonelectrostatic. Faraday caging becomes severe at high voltages. Some degree of Faraday caging is inherent in electrostatic application and cannot be avoided. For this reason, not all parts are suited for electrostatic application. This is particularly true for parts with complex geometry or parts that have sharp edges, points, deep ridges, or cavities.

Conductivity

In electrostatic paint application, the parts being coated must be somewhat conductive to complete the electrical circuit. The part need not be conductive throughout its construction; only the outer surface needs to be conductive. The surfaces of nonconductive materials, such as most plastics and wood, can be made conductive by the application (by dipping, usually, or spraying) of special conductive aqueous or alcohol solutions. These "prep" solutions contain ionic organic salts and alcohol, or water that contains a mix of potassium and calcium chlorides. The ionic material remaining on the parts after the alcohol or water evaporates provides the parts with sufficient surface conductivity for electrostatic coating. Applications of prep solutions for conductivity before painting are neither costly nor time-consuming. However, it is nevertheless an extra step in the painting process, and an operation to be avoided if possible.

Metallics

Metallic paint contains millions of tiny metal particles (usually aluminum flake or aluminum powder) that are mixed uniformly throughout. When metallic paint is atomized and applied without electrostatic charging, the particles tend to lie flat or "flop" on the surface being coated. This flat orientation provides the desired maximum metallic surface area for light reflection, giving the painted surface an attractive bright, sparkly effect.

When metallic paint is charged electrostatically, most of the charge is held by the metallic particles, since they are the most conductive components in the atomized droplets. This extra charge tends to make the metallic particles stand on edge (instead of lying flat as in nonelectrostatic application). Standing on edge helps the particles release their negative charge, but reduces the metallic effect. The "on edge" orientation reduces the surface area of the particles that can reflect light, causing a noticeable darkening compared to surfaces painted nonelectrostatically, as shown in **Figure 13-8**.

Metallic coatings are used extensively on automotive finishes—as high as 90% of particular car models receive metallic finishes. Automotive metallic topcoats are usually applied on the assembly line with nonelectrostatic spray for the final color coat because it gives a brighter, more appealing appearance to the finish. A less important reason that it is done nonelectrostatically is because almost all automotive repair shops tend to use only nonelectrostatic spray. It is easier for them to match repair paint when cars are given a metallic finish on an assembly line using nonelectrostatic guns. If vehicles were painted by the manufacturer with electrostatic guns and repaired using nonelectrostatic guns, the metallic sparkle of the repaired area would be noticeably different from the rest of the paint. This is true even if both use the exact same paint.

The electrostatic darkening effect of metal flake, and the increasing economic and environmental desirability of electrostatic application, have led some paint manufacturers to use mica flakes as a replacement for aluminum flake. Mica is a complex sodium aluminum silicate identical in chemical composition to asbestos but with none of the health hazards. It is a nonconductor and does not sink into the wet film or orient itself on edge during electrostatic application as do metal flakes. This prevents mica flakes from exhibiting unwanted darkening, and overcomes that difficulty when mica-containing paint is applied electrostatically.

A variety of pearlescent effects can also be achieved with coated mica flakes. Pearlescence refers to the multiple-color rainbow-like appearance noticed in a thin film of oil on water. This effect was often used in recent years, but its popularity has waned now.

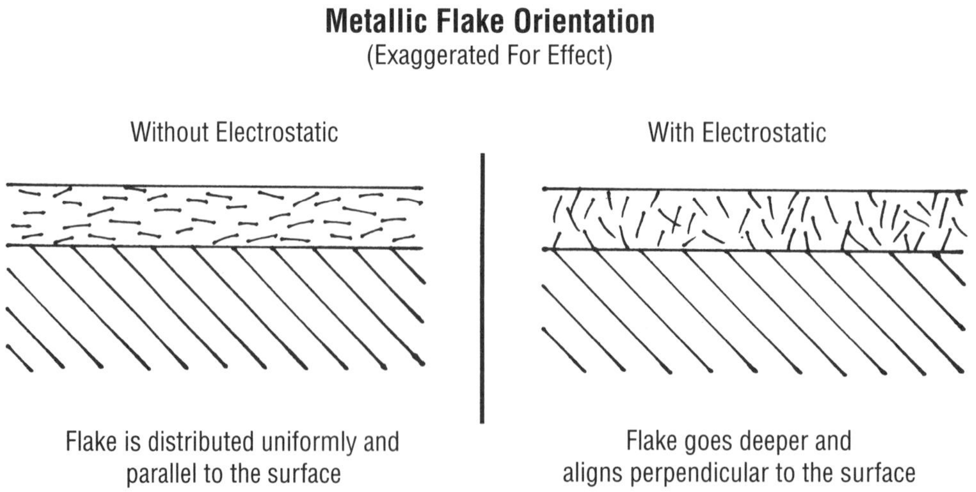

Figure 13-8. Comparison of flake orientation in nonelectrostatic and electrostatic application of metallic paint.

Solvent Selection

The selection of solvents is considerably more critical with electrostatic painting. The paint must be polar enough so that it will be able to accept electrons well. If the paint is too low in conductivity, the droplets are unable to accumulate electrons as they travel through the ionized air cloud, and the electrostatic effect is minimal. With "dead" paints, adding a minor amount of polar solvent can correct this shortcoming. However, this is rarely required. Nearly all coatings have no need for additional polarity. In fact, many high-solids paints are inherently almost too polar already due to the high reactivity of the resin molecules. For this reason, the preferred solvents for electrostatically applied paints are the nonpolar or slightly polar solvents. Paints that are too highly polar will accumulate so much charge while passing through the ionized air cloud that they cause excessive Faraday caging problems. Test meters are available to measure the conductivity of a paint to see if its polarity falls within an acceptable range, but they are seldom used.

Chapter 14

Paint Application VI: Rotary Atomizers

Introduction to Rotary Atomization

Rotary atomization involves the breaking up of paint into atomized droplets by delivering the coating onto a rapidly spinning circular "head." Depending on the shape of the rotating head, the application device is identified as either a *disc* or a *bell.* Instead of employing compressed air or fluid pressure to atomize the paint, as with spray guns, rotary atomizers utilize what may be referred to as centrifugal force. This force in a spinning object tends to project matter outward from the center and, as you can imagine, paint poured onto a rotating head will be hurled off as the result of this force.

Fewer than half of paint spray guns use electrostatic charging of the droplets; the majority does not. Rotary atomizers in use today, however, *always* use electrostatic charging. (I have encountered only one rare exception of nonelectrostatic rotary application being used for painting the insides of barrels.) With spray guns, the electrostatic charging is not believed to have much effect on the actual atomizing; but with rotary atomizers, electrostatic charging can play a key role in paint atomization. The older rotaries showed a sharp reduction in droplet size when no electrostatic charge was used. **Figure 14-1** shows a low-speed disc painting room air-conditioner housings. With the high speeds of the more modern rotary devices, however, almost all atomization is the

Figure 14-1. Low-speed disc.

result of rotation. This can be shown by the fact that droplet diameter size decreases only minimally when the electrostatic charge is turned off.

The rate of rotation of the older version low-speed discs was typically 900–8,000 rpm. **Figure 14-2** shows low-speed bells, which feature containment shrouds to catch the paint and solvent flushes during color changes and rotary head cleaning. The shrouds drain these fluids to a containment vessel and thereby reduce plant VOC emissions. The rotation of today's high-speed discs and bells can be varied from about 10,000–60,000 rpm. **Figure 14-3** shows three different models of high-speed bells. Summing up the difference between the low- and high-speed rotary atomizers: the low-speed units tend to be rather large and rotate relatively slowly; the high-speed rotaries tend to be small and spin very fast.

In 1956, Ransburg began offering a rotary disc they called a "number 2 process disc." Shortly thereafter, General Electric became the first company to use this then-new device in manufacturing for painting refrigerator shells. The old, slower rotaries are no longer used much for liquid coatings because they only give fair atomization even with low-viscosity (low-solids) paints at moderate flow rates. They were not able to handle medium-viscosity paints at normal flow rates; and with high-solids coatings the low-speed rotaries were totally ineffective. The newer high-speed discs were developed because they are able to apply just about any high-viscosity coating with ease, irrespective of flow rates. Slow rotation is used for powder application, however.

All modern rotary atomizers are alike in that they feed paint to the center of the spinning head and hurl paint droplets off its perimeter. However, rotary paint application devices differ in two principal ways:

- Shape of the atomizing head
- Mounting configuration.

Figure 14-2. Low-speed bells in a paint line.

Figure 14-3. Three different models of high-speed bells.

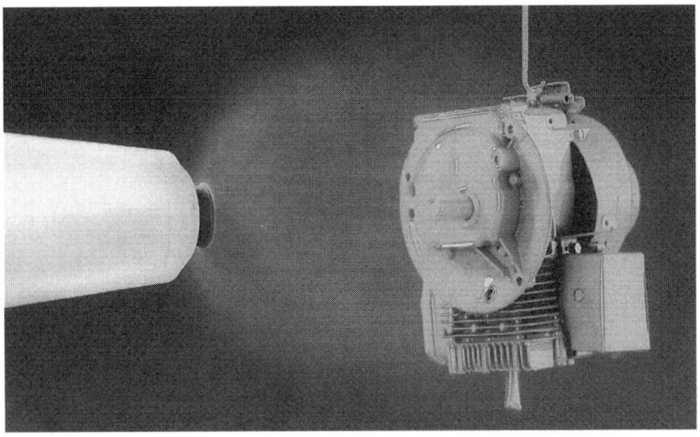

Figure 14-4. Rotary bells use centrifugal force, rather than air or fluid pressure, to break up the paint. High-speed electrostatic bells, operating at 20,000 rpm or more, emit a fine mist that produces a high-quality finish.

Shape of the Atomizing Head

Rotary atomizer heads are often made from a quality aluminum alloy and formed in one of two basic shapes: discs or bells (cups). The discs, as their name suggests, are thin, relatively flat, and round. "Dish-shaped" is a good description. Discs come in diameter ranges from about 4–8 inches (10–20 cm), but most have diameters of about 5–6 inches (13–15 cm).

The modern so-called bells are not, however, shaped very much like a bell. Depending on the manufacturer, they now more closely resemble a cup, a truncated cone, or a shallow sauce-dish. **Figure 14-4** shows a bell rotary atomizer in use; **Figure 14-5** illustrates how it

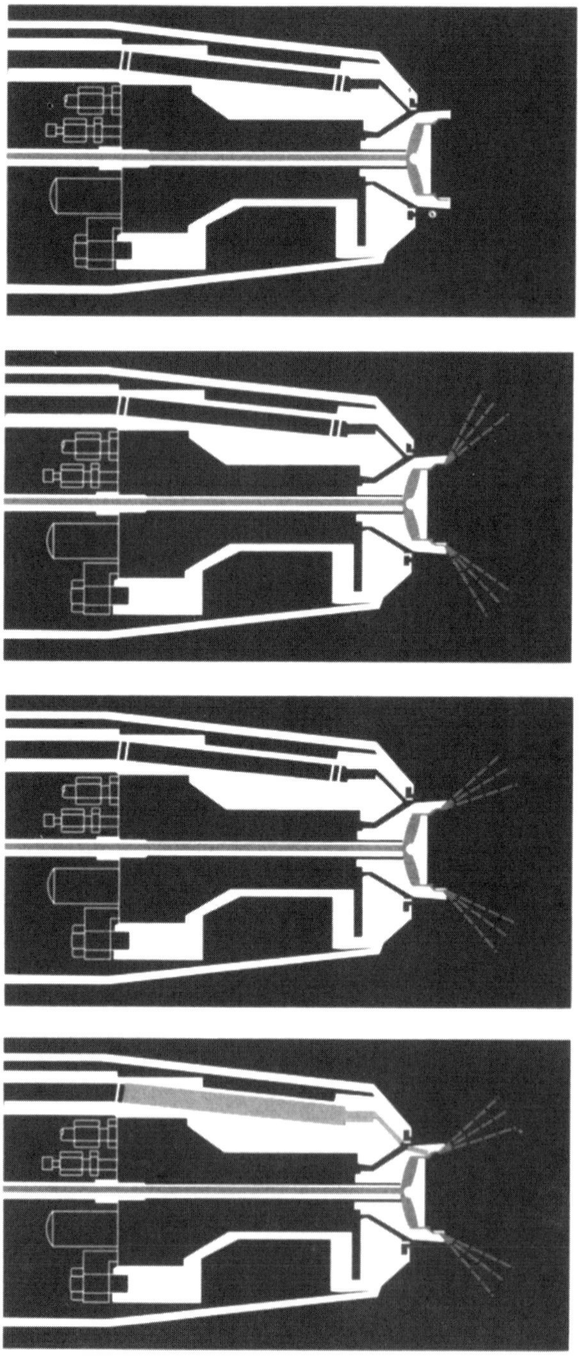

1. Coating material is fed through a tube in the center of the turbine shaft and flows into the rear of the cup.

2. Spinning at speeds up to 40,000 RPM, the coating material exits through small holes in the cup and is centrifugally atomized into very fine particles.

3. Shaping air is directed from behind the cup to provide excellent pattern control. It is also used to increase penetration into corners and deep recesses.

4. At the cup, an electrostatic charge is supplied to the coating material, creating a strong attraction between the paint particles and the grounded part being coated. This provides maximum transfer efficiency, uniform coverage, and excellent wrap onto part surfaces not directly in the spray path.

Figure 14-5. Bell rotary atomizer operation.

works. The bells range in diameter from about 1–5 inches (2.5–13 cm) but sizes below 4 inches (10 cm) are rare. Small bells are primarily used for robotic touch-up painting.

Mounting Configuration

Discs are mounted onto the lower portion of a vertical or near-vertical shaft so that the disc lies in a horizontal plane. The shaft and the air turbine that rotates it are enclosed in a protective housing. This assembly is generally attached to a ceiling-mounted *reciprocator* that moves the assembly slowly up and down to achieve total part coverage as shown in **Figure 14-6**. The vertical reciprocating stroke distance is determined by the length of the parts to be painted. Extremely long vertically hung parts, such as extrusions, may incorporate tandem floor- and ceiling-mounted discs, each with a reciprocating stroke slightly greater than half the length of the parts being painted. Very small parts may require no reciprocation, or may be painted with a disc tipped slightly away from true vertical. Tipping the disc widens the paint pattern "seen" by parts as they travel through the paint booth. Since discs deliver paint in a 360° pattern, parts to be painted by a rotary disc are hung from an overhead conveyor track that is looped around the disc. The parts must be positioned about 16–20 inches (41–51 cm) from the disc perimeter (target distance) to prevent sparks jumping from the charged aluminum disc head to the grounded parts.

Figure 14-6. A reciprocator moves the rotating atomizer up and down while painting.

Bells have a primarily unidirectional paint delivery pattern and so can be mounted in any orientation: vertical (see **Figure 14-7**), horizontal, or any angle in between. They can be mounted in fixed positions, onto vertically or horizontally moving positioners, or onto robot arms. If a positioner is used with a painting bell, it will move the bell to the correct target distance for each different part being painted.

Whereas a disc requires parts to be conveyed around it in a loop, a bell can be used much like an automatic spray gun in that it can be aimed at the part being painted. This allows bells to be used much like automatic spray guns (manual bells are not used in production painting). The typical distance from the bell to the part being painted is about the same as with discs. With newer heads made of nonconductive plastic coated with a conductive material, the target distance for discs and bells can be considerably reduced. These coated plastic heads are not only safer, they enable better paint coverage in Faraday cage areas.

Rotational Speed and the Degree of Atomization

As a rule of thumb, the faster the rotational speed, the greater the centrifugal force and the finer the atomization. However, two exceptions stand out. Rotational speeds below 2,500 rpm and between about 5,000 and 15,000 rpm exhibit anomalous behavior. **Figure 14-8** is a graph of atomized particle size as a function of disc or bell rotational speed. Atomization occurring with low-speed rotaries is due both to centrifugal force and the "pushing apart" of paint particles by electrostatic charge repulsion. The low-speed centrifugal force isn't great enough to hurl the paint off the perimeter in finely atomized droplets, but only as coarse particles of paint. It is the electrostatic charge that atomizes

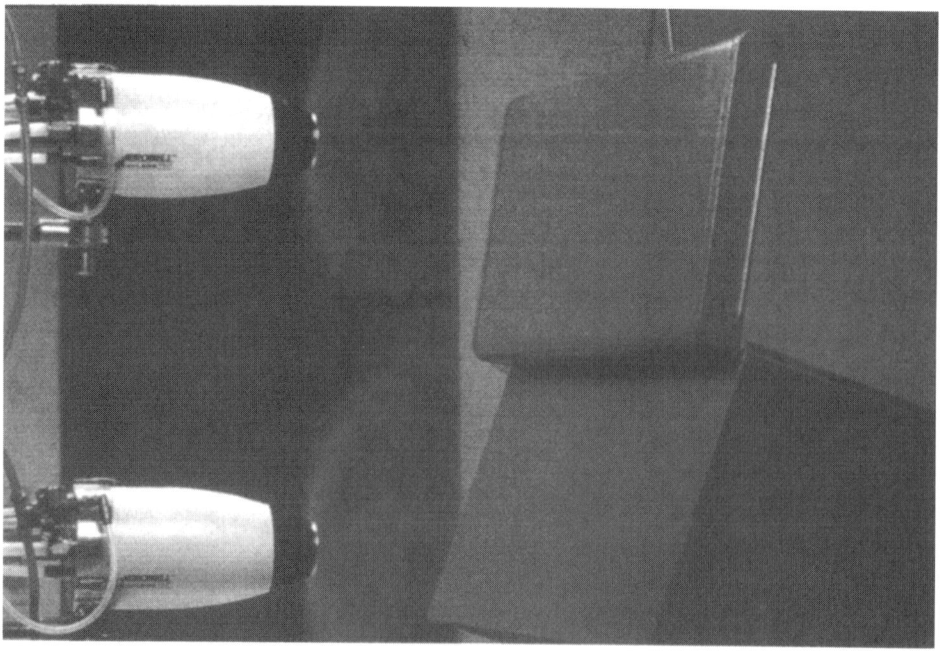

Figure 14-7. Vertically mounted bells apply a high-solids singlecoat epoxy to a metal tool chest.

the paint to a finer droplet size. This is termed *electrostatic atomization*. Without the electrostatic charging, atomization is not effective enough to use low-speed rotation to apply coatings.

Atomization quality on low-solids paints with low-speed electrostatic discs and bells is intermediate between the extremely fine atomization of air spray guns and that of airless spray guns. **Figure 14-9** compares atomized particle delivery velocities and particle sizes for air spray guns and bells. The low-speed rotary atomization was used very successfully on solventborne low-solids coatings for some years before EPA regulations brought about the current development in waterborne and high-solids coatings. Now, low-speed rotaries have largely disappeared, having been supplanted by the high-speed versions.

Refer again to **Figure 14-8**, and note that atomization is fair at speeds of 2,500–6,000 rpm (low-speed discs), and excellent at roughly 15,000 rpm and higher (high-speed discs). Strangely, atomization occurring between about 5,000 and 15,000 rpm tends to be rather poor. This is because the centrifugal force hurling paint off the disc forms string-like paint filaments instead of discrete droplets. Two or more adjacent filaments frequently tend to tangle and be flung from the rotating head as large sized "blobs" rather than droplets. As spin speed increases to above 15,000 rpm, the long filament formation ceases. Atomization is accomplished almost totally by centrifugal force at these speeds; droplet size decreases steadily as rpm's increase. The function of the electrostatic charge at these high rotational speeds is no longer helping atomization, but only placing a charge on the atomized droplets so they will be attracted to the grounded parts being painted.

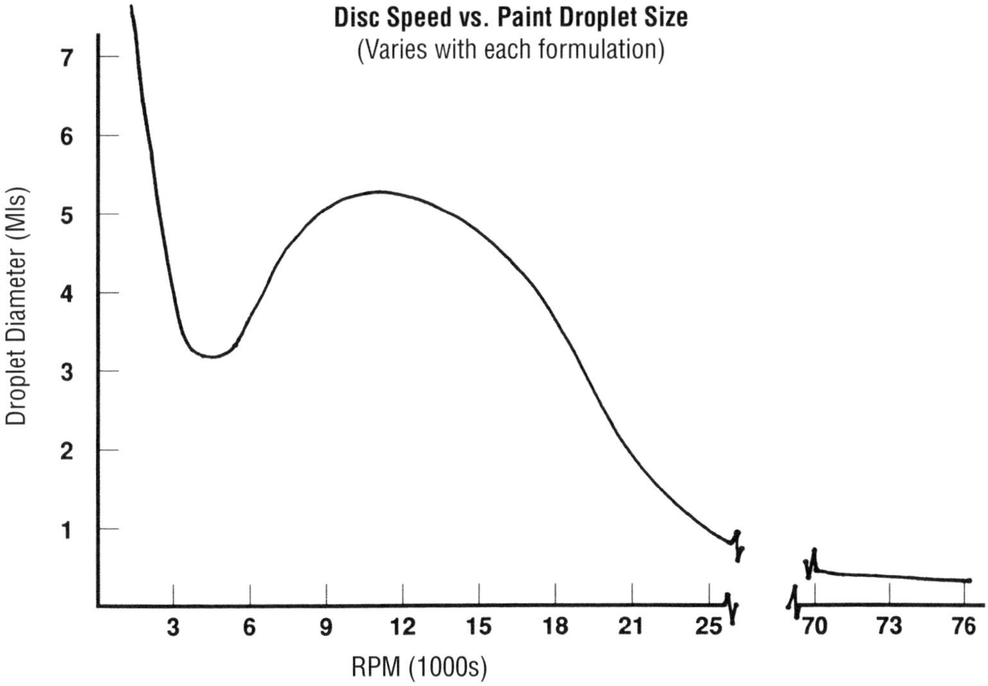

Figure 14-8. Droplet size as a function of disc speed.

Method	Typical Particle Delivery Velocity	Average Particle Diameter	Particle Size Range (Diameter)
Airspray	33 ft/s (10 m/s)	3 mil (75 μm)	0.5–5 mil (13–128 μm)
Bell	2.3 ft/s (0.7 m/s)	0.8 mil (20 μm)	0.6–1 mil (15–25 μm)

Figure 14-9. Comparison of particle delivery velocity, average particle size, and size range for air spray guns and bells.

High-speed discs and bells are almost always operated at above 20,000 rpm to ensure a fine atomization. Speeds up to 60,000 rpm are common. The particular operating speed is selected for the degree of atomization desired, the product being coated, the characteristics of the paint being applied, and the rate at which paint is fed to the rotor. As expected, the higher the viscosity and the flow rate, the faster the speed necessary for atomization. High-solids and waterborne paints are difficult to atomize at slow rotational speeds because they have a high surface tension, but application of both types works extremely well with high-speed discs and bells.

Paint Application

The atomized paint leaving a rotating disc is hurled outward in a 360° pattern. The droplets are guided to the parts to be painted mainly by electrostatic attraction. No directional air other than possibly "shaping air" is forthcoming from the disc head to give impetus to the atomized particles. With electrostatic attraction as the principal propelling force, the exhaust airflow in disc booths must be relatively calm. Strong air currents would distort the spray droplet pattern and tend to sweep atomized paint droplets away from parts being painted.

With the bells, directional "shaping air" concentric with the bell is sometimes used to reduce the circular size of the atomized paint particle cloud, helping the electrostatic attraction guide the particles to the parts to be painted. The air, varying in pressure from about 10–40 PSI (69–276 centibar), is directed through a circular groove or a series of small, closely spaced holes in the housing behind the bell's perimeter. Whenever a reduced size pattern paint delivery is needed, shaping air can be used to force the atomized particle cloud into a smaller "doughnut" of droplets. With small parts, the shaping air helps considerably to reduce the atomized droplet pattern diameter and thereby minimize paint overspray waste. The shaping air does not assist in atomization or the delivery of paint, but only helps to control the size of the droplet spray pattern.

The discs and bells are insulated from ground to accept the electrostatic charging voltage (about 100 kV) and transfer these charges to the paint as it travels across the rotating head. As with electrostatic spray guns, the parts to be painted need to be electrically grounded to establish electrostatic attraction. The parts to be coated, being the nearest electrical ground, attract the electrostatically charged particles. The electrical circuit is completed as with electrostatic spray: the paint particles give up their charge to the grounded parts on contact, and the negative charges flow to ground through the grounded hooks and conveyor.

Conveyor and Booth Configurations with Disc Painting

Various conveyor and booth arrangements are possible with disc systems to bring about painting of the total part. Unless the part is narrower than about 1 inch (2.5 cm), only the side of a part facing the disc will be completely painted as it is conveyed around the disc. The back will get some paint on it, as we have seen, from the electrostatic wrap-around. Both sides of flat parts can be completely painted by incorporating two booths with reciprocating discs and by having the conveyor line form a loop around each in a "figure S" configuration. For best appearance, the "S" arrangement is preferred, although a small separation between the loops of the "S" is acceptable. Too long a wait between loops will let the first application dry enough that a line will show where the two coats meet. Alternatively, parts can be painted, then rotated 180° and passed through a second disc booth. A third method can be used for circular parts, such as water heater jackets or propane tanks. These and similarly symmetrical parts can be entirely painted in one booth by automatically rotating the part slowly on its hanger as it is conveyed around a single reciprocating disc.

Flat parts to be coated by being conveyed around a disc must not be excessively wide or else the leading and trailing vertical portions of the part will receive less coating than the center. This is because the part must be conveyed around a circular path, and the ends are therefore much farther from the disc head than the center of the part. For example, if a 5 foot (1.5 m) wide part is 12 inches (30.5 cm) from the head at its center, the ends will be over 32 inches (81 cm) from the head. The maximum width of a part that should be painted by being conveyed around a disc is about 3 or 4 feet (0.9 or 1.2 m).

Disc application booths are shaped like a letter "C," that is, circular in shape with a notch out. Up to about one-third of the circular booth is open for the conveyor entry and exit. The conveyor to and from the entry and exit portions is linear, but the conveyor curves in, then circles in the opposite direction around the disc. This configuration of the conveyor in a top view yields a shape resembling the capitalized Greek letter omega (Ω), and so it is often called an *omega loop*.

The downdraft air in disc booths must be at a low velocity to prevent generating strong airflow within the booth that could interfere with the attraction of the charged atomized paint particles to the parts being conveyed around the disc. The booths are always the dry-filter types; the lower portion of the circular wall of the booth consists of paint filters. Air is drawn in through the open ceiling and booth opening, and it is exhausted through the filters and outside up through a stack.

Disc and Bell Rotation

Low-speed discs and bells were spun by electric motors. An air-driven turbine, however, rotates the high-speed discs and bells. They operate on the same principle as a high-speed air-driven dental drill. Much higher speeds can be achieved with turbines than with an electric motor drive. The high-speed rotary units require disc and bell turbine drive systems of exceptionally fine quality. Bearings that support the rotation are of two types: mechanical and air. The mechanical bearings must be precision-machined and lubricated to withstand the high rotational speeds. Some manufacturers still sell mechanical bearing disc and bell rotary applicators, but the air bearing type turbines are preferred. Air bearings substitute a steady stream of air for the mechanical bearing section. The air prevents the rotating element from touching the rotor housing. Some manufacturers of air bearings insist that air be turned on to support the shaft at all times, even when the device is not in use. This prevents the weight of the shaft from causing an indentation deformation that would adversely affect the unit for obvious reasons. If operated and maintained properly, air-bearing turbines have been shown to provide superior bearing life.

A frequent cause of high-speed rotating system failure is poor air quality and dirty bearings. Insufficient filtration of the air and failure to replace filter elements on schedule will permit oil, moisture, or particulate to gel or harden the lubricants in mechanical bearings. These contaminants can build up in air bearings as well as in mechanical bearings and eliminate the thin cushion of air between the sleeve and the rotating shaft.

The dynamic balance of high-speed bells with mechanical bearings is of crucial importance. Any abnormality can disturb the balance at high speeds and strain the bearings, shortening their life. These defects include nicks and dents in the disc or bell heads caused by improper handling and uneven paint buildup. An imbalance with air bearings generally has no adverse effect until it causes the rotating elements to contact the housing. Then bearing wear becomes rapid.

To preserve rotational balance, exercise care in handling and cleaning disc and bell heads. In some plants normal booth maintenance includes spraying most of the booth equipment with cleaning solvent. Other plants use a cloth or brush and a pail of solvent to remove overspray from rotary units. Either of these practices can cause problems by forcing paint buildup into the bearings. Solvent getting into mechanical bearings will wash away the lubrication and also can carry paint solids into the bearings. Nearly all rotaries have an air barrier to protect the bearings from paint or solvent incursion. This air should always be turned on during cleaning operations, but even this may not provide complete protection from pressurized solvent spray cleaning.

Automatic cleaning machines for bell heads have been designed by several automobile assembly plants to avoid rotary atomizer head damage. Nicks are easily made in a delicate bell or disc edge if it is hit by a part on the conveyor, or by a metal tool when the rotary atomizer head is being replaced or removed during cleaning.

Rotary System Operation

Paint is delivered to the surface of rotary disc and bell heads via delivery holes around the periphery of a smaller concentric inner cup. Center delivery feed and uniform distribution are needed to maintain precise head balance. Solvent is fed through this line to purge paint for color changes, and also to flush clean the head whenever it becomes necessary.

On older model bells, retractable flush shrouds were used to collect color-change purges and solvent rinses. Collection shrouds for side-mounted bells were fitted with gravity drain lines; overhead-mounted bells used siphon tubes. Capture of paint and solvent during these operations reduced the plant's overall VOC output and in many instances lowered maintenance costs significantly. Internal dump valves are now used for line purges and flush outs with solvent or air/solvent mixtures. Both alternating air/solvent pulses and integrally "foamed" air/solvent mixture systems are available for flushing. Mixing or pulsing air with solvent not only sharply reduces solvent usage, it also scours the fluid lines more effectively.

The rotational speed of a high-speed bell slows down when paint flows into it. This slowing is the result of the increased weight of rotating material. At a typical fluid flow of 10 oz/min (0.28 l/min), the reduction in spin speed is roughly 10%. For a fairly high fluid flow of 20 oz/min (0.57 l/min), the speed drop-off is about 20%. If one is near the bell, one can even hear the abrupt lowering of the pitch emitted by the spinning head when this occurs. Depending on the paint being used and the particular application, this slowing could be a significant problem. Therefore, the bell spin rate is preset high enough so that the spin-speed reduction when paint starts flowing does not become a negative factor in atomization. Some rotary manufacturers have used spin-speed governing controls. Although speed control governors maintain rotational speed irrespective of fluid loading, such devices have not been universally used. A third method uses a rather simple dual-air pressure system with a "high/low" setup. In this type, two different air pressures are applied to the turbine. A high pressure is used when paint is being applied; a low pressure is used during nonpainting or bell "idle-speed" periods. The dual-pressure system only works well with stable fluid-flow rates. Electronic speed control is preferred for painting operations that require frequent on/off triggering, those demanding extremely fine finish quality, or whenever fluid flow rates are varied.

Early discs and bells had no convenient, fast method to monitor rotational speeds. Systems that permit digital rpm readouts of spin speed are now almost standard. The spin rate of the rotaries can be continuously monitored using one of two similar systems. One system involves electronic speed regulation with a fiber-optic pickup, and the other with a magnetic impulse pickup. The magnetic impulse regulator offers the advantages of greater durability and lower cost for cable replacement.

Serrated (mechanically grooved) edges on some bells eliminate air entrapment "microbubbles" in the wet paint film. Entrapment of air is most likely to occur with high-surface-tension coatings, including both waterbornes and high-solids. Serrated edges enhance atomization so that bells can be operated at slightly reduced rotational spin rates. Each closely spaced serration line acts as a site that forms another paint filament to enhance atomization.

The fineness of atomization from a high-speed rotary is virtually unaffected by the electrostatic charging voltage, but it varies with the paint viscosity, the paint flow rate, and the head rotational speed. Bell spin speed can affect not only atomization and pattern size, but also the distribution of mica and metallic flakes in the film. Rotational speed may affect the paint color until a certain maximum speed is reached, above which virtually no color difference is noticed. While they do occur, fortunately visual color shifts are seldom detectable.

If different coating application methods are used on separate parts of an assembly, it can cause appearance mismatches from the texture or color variations they produce. Paint films applied by spray guns can take on a slightly darker shade of color from those applied by high-speed rotary atomizers. This is due to the differences in droplet particle size and the velocities at which droplets are delivered to the part. This color variation possibility should be investigated when interplant painting is introduced to avoid color matching difficulties later.

Hand-Held Bells

The bell and disc rotary atomizers described so far in this chapter have wide use in applying finely atomized paint on industrial painting lines. These bell systems are mounted as permanent capital equipment; thus, the parts to be painted are brought to them, usually on a conveyor. A portable hand-held bell rotary applicator is used, however, for low overspray finishing of selected immobile items such as fencing and pad-mounted electrical switch and transformer housings. Another widely used application is in repainting metal office furniture, as shown in **Figure 14-10**. It produces so little overspray that items to be painted need not be transferred to another location; they can be refinished right in their normal location. Another common use is to repaint installed decorative grillwork at a fast rate with minimal mess and little paint waste.

The portable unit has a rather long configuration but is surprisingly well balanced and easily maneuvered. The operator holds the unit roughly 6 inches (15 cm) away from the part being painted and moves the unit slowly back and forth or up and down in even strokes, or in circular strokes of 4–6 inches (10–15 cm) in diameter.

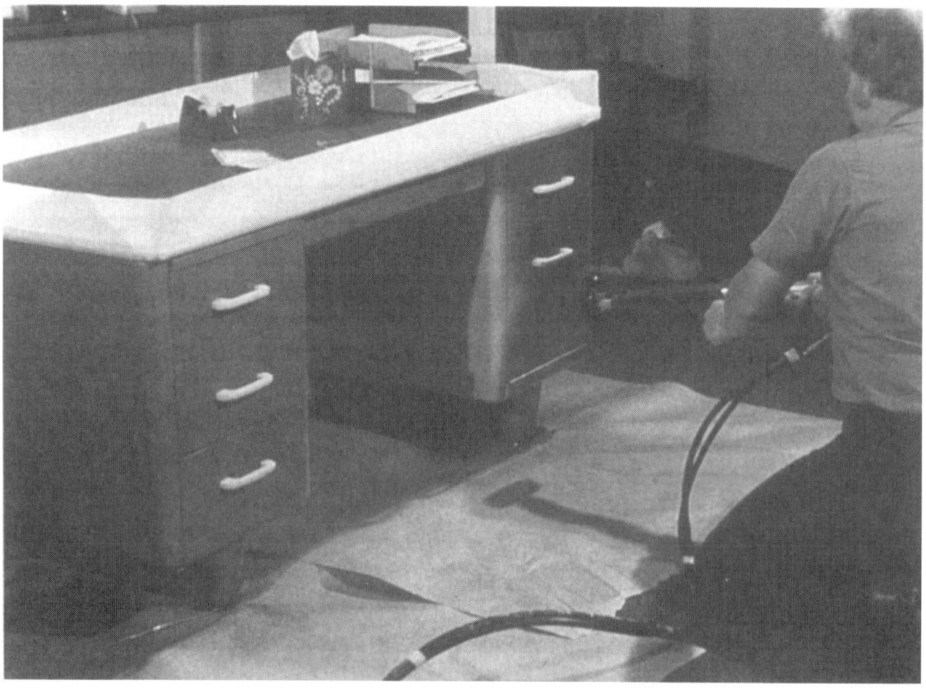

Figure 14-10. Portable bell rotary atomizer.

The bell electrostatically charges the atomized paint particles as they travel across it. An electrical grounding wire is affixed with a metallic clip onto office furniture being painted, which enables the part to attract charged paint droplets. Protective canvas or plastic sheets can be positioned around other objects in the vicinity and on the floor to catch any overspray. The painting is usually done after business hours. Air-dry paint allows the parts to be ready for use when business begins the next day. The bell can only apply up to 5–6 oz/min (142–170 cm^3/min) of paint. This rate of flow, while severely limited, is still satisfactory for applications such as in-office furniture repainting, but is just not adequate for industrial production line painting.

(It is not completely accurate to state that there are no manual disc devices. Of perhaps historical interest only is the hand-held rotating disc gun that was used in just a few places about 30 years ago. It had a shielded construction that blocked the paint thrown off the spinning disc except for moveable "doors" that controlled the width of the delivered spray pattern in the direction the device was pointed. It was not electrostatic. If anyone has a photo of such a gun, please let me know.)

High-Speed Rotary Atomizer Advantages

High-speed rotary disc and bell atomizers have several important advantages:

- Fine atomization
- High-solids, low-solids, metallics, two-component, and waterborne versatility
- Viscosity flexibility on both high and low VOC coatings
- High transfer efficiency.

Fine Atomization

The high rotational speeds and electrostatic charging voltage team up to do an exceptional job of atomizing paint into extremely fine particles to ensure a high-quality finish.

High-solids, Low-solids, Metallics, Two-component, and Waterborne Versatility

High-speed rotary atomizers can apply high-solids, metallic, two-component coatings, and waterborne coatings with the same ease as low-solids. Virtually all coatings can be applied with rotary atomization, even at high flow rates.

Viscosity Flexibility on Both High and Low VOC Coatings

The speed of rotation can be adjusted to compensate for coatings with varying viscosities, which eliminates the need to add solvent to adjust viscosity in many cases.

High Transfer Efficiency

Fine atomization, electrostatic charging, and minimal air turbulence in the vicinity of the rotary atomizer and part being painted ensure high transfer efficiency. **Figure 14-11** compares typical transfer efficiencies for the various types of spray guns with the rotary atomizers.

APPLICATION TYPE	TYPICAL TRANSFER EFFICIENCY RANGES (10 – 30 – 50 – 70 – 90)
AIR SPRAY High Pressure	25 ⊢⊣ 50
AIR SPRAY Low Pressure	40 ⊢⊣ 70
AIRLESS	35 ⊢⊣ 65
ELECTROSTATIC AIR SPRAY	35 ⊢⊣ 60
ELECTROSTATIC AIRLESS	40 ⊢⊣ 70
AIR-ASSISTED AIRLESS	40 ⊢⊣ 70
ELECTROSTATIC AIR-ASSISTED AIRLESS	45 ⊢⊣ 75
ROTARY (All are Electrostatic)	60 ⊢⊣ 90

Figure 14-11. Typical transfer efficiency ranges for various application methods.

High-Speed Rotary Atomizer Disadvantage

High-speed rotary atomizers have only one noted disadvantage. Although their capability to paint broad open surfaces is outstanding, the electrostatic attraction coupled with the absence of air or fluid pressure to push paint droplets onto the target limits their ability to paint into Faraday cage areas. Manual touch-up in these hard-to-paint areas is usually required. However, the use of nonmetallic plastic or ceramic discs and bell heads coated with an electrically conductive material almost eliminates any possibility of arcing from the bell to a grounded part. This permits the operator to safely position the rotary applicator head much closer to the part, which overcomes much of the Faraday cage problem.

Chapter 15

Coating Types and Curing Methods

The Curing of Coatings

The curing of a coating is the process whereby the freshly applied liquid or powder coating is transformed to a finished solid paint film. Coatings are generally of little or no value as they are applied; any simple touch will mar them. The raw coating must be converted to a more durable paint film to perform well, either for appearance or function, or both. Liquid coatings are converted to their cured state either by solvent evaporation only, or by solvent evaporation plus resin cross-linking.

Heating to melt the individual powder particles, and thus form a continuous film on coated parts, cures powder coatings. In roughly 98% of powder coating formulations, the heat also initiates chemical cross-linking of the resin molecules, although a few powder coatings simply melt without any cross-linking taking place. On cooling, the molten film becomes a solid once again, but now it forms a continuous film of paint instead of a layer made up of separate powder particles.

Coatings that Cure Without Resin Cross-Linking

A liquid coating that cures by solvent evaporation only is called a *lacquer*. Lacquers usually contain one or more resins with long-chain polar molecules dissolved in a solvent. As the solvent evaporates, the molecules are drawn together by polar attractive forces at numerous sites along their chain lengths; no chemical reaction takes place, however. The strong polar attractions produce a hard film once all the solvent has evaporated. In a similar way, lacquer powder coatings do not undergo any cross-linking during curing; they only melt into a liquid that solidifies upon cooling.

It is worth repeating the technical definition of "lacquer." Any liquid coating that cures by solvent evaporation only, and any powder coating that cures only by melting and resolidifying, is designated *lacquer*. Liquid lacquer paints can be made from any soluble resin (polymer). Common ones include chlorinated rubber, cellulose, phenolic, and acrylic resins. The resin molecules in lacquers characteristically have a high molecular weight, which gives good paint film hardness and water resistance. The main weakness of lacquer paints is a lack of resistance to the types of polar solvents originally used to formulate the liquid coating. This so-called weakness can also be an asset, however. In the refinishing of a lacquer film, solvent can be used to reflow the paint and thereby increase its surface smoothness.

Although liquid lacquers can often be air dried, they can also be processed (cured) in an oven. The only function of the oven would be to speed the solvent evaporation. Because of the high molecular weight resins of liquid lacquers, they are hard to dissolve and so their solvent content is necessarily high. A lacquer may be as much as 90% solvent and only 10% paint solids. You can understand, then, that with EPA limitations on VOC emissions, industrial lacquer usage has declined considerably since the early 1970s.

If the correct solvent is added to the dry (cured) paint film, the resin molecules can again go into solution. The solvent molecules will interfere with the polar attractive forces, which allow the resin to absorb some of the solvent and begin to dissolve. Lacquer resins can also be heated to reduce the polar attractive forces, which cause the resin to soften. The term *thermoplastic* is used to describe this thermal softening behavior of materials.

Coatings that Cure by Cross-Linking

All coatings that cure by cross-linking are categorized as *enamels*. This applies to both powder and liquid coatings. A liquid coating that cures by cross-linking normally contains one or more partially polymerized resins. Although there are 100% solids liquid enamels, most enamel paints, before they are applied, contain resins that are dissolved in a solvent. As the solvent evaporates or as heat is applied, a chemical reaction occurs either among the resin molecules or with molecules of another cross-linking resin so that chemical bonds are formed. Adding solvent cannot dissolve the newly formed chemical bonds. But although enamels will tend to be less solvent-sensitive than lacquer paints, this does not mean that solvent cannot soften enamel paints.

Coatings that cure by cross-linking can be divided into at least six categories:

- Oxidizing coatings
- Moisture-cure coatings
- Heat cross-linking coatings
- Reactive catalytic cross-linking coatings
- Catalyst vapor coatings
- Radiation-cure coatings.

Oxidizing Coatings

These liquid coatings use resins based on drying oils. The resins may be alkyds, phenolics, oil-modified urethanes, epoxy esters, or various oleoresinous systems. Some have drying oils added expressly to facilitate air-dry curing or force-drying (mild heating). After the coatings are applied, the solvent begins to evaporate, which exposes more and more of the unsaturated resin to the air. This air contact triggers a cross-linking reaction between the resin and atmospheric oxygen. These resins contain molecules with carbon-to-carbon double bonds that react with oxygen to form peroxy bonds. These bonds subsequently react further to form ether cross-linkages that comprise the final oxidative cross-linkages. This cross-linking occurs slowly at room temperatures but will progress rapidly at bake oven temperatures.

Moisture-cure Coatings

This category of liquid coatings functions in a manner similar to the oxidizing systems. After the paint is applied, moisture in the air will begin to react with the coating resin, resulting in resin cross-linking. An example of this type of coating is a moisture-cured urethane in which isocyanate groups react with atmospheric moisture to form urea linkages. As with oxidizing systems, elevated temperatures can speed the curing process.

Heat Cross-linking Coatings

Some liquid coatings will either cure extremely slowly or not cure at all by air-drying, and therefore require heat for curing to take place. Their essential composition consists of an uncross-linked primary resin, such as acrylic, plus a secondary cross-linking resin, such as melamine. During manufacture, the two resins are blended together with solvent, other additives, and possibly pigments as well. In this example, the resultant cured paint film is called a "melamine-cured acrylic." Once the container is opened and the coating is applied, the solvents will begin to evaporate, but little or no curing will occur at room temperature. The applied coating will merely remain tacky (sticky) and uncured unless heat is applied. The paint temperature must reach or exceed a level that triggers cross-linking of the resins. This is called the minimum cross-link temperature.

Heat cross-linking coatings can, of course, have various types of primary resins and a variety of cross-linking resins. Whatever the resin types, the basic principle is still the same: a certain minimum temperature must be reached to trigger the reactions that cause cross-linking and paint curing. For most high-solids coatings, the minimum cross-link temperature is well above room temperature. Such paints can remain tacky almost indefinitely without curing unless heat is applied. Heat can be introduced several ways including hot air ovens, induction heating, and infrared or microwave radiation.

All enamel-type powder coatings, which includes nearly 99% of them, also fall into the heat cross-linking category.

Reactive Catalytic Cross-linking Coatings

The coatings in this category cross-link cure. Each of these coatings consists of two separate parts: the major coating component and a lower-volume catalyst component. The major coating component contains mostly resin, various additives, solvents, and perhaps pigments. The catalyst component contains essentially only solvents plus the catalyst, or it may contain solvents, a catalyst, and a cross-linking resin. As long as the two paint parts are kept separate, each remains essentially unchanged in a fluid state. Once the two components come together, however, cross-linking reactions begin between the resin and cross-linking agent. The catalyst increases the rate of cross-linking and more rapid film curing is the desired result. The applied coating will cure just after or even during application without applied heat. However, elevated temperatures will speed the rate of cross-linking and solvent evaporation.

An example of a catalytic coating is a two-component urethane consisting of a polyester resin as one part and an isocyanate cross-linker plus amine catalyst as the second part. When the two are intimately mixed, a cross-linking occurs that forms a urethane linkage, thus the resultant coating is called a urethane. The coating can be mixed in a single container and then applied, or the two components can be brought together in the atomizing portion of a spray gun and mixed there (see **Figure 15-1**).

After the two components are mixed together, the resultant mixture must be used within

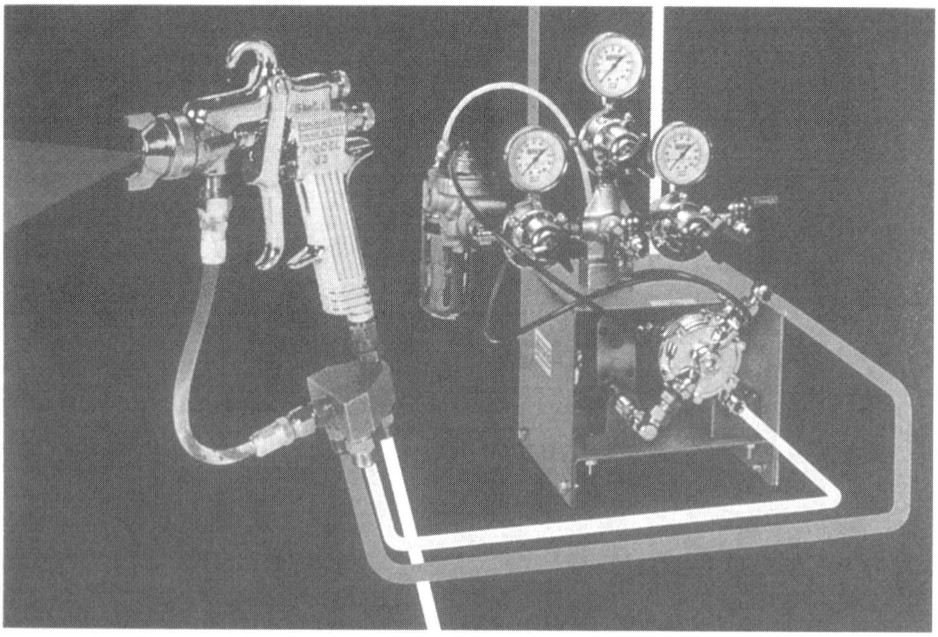

Figure 15-1. Spray gun mixing metered ratios of each of two components.

a limited amount of time. The maximum post-mix time within which the coating material must be used is termed the *pot life*. If the mixture is allowed to remain in the container or the paint lines, complete cross-linking will occur over a predetermined period of time, and the mixture will become too hardened to pour out. Paint formulation chemists are able to control the rate of cross-linking and the resultant pot life to various degrees. The pot life can be controlled to be as brief as a few minutes or as long as 16 hours. If the pot life is exceeded, excessive cross-linking lowers the film quality even though the paint may still appear to be applied properly and look normal.

A word of caution must be given here. The terms *catalyst* and *cross-linker* are often replaced with the term *accelerator* and *activator* by paint manufacturers, so the true catalyst and cross-linker are not always obvious. However, most manufacturers use terms consistently within their literature, so you should be able to follow their instructions and achieve correct results.

Catalyst Vapor Coatings

Known as *vapor curing*, this interesting but little-used process uses a room-temperature vapor catalyst to speed the curing of certain two-component liquid formulations. After the two-component paint is applied, the coated parts are conveyed into a chamber filled with the vapor of a catalyst such as diethanolamine. The vapor and liquid of diethanolamine are safe and nontoxic, but they have an unpleasant fishlike odor. The catalyst concentration is held at 1,200–1,500 ppm. Air seals at the entrance and exit of the curing chamber are used to prevent loss of the catalyst. The vapor accelerates the ambient-temperature cross-linking reaction rate of the isocyanate and hydroxyl groups, which forms the urethane bonding. The catalyst reduces the cure time from roughly 40–60 minutes down to about 5–10 minutes. Catalyst must be scrubbed from the oven exhaust with water to prevent its release into the environment.

A low amount of heating is usually employed on parts to drive off solvents after they leave the vapor chamber. That only low heat is required can be an advantage when painting large or thick castings and similar parts. It avoids the long, slow heat-up and cool-down times that are needed when oven-baking coatings on massive metallic substrates or on similar parts with large heat capacities. Catalyst vapor curing can also be used to advantage for curing coatings on heat-sensitive substrates such as paper, cloth, wood, and plastics.

A variation of the catalyst chamber process, called catalyst vapor injection curing (VIC), injects catalyst vapors into the atomizing air supply of air spray and air-assisted airless spray guns, as shown in **Figure 15-2**. It can also be introduced into the shaping air (air shroud) of rotary bell paint applicators. In this way, a separate catalyst chamber stage can be eliminated from the coating process. This variant is not used much either.

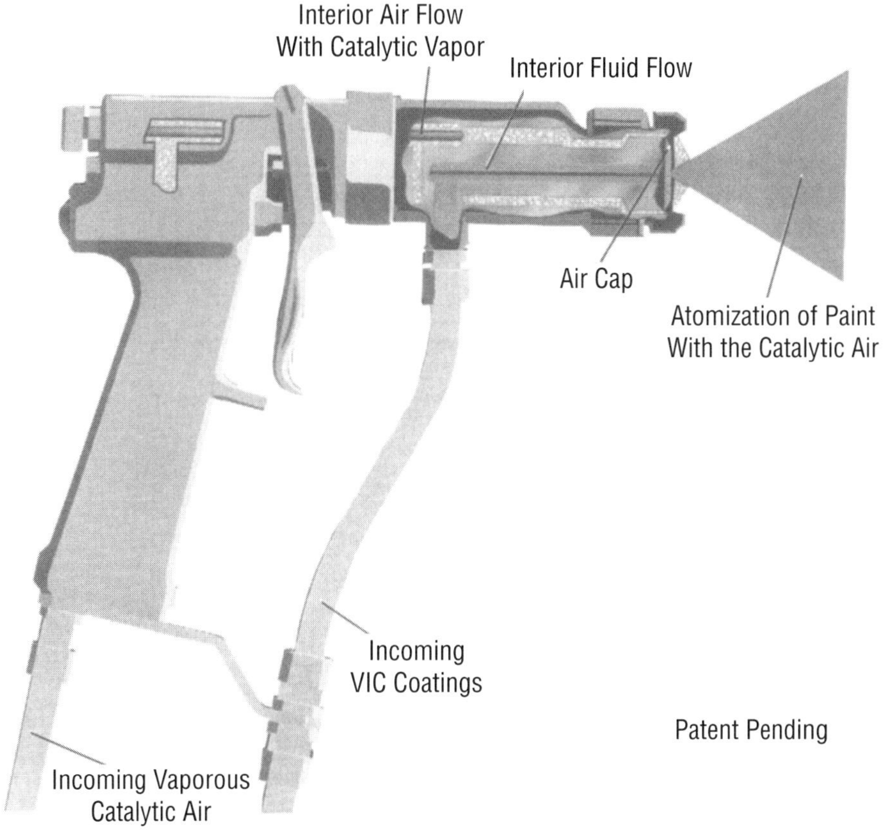

Figure 15-2. Vapor injection curing (VIC) coating/catalyst mixing.

Radiation-cure Coatings

Radiation-cure coatings contain various catalysts (or accelerators) that are dormant until acted upon by either ultraviolet (UV) light or electron beam (EB) radiation. The UV light or electron bombardment triggers a free-radical reaction among chemical groups that results in swift cross-linking (curing) among molecules of the paint resin(s).

UV-cure coatings. UV-cure coatings contain chemical photoinitiators that are sensitive to UV light. This radiation causes homolytic (equal) cleavage of certain chemical bonds (often peroxy bonds) of the photoinitiators to form free-radical groups. Reaction of the free radicals with the resin in turn triggers resin cross-linking. Curing happens in a multistep sequence. First, photoinitiator molecules absorb UV rays and break apart into free radicals, a step called *initiation*. Initiator free radicals react with resin molecules to form resin free radicals; this is termed *propagation*. Then combination of the various resin free radicals, a process labeled *termination,* forms multitudinous cross-linked resin bonds. This gives a cured coating film. Some radicals often remain for a brief time after UV exposure, and this gives a small degree of postcuring to the film. Heat is not necessary, although any heat absorbed from the UV lamps will accelerate the polymerization cross-linking of resin free radicals.

UV coatings undergo extensive cross-linking in just a very short time, often merely a few seconds. As a result, coatings for UV-curing are able to utilize low viscosity resins. The value of this is that they require little or no solvent for formulation. If any solvent is

Figure 15-3. Microwave powered UV lamps are used to cure powder coatings applied to electric motors and automotive radiators. UV curable powder coatings are often selected for products that contain temperature sensitive components.

used (and at least a small amount often is), a brief flash-off time must be allowed after application prior to parts entering the UV cure section. Frequently the fluidizing media is also a cross-linker, in which case it is called a *reactive diluent.* For reactive diluents, no flash-off time is required because they do not evaporate but instead become part of the cured film resin system. Phenolic alcohol-ester reactive diluent compounds, for example, are used as reactive diluents and thereby reduce solvent VOC. While that is their primary purpose, reactive diluents can also improve hardness and durability in some formulations. Because rapid, extensive resin cross-linking can be initiated with UV light, so that often extremely low-molecular weight resins with very low viscosities are possible in the coating formulation, these resins atomize and flow out so well by themselves. For a number of UV coatings in use today, no solvent content is required, which allows a zero VOC coating with its obvious advantages.

Cure by UV is accomplished in enclosed chambers saturated with high-intensity electrically generated UV light. These must be sight-shielded to prevent damage to people's eyes. For total curing to take place, the UV light must activate all of the photoinitiator molecules, which means that the light must shine on all areas of the parts. It must also shine all the way through the entire thickness of the paint film. This is fine for unpigmented coatings, but only about 1 mil dry-film thicknesses of pigmented coatings can be UV cured. With thicker films, the pigment molecules may block much or all of the UV light rays from some of the photoinitiator and inhibit full curing. In this case, the outer portion of the film will cure but the underlying bottom part of the film will remain uncured and soft.

An important factor to remember is that the energy of any light decreases with the square of the distance between the light source and the surface receiving the light. Doubling the distance of the lamp to the uncured paint surface will lower the UV light intensity to only 25%; tripling the distance reduces light intensity to less than 12%. This is a sharp curtailment, so the UV light source must be kept as close to the painted part as possible. Consequently, UV cure is used mostly on flat surfaces, or on surfaces that can be passed very near to the UV light source. Highly polished parabolic reflectors will reflect UV light and enable certain types of three-dimensional items to be UV-coated. These include clear coatings for golf balls, wall plaques, guitar bodies, and plastic or wooden pieces (grips, stocks) of firearms.

We have learned that UV curing is fast: usually it can be completed in 5–45 seconds. This permits UV ovens to be small and compact units compared to thermal ovens. The quick cure minimizes substrate heating, which is a great advantage when curing films on heat-sensitive substrates such as printed circuit boards, wood, and thermoplastics. UV lamp bulbs are sealed quartz tubes containing an inert gas and a small quantity of mercury. To alter the emitted UV wavelength frequency pattern (spectrum), the mercury may be doped with tiny amounts of metals such as gallium or iron. Two types of UV bulbs are employed. The lamps with electrodes are called arc type bulbs; they are less expensive and operate with less electrical energy, but they only last about 25% as long as electrodeless bulbs due to electrode burn-off wear. Wear on electrodes is particularly great when bulbs are started. Bulbs powered by microwave emitters are more complex. Their advantage is that they have no electrodes to wear down, and so they last far longer, even if they are turned off and on repeatedly. Since all UV lamps become rather hot, bulb cooling is often required. Even so, it may be necessary that lamps be turned off or shuttered temporarily whenever the production line is stopped. This becomes necessary in some cases to avoid scorching or causing heat-distortion on the product being coated. Air-cooled lamps are suitable if the substrate is not overly heat-sensitive, and these are more economical than fluid-cooled types. Water-cooled lamps should be considered for coatings on paper or plastic films. In the past, many UV lamps could not be restarted quickly once they had been turned off, but now fast "on/off" electrodeless UV lamps are used. These lamps cool rapidly so the coating line can be started and stopped quickly without ill effect on parts.

EB-cure (electron beam) coatings. EB-cure coatings are formulated with resins that undergo cross-linking when radiated with beams of electrons. As with UV curing, only a limited variety of EB-cure coatings is available. Electron radiation is the weakest of the radioactive emission types, but it is still potentially hazardous. The electron beams emitted by an electron radiation cure source are sufficiently strong that they can be dangerous to human tissues. Therefore, the EB radiation source must be completely shielded. The electron emission area is designed to be inaccessible to operators, and fitted with safety interlocks on doors. The EB source shuts off and becomes inoperative when system doors are open in order to prevent humans from accidentally getting close enough to be harmed.

EB systems require an inert gas flush, such as nitrogen or argon, to keep air out of the electron radiation zone. This is necessary because electron radiation passing through air

Figure 15-4. Testing clearcoats that are UV cured. On flat surfaces, the cure is practically instantaneous.

(which you may recall contains approximately 20% oxygen and 80% nitrogen) would create ozone and mixed oxides of nitrogen. These are harsh irritant gases that present severe health hazards for humans even at low concentrations. People must also be excluded from the inert gas atmosphere as a lack of oxygen presents a danger of asphyxiation.

The method of initiating chemical cross-linking by electron beams is in many ways similar to UV curing. Not surprisingly, therefore, some of the same advantages accrue with both UV and EB curing. The quick line start and stop capability, the low floor space requirements, the minimal substrate heating, the low or zero VOC emissions, and the rapid production rates possible for UV curing are also advantages of EB curing. While both are fast curing methods, EB curing is even faster than UV curing. Completely film curing by EB is often possible in less than a second, for example. The thickness of the coating is not a problem in EB curing as it is in UV curing; electron beams are far more penetrating than UV light.

The electrons collide with the atoms or molecules of the gaseous atmosphere around the source. As a result, the electron beams lose their energy rapidly as they travel away from the generating source. A 6- to 8-inch (15–20 cm) target distance is about the maximum possible. With extremely thin nonmetallic substrates some "through curing" may occur, but it is principally a "line-of-sight" curing process. No through curing is possible when metallic substrates are involved. Electron beams will not even penetrate metal foil. A slight amount of reflected beam curing can occur, but that is ordinarily not significant in the EB curing process.

At first it might be expected that EB curing would be ideal for curing paints on heat-sensitive substrates such as wood, paper, and plastics. This is only somewhat true. EB curing is possible for some papers, cloth, or wooden materials, but it is not well suited for the general curing of coatings on plastic parts. Polymeric substrates do not readily block the passage of electron beams. In addition to curing the paint, the penetrating electron beams tend to over cross-link the underlying plastic substrate as well. Substrate cross-linking can dramatically reduce the plastic's impact strength by excessive embrittlement.

Types of Thermal Ovens

The most commonly used coating curing ovens are the hot air convection and infrared types. Both have the same goal: to heat the applied coating to its curing temperature. The heat can cause both solvent evaporation and resin cross-linking reactions to occur. Either gas or electric energy could be employed for convection heating, or for creating infrared radiation. Natural gas and propane gas are the fuels most frequently used to heat convection ovens, because in most locations electricity is more costly for convection heating than gas. Electric convection bake ovens for coatings are very rare. On the other hand, electricity is a more common energy source for infrared radiation ovens. Electric infrared ovens permit considerably easier temperature regulation, and they also have a much faster start-stop response (faster heat-up and cool-down) than gas infrared ovens. Nonetheless, gas infrared ovens are used in a large number of plants. They tend to be more economical to run, particularly for bake ovens that each day operate at the same temperature for an entire shift or longer.

Convection Ovens

Convection ovens heat the applied coating by first heating the oven air, which in turn transfers the heat to the coating. Convection ovens are built in two distinct styles: *direct-fired* and *indirect-fired*. In both types, the hot air is normally moved around within the oven by fans and directed through an arrangement of vents. Some hot air is lost through the parts entry and exit openings, and a portion of the oven atmosphere is exhausted with a fan. Fresh air makeup into the oven, normally filtered to remove contaminants, is also supplied by a fan.

Direct-fired ovens warm the oven air directly with a source of heat, generally a gas flame. This flame's products of combustion are therefore also present throughout the oven, including where the coated parts are being conveyed through the oven. The combustion products may discolor or wrinkle some types of coatings, but in most cases they cause no harm. However, for fine finishes it may be necessary to separate the oven burner combustion products from the curing chamber. Since it is more efficient to heat air directly than indirectly, direct-fired ovens tend to be preferred if their use causes no problems with curing or paint appearance.

Critical appearance requirements, such as those on automobiles, necessitate the exclusion of oven combustion gases from the areas in which coated products are being cured. Indirect-fired ovens confine fuel combustion—the source of heat—to a separate enclosure. The products of fuel combustion are exhausted outside the plant by a vent or stack. The walls of the "fire box" get hot and warm the enclosed air contained in the plenum section of the oven. A blower circulates the hot air out of the plenum and into the oven's curing section. Thermostatic controls can be used to maintain a constant temperature

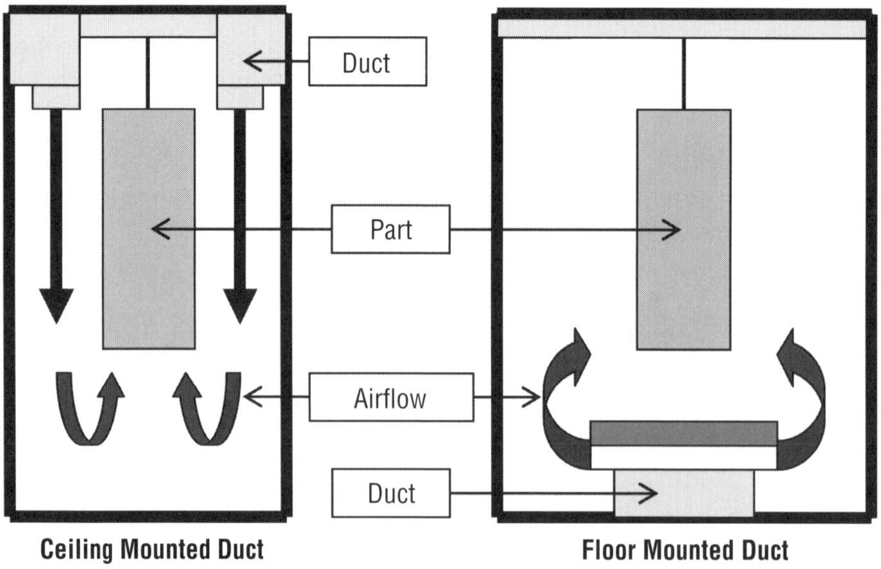

Figure 15-5. Both designs shown above are in use, but ceiling mounted ducts for air discharge allow for easy oven floor cleaning.

throughout the cure zone. A constant measured amount of fresh air makeup is introduced into the curing section, and an equivalent amount of oven air escapes through the oven openings or is blown out through the oven exhaust stack. This air exchange prevents a potentially dangerous buildup of solvents within the oven, which could otherwise cause an explosion.

Normally the maximum allowable concentration of flammable vapors inside an oven is limited by fire prevention codes to about 25% of their lower explosive limit (LEL). LEL refers to the leanest possible combustible air/solvent mixture expressed in percent vapor by volume. In actual practice, the volume of solvent vapor in most industrial paint ovens is generally far lower than 25%. Many ovens are found to be operated at only 2–7% of their LEL. Oven air volume turnovers per minute generally range from 2 to 10, depending on solvent loading levels in the oven.

To save energy, some plants have installed monitoring systems to minimize the fresh air makeup needed in the oven. Insurance companies may permit ovens so equipped to be operated as high as 50% of the LEL value. The problem with such high LEL levels is that paint curing is not always efficient even at 20%, not to mention at 50% of the LEL. Nevertheless, hot air dilution of oven solvent vapor to lower concentrations than necessary is an outright waste of fuel. A study by a U.S. automaker revealed corporate-wide energy waste in this regard. The company's paint ovens were operated at an average of merely 2% of the LEL; the highest single vapor concentration found in any of their 58 paint ovens was only 8% of the LEL. Fresh air turnover rates corporation-wide were reduced sharply, producing an enormous annual dollar savings.

Many other techniques can be used with curing ovens to conserve energy. One consideration when ovens are designed is to use air seals or "bottom entry and bottom exit," which confines the oven's hot air inside where it belongs because hot air rises. Bottom entry/exit is better than air seals, but a combination of both is the most energy efficient design. Another help is to fully insulate the oven. A commonly overlooked measure is to operate the oven at no higher temperature than what is required to elevate the applied coating to its curing temperature for the specified amount of time that parts remain in the oven. Excess oven temperatures may not overbake the coating, but it may very well be wasting heat. Of course, it is likewise important that underbaking be avoided; too low an oven temperature will cause even greater problems and expense. Two or three oven zones at increasingly higher temperatures can be effective in preventing solvent boiling (popping) with both solventborne and waterborne paints if the oven is constructed to make this possible.

Indirect ovens with a full top or bottom plenum are best for efficiency and temperature uniformity. A variation on conventional indirect-fired ovens uses high-velocity recirculating hot oven air that is directed by nozzles or vents to impinge rather directly onto the applied coating. These designs are termed "high-velocity ovens." The rapid exchange of air across the part surfaces heats the coating more quickly and sweeps away solvent vapors at a faster rate. This causes fast curing and allows the conveyor speeds to be increased by as much as 50%.

Gas is usually the source of heat for indirect-fired convection ovens. Oil or electricity can be used, just as with direct-fired ovens. The ideal oven type and shape varies with parts volume and the specific product or products being coated. The oven may be square or long and narrow, and it may be conveyorized to handle any number of parts. Square ovens use fewer panels and are the easiest configuration to balance. Low-volume finishers often use a batch (usually unconveyorized) oven. These may have a single door for in/out access, or they may be fitted with doors at both ends for flow-through of parts.

New ovens must be cleaned thoroughly during construction and at completion to avoid built-in “dirt” problems. Paint on some new enclosed light gauge conveyor tracks can degrade to the extent that it also contributes to “dirt” defects.

Infrared Ovens

Infrared (IR) ovens use radiant energy rays that pass freely through the oven air, but without heating the air. Some air heating occurs in infrared ovens, but this is because heat-up of the coated parts and the oven walls generates a modest amount of convection heating of the oven air. The IR energy is converted to heat when its rays strike a surface, such as the coated part and oven walls. The source of infrared rays can be electrical heating elements of various types and design configurations, but it also can be surfaces heated by combustion of fuels such as natural gas. Hot surfaces of many types will emit infrared rays. Gas-fired black-surface infrared ovens are not as common and are less easy to control but can be just as effective as electrical infrared ovens in curing paint films. IR ovens have been shown to have up to 90% efficiency if both heat and IR emissions are included in the calculations.

A properly designed infrared oven directs the radiant energy as efficiently as possible onto the coated surfaces. Most IR ovens use either tubular lamps, quartz glow bars, or

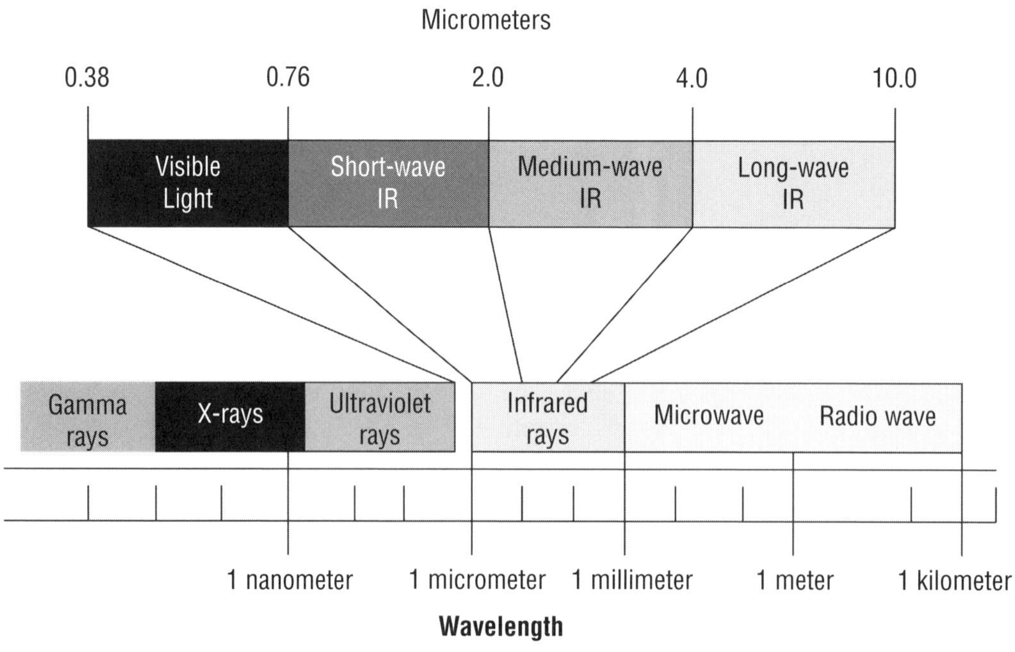

Figure 15-6. Electromagnetic spectrum of IR wavelengths.

filament-wound ceramic cones to emit the energy. The wavelength output of the radiant energy varies slightly with each source as shown in **Figure 15-6**.

Short-wave infrared rays, having wavelengths of 1.0–2.3 microns, appear to the eye to have a white or yellowish color. These waves are absorbed well by black and dark colors but are reflected or only absorbed poorly by white and other light colors. Short-wave infrared rays heat up bare metals, unlike the medium- and long-wave IR.

Rays with 2.3–4.5 micron wavelengths are classed as medium-wave infrared. They have a distinctively red color. Most colors absorb medium-wave IR so it can be used to cure a variety of paints; however, dark colors are able to be cured more readily than paler shades. Medium-wave infrared is used to good effect with waterborne paints, which absorb these wavelengths strongly. Spot or panel repair paints on motor vehicles are frequently cured with portable medium wavelength IR lamps.

"Black wall" radiation, named for its visual appearance, has wavelengths greater than 4.5 microns on up to several hundred microns and is therefore termed long-wave infrared. This is the most colorblind region of IR radiation and will be absorbed rather completely by all hues of paint.

Tuned infrared ovens utilize mainly the IR wavelengths that can be most fully absorbed by the paints being cured. This increases efficiency and thus lowers cost. The energy of the waves decreases as their wavelength lengthens, but any difference in safety among the three categories is insignificant.

Each of the three types commonly will use polished reflectors behind them to direct the reverse-side radiation toward the painted parts being cured. Studies have shown that gold-

Figure 15-7. Infrared oven equipped with gold-coated reflectors.

coated reflectors, although more costly initially, surpass the others in infrared reflecting efficiency and are ultimately the most economical choice. **Figure 15-7** shows some painted parts being conveyed through an infrared oven equipped with gold-coated reflectors.

Infrared radiation is often the best method to cure powder coatings because no air blows through the oven to disturb the delicate powder layer before it melts and cures. Fast heat-up rates can be achieved with infrared, a highly desirable procedure for generating optimum powder coat appearance. Ideally, powder coatings should melt and flow out completely before any cross-linking takes place. If heated slowly during curing, rather than melting, powder coatings may undergo considerable cross-linking. Complete melting and full flowout may never occur. Fast initial heating of powder coatings gives a noticeably smoother and more uniform paint finish.

Other Curing Methods

Various other curing methods have been devised and have been used. For that reason they are explained here, but none of these has anything close to widespread use. These minor paint cure processes include:

- Induction heating
- Microwave and radio frequency (RF) curing
- Heat-of-condensation curing.

Induction Heating

For induction heating, a source of low-frequency alternating current is placed close to the freshly coated metal. The rising and falling electromagnetic lines of force from the alternating current induce powerful circulating currents in the metal, which creates a rapid temperature rise both in the metal and coating. Unless the solvent content of the paint at this point is quite low, popping and bubbling could occur in liquid paints. Carefully regulating the alternating current can control the induced current and the resulting temperature rise in the metal and coating. To be successful, the source of alternating current must be positioned very close to the coated metal.

Induction heating has been used successfully to cure both powder and liquid coatings applied to a fast-moving strip of metal (coil coating). An advantage to induction heating is that metal can be heated almost instantaneously, eliminating the need to have long extended convection ovens for high-speed coil lines.

Microwave and Radio Frequency (RF) Curing

These radiation types of curing are actually similar in nature to induction heating. The only difference is in the higher frequencies used, which are in the thousands of cycles per second. Both microwave and radio frequency curing require safety shielding and are far too slow and/or too costly for nearly all applications. The further disadvantage of RF curing is that the source of RF energy is a radio transmitter, creating a powerful radio signal at the particular frequency being used. The signal must somehow be totally squelched to satisfy Federal Communications Commission (FCC) regulations.

Some types of ultraviolet lamps are powered by microwave energy sources, but the actual method of cure in that case involves ultraviolet light absorption rather than microwave absorption.

Heat-of-Condensation Curing

Heat of condensation is based on this scientific principle: when vapor is allowed to condense to the liquid state, a large quantity of heat is given off, as shown in **Figure 15-8**. At least two companies reported exploring the possibility of using the heat-of-condensation of a nonflammable vapor converting back to the liquid state in order to generate heat for curing paints. In practice, freshly painted parts would be placed into a chamber containing hot liquid vapors. These vapors would then condense on the cool parts and give off heat to cure the paint. The problem is that the condensed liquid material runs down the parts, creating somewhat unsightly lines and grooves in the paint.

One possible solution to this problem is another curing method—a heat pump oven—which to my knowledge nobody has ever tried. But it would certainly work. In a refrigerator, the evaporator allows a liquid compressed refrigerant gas to expand. This process absorbs heat, keeping the inside of the refrigerator cold. The compressor forces the expanded gas vapor back to a liquid, giving off heat. This is dispersed through coils behind or under the refrigerator. Usually a fan blows ambient air across the condenser coils to help remove this heat. On a large scale, such heat could undoubtedly run a curing oven, much as a heat pump is used for residential heating. The economics of a heat pump oven, however, might be another story.

These are rare curing methods, but the future is something at which we can only guess. Some years before vapor curing was a small commercial reality, opinions were expressed by painting "experts" that catalyst vapors to cure paint would never be used outside a test laboratory. They were wrong. (But not by much!) Skeptics doubted ecoating would ever be economically feasible, too, so who knows what the 21st century will bring?

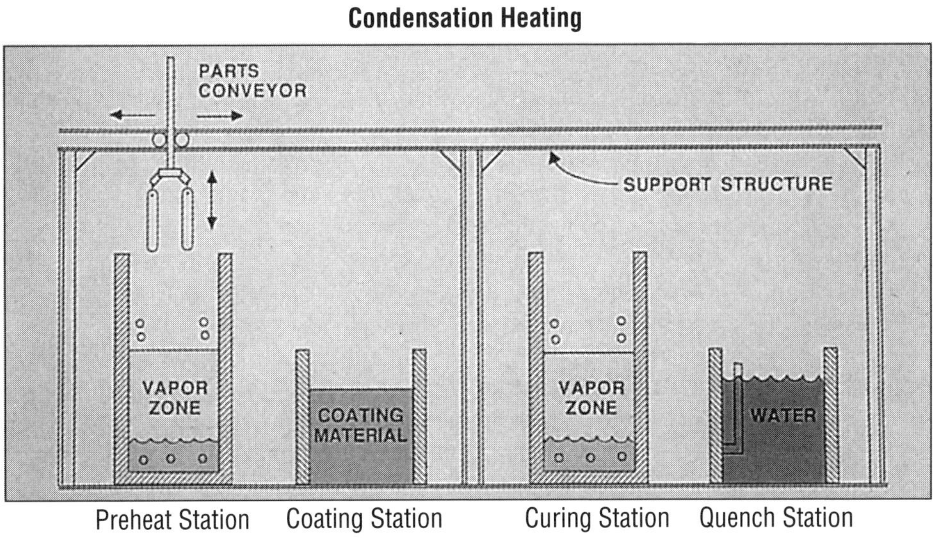

Figure 15-8. Heat-of-condensation curing.

Infrared Oven Heat Source Properties

Source	Type	Wave Lengths	Surface Temp °F	Uniformity of Temp	% Max Efficiency	Time to Warm-up
Quartz panel	electric	medium-long	1800	excellent	85	5 min.
Quartz tube	electric	medium-long	1800	very good	80	35 sec.
Radiant tube	gas	medium or long	1200	fair	40	10 min.
Porous ceramic	gas	medium	1875	good	45	90 sec.
Glassfaced panel	electric	medium-long	1725	excellent	80	5 min.
T-3 lamp	electric	short	4000	very good	85	2 sec.
Catalyst wire (mesh)	gas	long	1000	good	50	15 min.
Calrod	electric	medium-long	1400	good	65	5 min.
Metal coil with reflector	electric	long	1100	very good	70	5 min.
Metal ribbon	electric	medium-long	1850	very good	85	45 sec.
Ceramic tile	gas	medium	1900	good	45	90 sec.

Final thoughts: WHAT HAPPENS IN ULTRAVIOLET LIGHT CURING?

UV light of designated wavelengths is focused onto special coatings that are specially formulated with agents called "photoinitiators." A photoinitiator reacts with discrete UV light wavelengths and becomes hyperactive, exciting resin molecules and causing them to polymerize together rapidly. The extent of cure is dependent upon several factors: film thickness, the amount of photoinitiator, the intensity and wavelengths of the UV light, and the duration of exposure to the UV source. To properly cross-link a film of UV cure paint, the energy intensity and the total UV dosage must be sufficient. UV intensity, the amount delivered per unit of time, is usually given in watts per square centimeter, while dosage is expressed in joules per square centimeter.

Several approximately contiguous ranges of UV wavelength, from weakest to strongest, are commonly identified: 1) UVV – 450 to 400 nanometers (nm) is the visible UV portion. 2) UVA – 400 to 315 nm represents the largest range of UV energies. It is present in sunlight and is responsible for skin wrinkles and tanning of the skin by inducing melanin production. 3) UVB – 315 to 280 nm is the range associated with sunburn on skin and potential retinal damage to eyes. 4) UVC – 280 to 200 nm is the most powerful portion of the UV spectrum, called the extreme UV.

Chapter 16

Film Defects in Liquid Coatings

The Root Cause of Defects

The types of defects are various and sundry. However, all paint defects have one thing in common: a cause. The cause is usually due to the breakdown of good painting practice, somewhere along the line, from cleaning and pretreatment to paint application and curing, although on rare occasions it can be the result of improperly manufactured coating.

Tracking down the cause or causes of a paint defect ought to be pursued in an organized and straightforward manner. A painting defect is like a warning sign along a highway—it gives cautionary information that alerts one to potential hazards and situations that may cause problems. The defect alerts painters to a violation of good painting practice. Therefore, the secret of good "detective work" to remedy factors causing paint defects is to build solid foundations of knowledge of good painting practices and to be aware of frequent culprits in painting defects.

Types of Defects

A comprehensive list of painting defects, if it were to include even those that occur rarely, would encompass perhaps several hundred paint flaws. Such a lengthy list is abbreviated considerably in this chapter by limiting it to the more frequently encountered types of defects. By and large, these major defects come close to covering all types of defects experienced by most painters. The defects that will be covered are:

- Blisters
- Bubbles and craters
- Color mismatch
- Dirt
- Fisheyes
- Gloss variations
- Mottle
- Orange peel
- Runs, sags, and curtains
- Paint adhesion loss
- Soft paint films
- Solvent pops, boils, and pinholes
- Solvent wash.

Blisters

As shown in **Figure 16-1**, a blister in a paint film is a small dome-like raised area that contains, or once contained, moisture (water, water vapor, or both). The surface may contain a single isolated blister or have numerous blisters scattered over the painted surface.

Causes on metal. Blisters on painted metal are traceable to contamination left on a surface prior to painting, either from water, residual soils, poor rinsing, or incomplete dry-off of rinse water. Fingerprints may contribute salts or skin oils that can lead to blis-

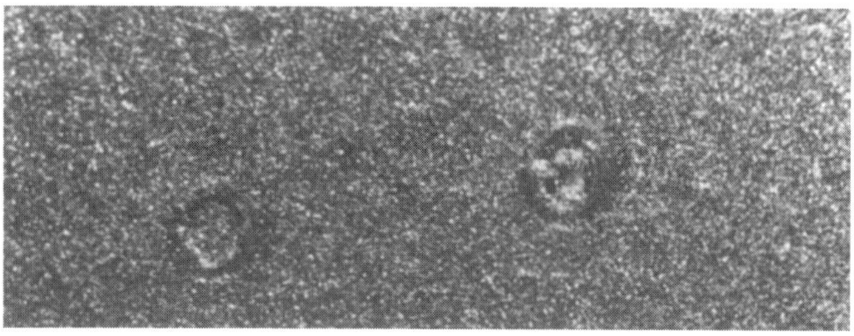

Figure 16-1. Blisters.

ters. Any unrinsed cleaners, pretreatment chemicals, or unremoved greases and oils can permit moisture to get under a paint film in a matter of days and form a blister. In chilly weather it sometimes happens that parts from a cold area are brought into the paint shop and painted before they warm up. But the cold parts may condense a film of moisture that is not seen, and this will cause blisters when it is painted over. Moisture can also come from poor water removal in a compressed air supply; however, when that occurs, it tends to cause large, obvious flaws in a paint finish.

Moisture in or under a paint film expands and contracts with temperature change, thereby producing rapid blistering and lifting of the film. The expanding moisture strains the paint film, which is unable to resist this movement because of reduced adhesion at the part surface. Once the film expands into a blister or even forms a crack, moisture easily reaches the underlying surface, and corrosion can begin.

Inadequate rinsing or, more commonly, rinsing with water containing high amounts of dissolved solids, frequently causes blistering. Unless the rinse water is quite free of dissolved salts, it may be necessary to use an additional postrinse with ultrapure rinse water. Some plants use postrinses containing as little as 10 ppm total dissolved solids. A brief final rinse with deionized water is common practice in the auto industry, for example, to eliminate blisters from regular tap water rinses. The solids level that can be tolerated in rinse water varies from case to case. In some applications dissolved solids up to 200–300 ppm may be adequate to avoid rinse water deposits, but above this level the adhesion may suffer under moist conditions. This is because water is drawn through the paint film by dried-on residual salts that remain after the rinse water is evaporated in the dry-off oven.

Water blisters are produced on such parts by a process called "osmosis." When a semipermeable membrane separates aqueous solutions of different concentrations, the solution with the lesser concentration will seep into the solution with the higher concentration because of osmotic pressure. Such a scenario is formed with a paint film when solid residues remain on the surface of a substrate beneath the paint film. When the paint film gets wet, moisture will seep through the paint film, contacting the residues and forming pockets of concentrated solutions. Osmotic pressure will draw additional moisture from the outside of the paint film through the paint film to the concentrated solution sites under the film. When a sufficient amount of moisture has been drawn through the paint film to the pockets of solutions below, the paint film will be lifted from the substrate at the solution sites in the form of blisters. (Osmotic pressure can be amazingly high; osmotic pressure is what will force water up to the topmost leaves of tall trees.)

Causes on wood. Blisters on painted wood are usually caused by moisture escaping from the wood to the wood/paint film interface. When the moisture gets warm, it expands, which exerts enough pressure to raise the film into a crack or blister. If the expanding moisture cannot escape other ways, it will force itself up under the paint. The best achievable adhesion of paint on wood is totally insufficient to stop the pressures produced by expanding or evaporating water.

Prevention. To prevent blister formation on painted metal, it is vital that the surface be free of contaminants before paint is applied. Also, if air is used in the application, be sure that the gun or bell air supply is both clean and dry. To deter blister formation on wood (in addition to the cleanliness precaution for metal), make sure that moisture can find another exit route out of the wood instead of pushing up through the paint film.

Bubbles and Craters

A bubble in a paint film is a small dome-like raised area that contains, or once contained, solvent vapor. A bubble closely resembles a blister (which contains water and/or water vapor). A crater is a small, concave depressed area that formerly was covered by a bubble, as shown in **Figure 16-2**. The breaking of the bubble contributes to the crater's rounded bottom and built-up sides.

Causes. Bubbles tend to form during a bake cycle when the top layer of the paint film skins over before most of the solvent in the film has had a chance to escape. The rising solvent vapors accumulate and push up portions of the skinned-over top layer, forming bubbles. If the bubble breaks, the raised skinned-over area falls or contracts to the side, forming a depression, or crater. These defects are rarely seen on air-dry finishes because ambient conditions tend to allow most of the solvent to escape before the film skins over. Bubbles (and craters) are usually traceable to the following causes.

- Inadequate flash time before the bake cycle. This leaves an excessive amount of solvent in the paint film when it enters the oven. The high oven heat may produce a skin on the film before most of the solvent has escaped.
- An extra heavy wet film application. This can trap large amounts of solvent in the film. The bake oven heat may skin over the top layer of film before most of the solvent has escaped.
- An improper solvent blend that evaporates too slowly. This will invite skinning over of the top layer of film before most of the solvent escapes. Retarder solvent may be added to paint to eliminate blushing, but excess retarder can cause blisters, bubbles, or craters.
- Insufficient primer bake. This can leave an excessive amount of solvent in the primer, which may form bubbles in the subsequently applied topcoat during the topcoat bake cycle.
- Overly hot bake oven. Solvents

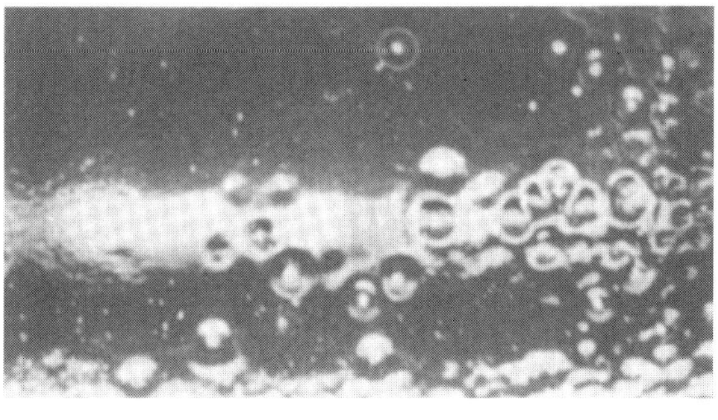

Figure 16-2. Cratering.

are evaporated so rapidly that all of it cannot dissipate from the film without forming vapor bubbles in the wet film.

Prevention. The best ways to prevent bubble and crater formation are: apply a coating in several thin layers (rather than one thick layer); use the proper solvents and use the proper proportions of each; allow enough flash time before a bake cycle; maintain specified primer bake parameters of time and temperature; and, finally, set all bake oven temperatures at their correct settings.

Color Mismatch

The color deviations from one part to another referred to here are shifts in the color of a paint film that occur when applying the same batch of paint. These color deviations are totally independent of *metamerism*, which is a color shift due to a change in the nature of the incident light. Color mismatch between separate batches of the "same" paint can also happen, of course, but these are the fault of the coating manufacturer.

Causes. The causes of color deviation can include the following.

- Variations in the degree of film wetness. The degree of film wetness can alter the shade of a paint; this is especially true of metallics. The wetness affects the relative distribution of color pigment particles and metallic flakes. In general, the drier a metallic paint is applied, the lighter it appears. In solid shades the pigment particles may float or sink in the film and thus shift the color.
- Inadequate agitation. Improper agitation of the paint can leave uneven pigment distribution, which will naturally cause color deviation. The paint may not have been thoroughly agitated in its container before being put into the paint application circulation system. The pigment portion of nearly all paints will settle during use if the paint is not agitated at least occasionally throughout use. Inadequate mixing will result in the alternate application of resin-rich paint and pigment-rich paint.
- Low film builds. Excessively thin films may fail to hide a substrate thoroughly. Some colors are far more effective at hiding than others. Light colorcoats often require a greater film thickness to hide a given substrate completely than is required for darker shades.
- Different application procedures. Various application procedures can produce different apparent colors with the same paint. Paint applied with air spray may have a different shade than the same paint applied by a centrifugal rotary applicator due to different degrees of atomization and particle delivery velocities. Electrostatic versus nonelectrostatic application can produce color deviations as well. This is particularly obvious with metallic paints but it can also happen with solid colors.
- Different substrates. Even if the same paint is applied with the same application device at the same time, the color may look different if the substrate material varies. If an assembly of plastic and metal is painted, a variation in the shades of color may occur. This is suggested to possibly be due to the different heat-up rates (heat capacities) of metals and plastics during the curing process. The paint on the plastic could remain wet for a longer period than the paint on the metal, or vice versa, allowing variations in the rates of pigment separation. This would less likely be a problem with an air-dry paint than with a baked finish.
- Different surface textures. Surface texture differences can cause perceived color

deviations. However, color detection instruments may not indicate a color difference. It can be a bit disconcerting to have a color-measuring instrument indicate two samples are identical in color when your eyes tell you they are very definitely NOT the same color. This is why both visual and instrumental color match tests should be performed. No correlation seems to exist in the perceived shift of light and dark colors on smooth and rough surfaces.

• Overbake. Baking either too long at normal temperatures or at overly high bake temperatures can darken light colors. Some paints seem to be much more sensitive to this than others. Oven loading rates can complicate the situation. Since the heat removal is less when only few parts are in the oven, the tendency to overbake is greater when the conveyor line is only lightly loaded.

Prevention. The prevention of color deviations requires absolute consistency in paint agitation, degree of film wetness, film thickness, and application procedures. Color shifts with widely different substrates such as metal and plastic can be prevented in some instances by baking each substrate separately with different bake cycles. Eliminating the differences in texture (changing the product design) can prevent perceived color shifts with different surface textures. The so-called "solutions" to the problems of color mismatch that I have offered here are often impractical, however, so take my advised instructions with a grain of salt.

Dirt (or Trash)

Dirt or trash in paint or in painted surfaces is defined as any and all contaminants, including lint, dust, small clusters of improperly mixed pigment, tiny particles of overspray paint debris, and oil mist.

Causes. The causes of dirt in paint or in painted surfaces can almost always be traced to inadequate facilities, to poor housekeeping, or to poor painting practices. Some of the causes of such dirt include the following.

• Lint. This type of dirt is one of the most frequently found paint contaminants. Lint can originate from pressed or corrugated cardboard packaging such as cartons, interleaf sheets, masking paper, shop cloths, clothing, boxes, etc., especially if the pieces are frayed and worn. Ripping apart these materials throws literally showers of lint particles into the air. A large plant conducted a detailed study on the nature of the contamination in its painted products and discovered that over 50% of the dirt was attributed to various types of lint fibers.

• Overspray. A common source of dirt is oversprayed paint that dries and accumulates on various objects in the spray booth. Overspray can disintegrate into tiny particles that can fall or float into the applied paint. Improperly balanced airflow in a spray booth may contribute to extensive overspray accumulation.

• Ruptured, loose, or missing filter elements. Poor filter maintenance, whether for paint or for air makeup/air exhaust, can contribute to dirt in paint. Clogged air filters, for example, can add dirt by reducing either makeup or exhaust air flow and can themselves be a source of dirt by having debris collect on and fall off the filters.

• Dust. Airborne dust can be generated in the painting area by many sources, which may include dirty floors, dirty conveyors, fans, forklifts, and sanding processes. Sanding rework is often performed just outside the paint booth because of the convenience of that location. Sanding should be done as far from the spray booth as possible, and if feasible kept in a confined area with separate filtered ventilation to contain the sanding dust. Sealing concrete floors with epoxy or urethane can help contain the fine dust and debris that otherwise will arise from abrasion of the surface.

- Inadequately cleaned paint delivery piping and tubing. Paint piping and hoses that are seldom cleaned can be a source of pigment cluster and resin globule formation. The outside is meant here, but the inside of the paint delivery system must also be properly cleaned to avoid pigment buildup that may flake off and be seen as dirt in paint.
- Reduction with improper solvents. Adding thinners of the wrong polarity can force resin out of solution. Soft lumps of the resin may readily be squeezed through filters and end up in the finish as visible clumps.
- Dirty paint from the supplier. Paint can be contaminated with dirt when delivered from the coating manufacturer. Occasionally, an uncleaned tote tank or dirty pails are inadvertently filled with paint. The steel balls used to disperse the ingredients into a paint formulation may give off small pieces of steel flake that do not get filtered out before the paint is packaged. Although it is sometimes suspected that the paint arrives at the plant with dirt in it, I have found this is not very common, but it does happen. Filtering will remove the foreign material.

Prevention. Eliminating dirt in paint or in painted surfaces should be a daily priority assignment for everyone in the painting department of a plant. Sharp vigilance must be maintained in the following areas.

- Lint. All possible sources of lint should be minimized in the paint department, which if possible should be a closed and restricted area. Only authorized persons properly attired with lint-free outerwear should be allowed to enter. Doors should be opened only for authorized personnel access. Do not clean painting areas with cotton mops; these are a rich source of lint fibers. Instead use mops with heads made of synthetic material.
- Overspray. Spray booth air balance should be checked regularly to confine overspray to the booth's interior. Booth doors should be opened only by authorized personnel and closed promptly, not propped open. Please read the previous sentence once again. Close the booth doors. Open booth doors are a common violation that I see far too often in my consulting work.
- Filters. Periodic cleaning or replacement of air filters in the compressed air line helps prevent particle accumulation and dirt. Filter booths that draw plant air into the spray zone must have tight-fitting intake filters with a pore size small enough to remove fine dirt. Flash zones need to receive the same timely maintenance.
- Dust. Converting the painting area into a "clean room" can control dust. Incoming air should be filtered; shoe baths and blow-off vestibules are helpful; sweeping with a broom should not be permitted—only wet mopping. Wiping products with a tack cloth prior to painting can be an asset. Ionized air blow-offs can be directed onto plastic parts, which tend to attract dust. It is prudent to cover pallets of parts that are awaiting painting to protect them from possible contamination by oil mist, dust, and particulate present in plant air. Sheets of inexpensive lint-free material such as plastic are normally suitable for this purpose.

• Inadequately cleaned paint piping and tubing. Simply flushing thinner through paint piping and tubing until the solvent is clear often does not do an adequate job of removing the old paint. Blends of aggressive solvents can be purchased that are specifically designed to clean paint lines before they are used for a new color. Plants that use such solvents for the first time are always amazed at the material removed from supposedly "clean" paint lines.

• Inadequately stirred paint. Paint should be thoroughly agitated before and during use. If minor amounts of pigment cannot be redispersed, it is satisfactory to filter the paint carefully before using. If appreciable quantities of pigment have settled out of the paint, the coating must be discarded or returned to the supplier. Coatings that are significantly past their normal shelf life may have formed irreversible pigment seediness.

• Reduction with improper solvents. The only solvents that should ever be used for reducing paint viscosity are those recommended by the paint suppliers. Add reducing solvents slowly with stirring to avoid resin kickout.

• Dirty paint from the supplier. Paint should always be carefully filtered immediately before use. The paint from the supplier may well be perfectly clean. Last minute filtering before painting eliminates any possible accidental contamination and can prevent untold quality problems.

Fisheyes

A fisheye defect in a paint film is a small depression (crater) with a mound (dome) in the center (see **Figure 16-3**). The resemblance to a fisheye is the roundness of the depression (the outer circle of the eye) and the central mound (pupil).

Causes. Residual oils or greases, especially the silicone-containing types, nearly always cause this defect. Silicone materials may be accidentally introduced onto surfaces of previously cleaned parts or may contaminate parts not completely washed free of silicones during prepaint cleaning. Silicone molecules are extremely polar at one end and thus act as highly tenacious lubricants for metal surfaces. However, this characteristic makes them difficult to remove as well. Tiny quantities of silicone in sanding dust, on contaminated paper or cloth towels, or from mold releases and polishes (even if sprayed at a considerable distance away from parts yet to be painted) can cause serious fisheye problems. Because the silicones are so highly polar, paint tends to "crawl" away from spots of silicone contamination. This leaves behinds a small dome of silicone barely covered with paint sitting in the center of a depression in the paint film.

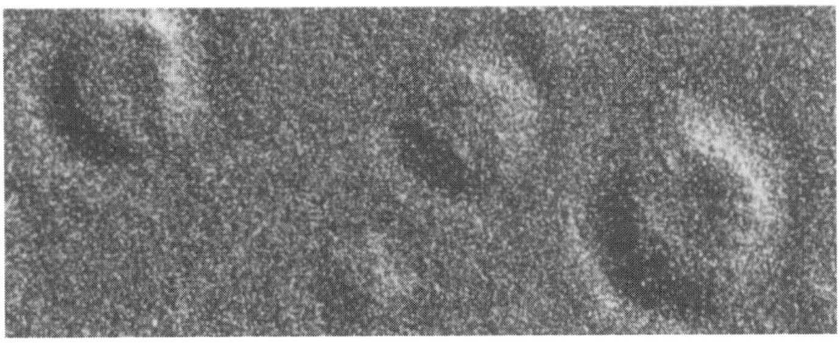

Figure 16-3. Fisheyes.

So many different silicone-containing materials are used that it is impossible to identify a single solvent that is able to dissolve whichever silicone contamination might be present. Various silicones are found in products such as hand creams, hair sprays, underarm deodorants or antiperspirants, all types of polishes and cleaners, mold releases, rubber seals, waxes, gaskets, and lubricants. One corporation traced the source of its daily fisheye problem, which always occurred briefly between 9 and 11 AM, to a silicone-containing glass cleaner being used by a vending machine attendant.

Prevention. To avoid silicone-caused fisheye problems, try to keep silicone products out of any plant that does painting. Employees should be instructed to avoid bringing silicone-containing products into the plant. Several companies have provided employees with lists of specific personal care products that do and do not cause fisheye problems, along with a polite request to use only the ones that are acceptable. Where fisheye problems already exist, the contamination must be removed. Mechanical abrasion may be effective, but care must be taken so that any sanding debris or cleaning solvents do not contaminate other parts. Solvents that have had good success in removing various silicone materials include butyl acetate; 1,1,1-trichloroethane; and trichlorotrifluorethane. Acetone has been said by several persons to be helpful in removing silicone and may be worth trying, but this author has not found acetone to be effective in this regard.

Materials sold as "fisheye eliminators" are designed to be added in small amounts to the paint and mixed in before application. Roughly 0.2–2.0 milliliters per gallon (one milliliter is equal to 0.0338 oz) are used to reduce the tendency for the paint to pull away from the oil remaining on surfaces to be painted. The fisheye eliminators tend to work well because they contain molecules that have a polar end and a nonpolar end. The polar portion is compatible with the highly polar silicone material, and the nonpolar part of the molecule mixes readily with the less polar paint molecules. (Soaps and detergents have similar molecules that are also co-soluble. Thus they can help to remove oil or grease that is nonpolar, and disperse it into the polar substance water.)

Although fisheye eliminators are effective, their routine use is not recommended. Fisheye eliminators can actually cause fisheyes by interfering with adhesion when a coating that has had fisheye eliminator added to it is repainted. This situation can be acute when touch-up is performed on rework parts. Fisheye eliminator can be nearly impossible to clean out of paint pots and lines. Sometimes continued use of the eliminator is required to prevent fisheyes, even though the silicone contamination on the parts has been corrected. In such situations, a possible solution is to wean the system from fisheye eliminator with gradually diminished amounts of the additive. My recommendation is that you not use fisheye eliminator on a constant basis as a means to circumvent inadequate cleaning of parts. Finding and eliminating the source of contamination is the best way to prevent the fisheye defect.

Gloss Variations

Gloss is a measure of the capability of a surface to reflect light. A painted surface that has a high gloss reflects considerable light; and therefore one with low gloss reflects little light. A painted part does not necessarily have the same gloss everywhere even though painted identically. Gloss-deficient patches of paint film can sometimes occur and are known as *flat spotting* and *striking in.* Some of the reasons for this variation are given below.

Causes. Possible causes of gloss-deficient areas can include the following.

- Wet spots in a basecoat. Wet spots in a basecoat prior to topcoating may cause point-to-point gloss variations. The wet spots arise from improper application techniques. Extra wetness at any area can allow minor pigment floating that can appear as decreased gloss, or in rare instances as increased gloss.
- Insufficient oven makeup air. Insufficient fresh makeup air in the oven may cause paint dulling. Few ovens are ever found to be operating at air turnover rates that are too low. However, occasionally solvents need to be swept out rapidly to prevent their dulling the paint. When direct-fired ovens are used for curing, the solvents in combination with oven combustion products can contribute to poor gloss formation. In fact, if the oven is fouled in any way, this can definitely produce a haze on the parts.
- Excessive humidity in the flash zone. Excessive humidity in the flash zone, particularly with paints that contain fast evaporating solvents, can add small amounts of fine water droplets to the paint film. The *blushing* is caused whenever so much evaporative cooling takes place that the surface temperature of the paint film drops below the dew point. Minor blushing may be perceived as low gloss because of its dulling effect on the paint film, as shown in **Figure 16-4**. Rarely do plants experience true blushing nowadays because so few low-solids lacquer paints are still used industrially. Blushing results in a distinct whitish haze due to condensation of water droplets in the wet film. Minor blushing can generally be buffed out, but in more pronounced cases repainting is necessary. Usually not enough water condenses into the film to cause a whitish blush, but only enough to lower the overall gloss levels.
- Insufficient film builds. Low film builds do not permit the paint to flow sufficiently for good gloss. Variations in film thickness can therefore produce nonuniform gloss.
- Excessive oven temperatures or dwell time. Excessive oven temperatures and extended oven dwell times cause overbake that can reduce overall gloss, sometime dramatically. Each paint is different in respect to gloss loss from being overbaked. Dark shades are most likely to exhibit reduced gloss if cure temperatures are too high, but light shades are certainly not immune.
- Molded plastic density variations. Point-to-point density variations can occur in molded plastic parts due to unequal mold filling and unequal mold pressures. These

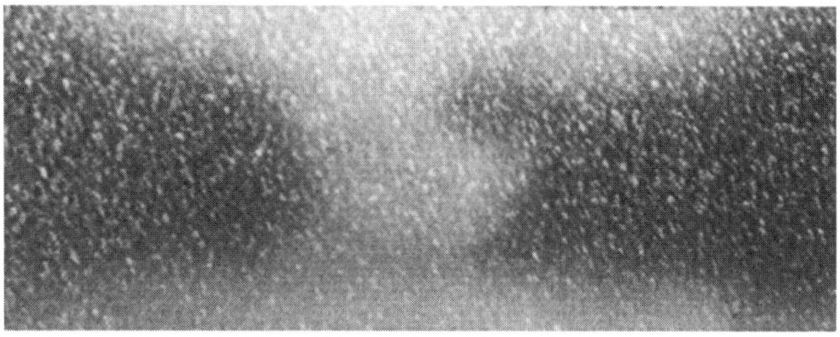

Figure 16-4. Blushing.

density differences cause varying degrees of solvent adsorption and evaporation with subsequent differences in gloss level from one spot to another. Gloss variations on painted plastics are not a common problem, however. In fact, gloss variations on "as molded" plastic parts are often so obvious and undesirable that plastic parts are painted in order to create a uniform gloss on all surfaces.

Prevention. Gloss-deficient areas can be prevented by a six-point program.

1. Use proper paint application techniques.
2. Maintain consistent paint film thickness.
3. Avoid excessive humidity in flash zones.
4. Make certain that oven makeup air is sufficient.
5. Bake painted parts at the specified time and temperature.
6. Make sure that plastic parts are free of density variations by priming them first and then topcoating.

Mottle

Mottle occurs when metallic paint is applied excessively wet and the color pigment particles separate from the metallic flakes. The separation of pigments creates darker "rings" of color that deviate from the overall color while the metallic flake forms light regions in the centers of the rings.

Causes. The cause of mottle is almost always traceable to applying paint extra wet (containing excessive solvent). Many colors are blends of two or more pigment materials whose densities may differ widely. When nonmetallic paints are applied overly wet (flooding), the pigments may tend to separate; one pigment may seem to float. The result is a dark visible appearance where color float occurs.

If metallic paints are applied extra wet, solvent evaporation cooling at the surface can cause hundreds of small (0.2–2.0 mm diameter) circulation swirls known as *Benard cells*. The swirling action sweeps the finely ground color pigment to the edges of the cells and leaves the metallic flake in the center. Because of their size and weight, the metallic flakes are not readily moved by these miniature circulation patterns in the drying paint film. This gives rise to dark rings of color pigment particles with light centers because of the concentration of aluminum flakes. When applied too wet, light-colored metallic paint is especially prone to this visible separation of metallic flakes and pigment particles.

Metallic paints sometimes undergo pigment clumping when applied electrostatically, but that is not related to mottle. Pigment clumping is caused by a sudden dissipation of the electrostatic charge accumulation on the metal flake pigments. The pigment clumping can usually be prevented by using grounded paint hoses.

A related defect found in both metallic and nonmetallic films is sometimes caused when extra paint, either due to surface tension or because of electrostatic attraction, accumulates on the edges of panels. This extra paint requires increased drying or cure time, occasionally allowing pigments to "float" to the surface and leave a dark line around all four edges of panels. For obvious reasons, this defect is frequently referred to as *picture framing*; preventing extra paint buildup on edges of parts avoids this defect.

Prevention. Applying the paint at the proper degree of wetness can usually prevent separation of color pigment particles and metal flakes. In efforts to prevent mottle, often the initial (wet-on-wet) metallic coats are applied at normal wetness, but then the final metallic coat is put down rather dry. This has been done for years and works well.

Orange Peel

Repetitive bumps and valleys of a size and distribution similar to those on the rind of an orange characterize orange peel in paint. An example of this defect is shown in **Figure 16-5.**

Causes. Orange peel results when the freshly applied paint film does not flow out completely and surface irregularities remain as a consequence. The causes of poor flowout are usually one or more of the following.

- Excessively dry spray. This occurs when excessive solvent evaporates from the atomized paint particles either en route to the target, or too quickly after reaching the target, so that satisfactory paint flowout is impossible. Excessive solvent evaporation from droplets en route to the target can be caused by overatomization, by use of a solvent that evaporates too fast, or by too much distance between the gun and paint target. Overatomization creates finer paint droplets and thus much greater particle surface area from which solvent can evaporate. As a result, the droplets are drier and do not flow as well. Excessive distance between the gun and target can occur from poor operator technique. For example, too wide a spray fan pattern can make the spray at the edges travel an excessive distance to the target and paint droplets will then dry out more on the way. Maintaining booth temperatures too high or having parts too hot when painted may cause excessive solvent evaporation once the paint is deposited on the target. Excess solvent evaporation can also result from utilizing an improper solvent blend that has too much fast-evaporating material in the mixture.
- Poorly atomized spray. Underatomization can be just as bad as overatomization. Inadequate atomization can be caused by a low setting on the atomizing air pressure, an overly high paint viscosity, and by conventional spray and HVLP air spray fluid pressures set too high. Fluid pressures set too low with airless and air-assisted airless application also cause orange peel. Low paint viscosity can be caused by cold paint that was not allowed to warm to ambient temperatures; or it may be from paint line heaters not working correctly, as well as from improper mixing of paint.
- Overly thin film builds. Films that are not thick enough can be caused by improper application parameters, and usually by insufficient paint flow from the application device (gun, disc, bell).

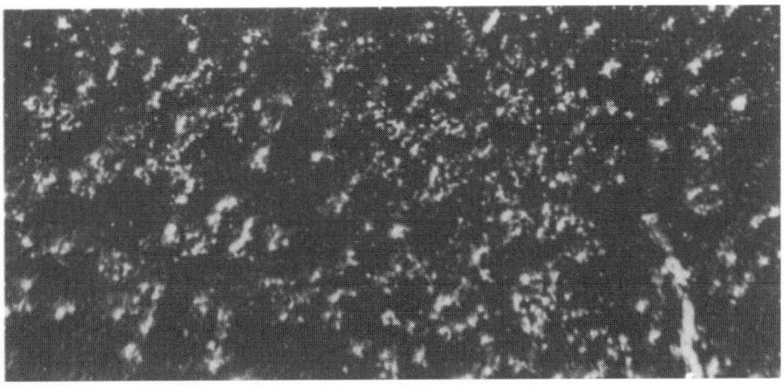

Figure 16-5. Orange peel.

• Rough substrates. Except in rare cases at high film build levels, paint films cannot level out rough or uneven substrates. It should be recognized that a rough substrate surface will automatically produce a degree of roughness in the topcoat.

Prevention. Taking these measures will ensure that paint flows out satisfactorily and will prevent orange peel.

• Eliminate dry spray. Atomizing air pressure should not be excessive. Solvents should not evaporate too fast. The distance between gun and target should be monitored. Proper spray procedures should be used.

• Attain proper paint atomization. Atomizing pressure and paint viscosity should be monitored closely. To rule out cold paint (as a cause of high viscosity) and poor atomization, bring paints inside at least 24–36 hours before they are to be applied.

• Monitor paint film thickness. A sufficient film build is necessary to ensure good paint flowout. Film thickness should be monitored frequently.

• Check substrate smoothness. The smoothness of the substrate should be achieved before you begin to paint. If the surface has been filler coated or primer painted, smoothing can be done by sanding.

Runs, Sags, and Curtains

Paint applied on vertical surfaces may flow downward in various amounts before the curing process hardens the film and stops the flow. All such downward flows are termed *runs* or *sags* (see **Figure 16-6**). The terms *curtains* or *curtaining* (and rarely *draping)* are sometimes used because the lower portion of wide runs and sags may resemble the curves in drapes or the scalloped lower edge of a curtain.

Causes. The origins of runs, sags, and curtains are almost always either that the paint was applied too thick, or the paint was applied too wet.

The reasons why a coating was applied too thick include the following.

• Dirty gun(s). A dirty gun, especially clogged air passages in the tip, can distort the normal spray pattern and make it apply too heavily and concentrated in some areas of the fan. Immersing a spray gun completely into a container of solvent for cleaning can cause this situation. If the entire gun is repeatedly put into solvent, the solvent (now with paint solids in it) may flow back into the air passages. The paint-laden solvent can dry in these

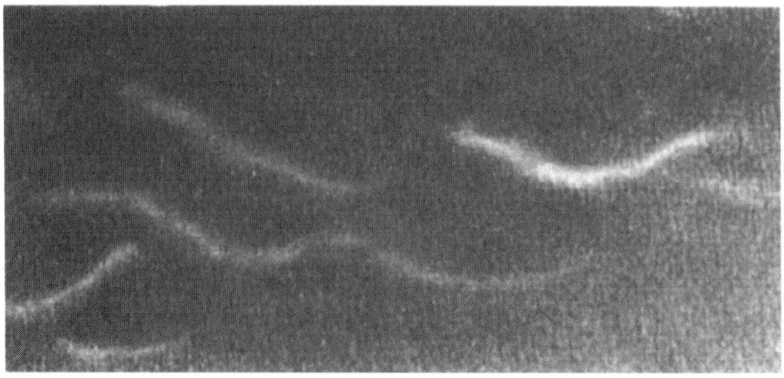

Figure 16-6. Runs and sags.

air passages, leaving behind more and more paint solids that eventually will constrict or clog them and reduce their operating effectiveness.

- Poor operator technique. Poor operator technique in painting complex parts may contribute to excessive application in certain areas of a part. Painters need training to be able to paint well, and, once trained, they need periodic retraining to maintain these skills.
- Paint may be applied too wet for the following reasons.

Gun was positioned too close to workpiece. If the spray gun is too close to the part being painted, an insufficient "flight time" for the atomized paint droplets will cause them to retain excess solvent. When they arrive at the part surface they are too wet.

Excessive solvent or too much slow-evaporating solvent (tail solvent) was added. Adding excessive solvent to a coating will allow too much solvent to remain in the newly deposited paint. The same is true if excess tail solvent is present in the blend.

Atomizing air pressure was set too low. An insufficient air pressure will result in poor atomization, forming larger paint droplets from which solvent evaporates less rapidly.

Flash time allowed was not long enough. An inadequate flash time between coats of paint applied wet-on-wet can leave a coating excessively wet. The problem of inadequate flash time often is encountered near the end of each month when line speed tends to be increased to get products out on time.

Cold parts were coated. This is not going to happen if parts are washed or primed just before being painted; it only applies to parts that are preprimed or which receive no pretreatment. Applying paint to cold parts may slow solvent evaporation enough to cause a wetness problem. This problem can result when parts are stored in an unheated warehouse and not taken in to warm long enough before being topcoated.

Fluid tip (nozzle) or fluid pressure used was too large. Always use a tip that gives the correct fluid flow at the correct fluid pressure. It can also occur from not partially triggering the gun for the areas of parts that are difficult to paint properly otherwise.

Prevention. Steps to take to prevent runs, sags, and curtains can include the following.

Cleaning dirty guns. A spray gun should be cleaned properly. It should never be totally immersed in solvent; only the spray tip should be cleaned in this fashion. Leave the air on slightly when cleaning to prevent paint and solvent from entering the air passages. The gun should be checked regularly to prevent defective spray patterns caused by plugged air passages.

- Training operators. Operators should be trained to develop correct and efficient spray procedures. They should periodically check the film thickness at different areas of a part to make sure they are not applying paint too thick (or thin) in spots.
- Operating guns properly. By moving the gun farther from the workpiece, the travel time for droplets increases, allowing additional time for solvent evaporation. This procedure reduces the possibility of wet paint and the resultant runs and sags. Naturally, excess target distance is likewise to be avoided.
- Using correct type and amount of solvent. If excess solvent has been added, a correction should be made. In some cases, particularly in cold weather, a fast-evaporating solvent blend should be used. This will increase solvent loss during application so the applied paint is drier. The converse is true in warm weather when slower solvents may be added.

• Correcting air pressure. A slight increase in air pressure will improve atomization and increase the surface area of the droplets. This will improve solvent evaporation and contribute toward a slightly drier paint.

• Providing adequate flash time. When applying coatings wet-on-wet, you must allow sufficient flash time after each application. If the flash area is colder than normal, flash times need to be extended; otherwise, excess solvent remains and the coating may run or sag.

• Warming parts. To prevent cold parts from slowing solvent evaporation and increasing wetness, always bring parts to room temperature before you paint them. Parts that go through a washer will be warmed in the process, but occasionally parts that have already been primed are repainted or topcoated. If these parts are stored in an unheated warehouse, they should be brought to room temperature before being hung on the line to be topcoated. A good practice is to require that such parts be delivered to the paint area at least 24 hours before being painted.

Paint Adhesion Loss (and Weak Adhesion)

Adhesion loss is the premature separation of a paint film from a substrate.

Causes. The causes of inadequate paint adhesion or complete adhesion loss can be one of the following.

• Contaminants under the applied paint. Almost 95% of the loss of large or small areas of paint due to poor adhesion is caused by some form of contamination under the paint. This may be oil, grease, sanding residues, water, or other foreign materials that should not be there.

• Excessive bake time or oven temperatures used for curing primers (or other previously applied coatings). This may render the primer film (or other auxiliary coat) so thoroughly cross-linked that solvents in the next paint layer cannot micro-etch the surface to attain good adhesion.

• Paints that differ widely in polarity. When an applied paint film differs substantially in polarity from the next applied paint, adhesion of the second coating may fail. Paints with widely differing polarities tend to repel each other. This is rarely a problem unless a customer requests that one special coating be used and it happens to be incompatible with the primer being used.

• Condensed moisture. High humidity in the paint shop may form condensation on parts to be painted, especially on cold parts. Applying paint over the moisture can cause fine blisters and even large-scale adhesion weakness and loss.

Prevention. To prevent weak adhesion and adhesion loss, the following steps should be taken.

• Clean parts thoroughly before painting. Obvious as this may be, many plants don't seem to recognize poor cleaning when it occurs with their parts. All washer stages should be monitored regularly to ensure that bath temperature, chemical concentrations, pH, and other critical factors such as spray pressures and nozzle openings are within specifications. Periodic tests should be conducted on cleaned parts to verify that all soils have been completely removed.

• Maintain correct bake oven parameters. Maintaining correct bake cycles and oven temperatures can prevent overbaking a coating, enabling the next applied paint to micro-etch the coating for a proper adhesion "bite." If a previous coating is overbaked, a light scuff sanding may be beneficial to provide an anchor pattern for the next layer of paint. Sanding will also cut through much of the hard outer glaze on the existing paint layer and allow solvents in the topcoat to "bite" better into the previous paint layer.

• Maintain similar polarities for applied paints. When feasible, use primer and subsequent coatings manufactured by the same supplier. Test primer-to-topcoat compatibility if different paint companies make them. Sealer coats of intermediate polarity must be used between paints that have widely differing polarities because they will never have satisfactory adhesion with each other. Sealers are designed so that the adhesion of primer to sealer, and sealer to topcoat, will be excellent.

• Prevent condensed moisture. Making sure that cold parts are warmed before they are painted can help prevent moisture from forming on parts. This occurs readily if the humidity is high, as is often the case around the pretreatment and paint area due to hot cleaner solutions and rinses.

Soft Paint Films

Soft paint films are incompletely cured coatings and have hardness levels below a designated specification. Soft paint films are easily marred or gouged (e.g., are readily penetrated with a fingernail) and will have inadequate solvent resistance. When parts with soft paint are stacked, the parts may "block" (meaning they leave marks in the paint or stick to each other); they may also adhere or rub off onto packaging materials such as paper or cardboard. Another term for blocking or sticking together is *printing*. It is not limited to baking paints. Soft paint can result with air-dry paint, especially when ambient humidity is high.

Causes. The causes of soft paint films may be one or more of the following.

• Low oven temperature. A faulty oven temperature controller or indicator system may result in oven temperatures running below those required. Inadequate cure may be taking place without any indication that the oven temperature is incorrect. Periodic oven temperature checks are necessary. Automotive plants normally check oven temperatures daily using a multi-point oven temperature probe to confirm readout gauge accuracy.

• Low oven air makeup. Insufficient fresh air makeup in the oven may retard the extent to which oxygen molecules react with paints that cure by oxidative cross-linking. Low air turnover may keep paint from developing full hardness by not allowing the removal of sufficient quantities of solvent during the bake cycle.

• Contaminant(s) acting as softening agents. Solventborne paint will usually dissolve minor amounts of residual substrate wax, oil, and grease that remain from marginal cleaning effectiveness. Paint will also readily absorb moderate amounts of oil from compressed air. However, if significant amounts are incorporated into the paint film, the oil and grease can act as plasticizers or softening agents.

• Excessive paint storage time. Paint that has exceeded its shelf life may not cure properly. This is especially true with all powdercoatings and some high-solids coatings, which have a shelf life considerably less than low-solids paints. High-solids paints contain small, reactive molecules, so shelf life inevitably is short. When paint exceeds its shelf life, cross-linking in the can has progressed so far that not enough cross-linkability remains to fully harden the paint when it is applied to the parts.

• Excessive retarder solvent. Excess retarder (tail) solvent can prevent paint from achieving normal hardness levels because it does not get fully expelled from the paint film.

• Excessive film builds. Excessive film builds may not permit full curing during the normal cure cycle. Thick paint films require more time and/or temperature in the curing oven than thinner coats.

• Insufficient cure time. Too brief a cure time at a given temperature may leave the paint soft. This sometimes occurs when plants increase the line speed to meet production schedules without raising the temperature of the oven at the same time.

Prevention. Soft paint may be prevented by taking the following steps.

• Check the oven temperature. Oven temperature gauges and control systems should be checked periodically to ensure that "cool zones" are not occurring and that the indicated oven temperature is the actual temperature throughout the inside of the oven.

• Maintain oven air makeup. Oven air makeup should be maintained at recommended levels to be certain that sufficient oxygen is being supplied for oxidative cross-linking. Oven air turnover must be high enough to remove evaporated solvent.

• Avoid residual softening agents. Check for proper cleaning of parts. Wax, oil, and grease should be kept off items to be painted. Compressed air supplies should be checked for oil content.

• Check paint storage time. Paint should not be stored beyond the shelf life recommended by the coating manufacturer. It would surprise readers to discover how frequently this author has found that FIFO (first in–first out) paint usage was not being followed.

• Use proper amounts of retarder solvent. When retarder solvents are added to improve flowout and gloss, or to avoid blushing, extreme care must be exercised to avoid soft paint. Retarders should be added with a light touch in small increments.

• Avoid excess film build. Operator care must be taken to avoid excessively thick paint films, for these will not cure satisfactorily during the normal oven cycle. If thick coatings are needed on some parts, their oven dwell time and/or the bake temperature should be increased.

• Cure for the correct time. If the conveyor line speed must be increased, thus decreasing time in the oven, then oven temperature must be increased to compensate. Operators should be made aware of the relation between oven time and temperature. The paint manufacturer can supply time-temperature curves to allow correct temperature compensation for cure time variations.

Solvent Pops, Boils, and Pinholes

Solvent pops, boils, and pinholes are tiny craters on the surface of paint films, as shown in **Figure 16-7**. They are small versions of bubbles and craters and are typically far more abundant and more widespread across the area of the paint film.

Causes. These defects are basically caused by overly rapid solvent loss from the wet paint. Solvent vapor may escape in small "bursts" from a paint film that has partially dried. The paint may be unable to flow back together to hide the escape ports, leaving

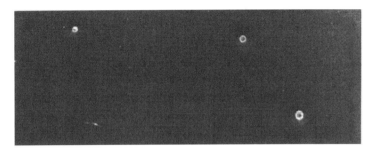

Figure 16-7. Solvent pinholes.

tiny craters. The causes are basically the same as for bubbles and craters, only to a lesser degree. Too little flash time and overly thick wet films are the two major causes.

Additional causes for solvent pops, boils, and pinholes include the following.

- Pigment clusters. Insufficiently agitated paint may form pigment clusters that can trap solvent and delay evaporation.
- Surface roughness. Some types of surface roughness can act as nucleation sites for vapor bubble formation.
- High solvent evaporation rates. High solvent evaporation rates can cause and exaggerate solvent pops, boils, and pinholes.
- Slow solvent evaporation rate blends. Slow solvent evaporation during flash periods leaves an excess amount of solvent to evaporate in the oven. The sudden rush of solvent vapor formation leaves behind these marks in the film.
- Drafts. Strong drafts can sometimes speed solvent evaporation to the point that solvent pops, boils, and pinholes are produced in the paint.
- High oven entry temperatures. Sudden exposure to a high heat at the oven entry can heat the paint film too rapidly. A warm flash zone or a lower temperature for the oven entry zone can prevent solvent popping and boiling.
- Bell-trapped air. This is identical to solvent popping and is caused by tiny nicks in the edge around the circumference of turbine-driven bells. Atomization at the nicks can be inefficient and form large droplet particles that contain more solvent than the remainder of the efficiently atomized paint droplets.

Prevention. Taking the following actions can help to prevent pinholes, solvent pops, and solvent boils.

- Avoid pigment clusters. Efficient agitation will prevent the formation of pigment clusters. Filtering the paint will remove the nonredispersible pigment clusters.
- Check surface roughness. If compatible with the product design, the problem-causing surface roughness should be eliminated by a smoothing step in the manufacturing process.
- Control solvent evaporation rates. Using only solvents recommended by the paint manufacturer can regulate solvent evaporation rates. Extra flash time is beneficial if slow evaporating solvents must be utilized for better film appearance.
- Avoid drafts. Direct impingement of blower-driven air onto freshly painted parts should be avoided.
- Use correct oven entry temperatures. The problem can sometimes be corrected by

zoning the oven. This permits somewhat lower temperatures at the oven entry to warm the part gradually, and thus avoid "heat shock" and overly rapid solvent evaporation. Subsequent zones must be set increasingly warm to cure the paint fully if the entry zone temperature is lowered.

- Prevent bell-trapped air. Replacing bell heads that have any nicks along their circumferential edges can eliminate this problem. Careful handling will help avoid costly repair or replacement of bell heads. Teach people not to drop or damage them; new rotary heads now cost around US$450–$1,100.

Solvent Wash

Solvent wash is the term used to describe what has happened in painted areas of parts that emerge from the oven with little or no paint on them. This is due to solvent condensation in cooler part areas. The solvent vapor evaporates from warmer part areas but then it condenses on cooler spots. The liquid solvent thus "washes" down that area on a part, resulting in the paint being dissolved and rinsed off by liquid solvent in that part location.

Causes. Solvent wash is caused by excessive solvent evaporation from a painted part in the entry area of an oven and the condensation of this solvent on cool areas of the part. The solvent condensation can wash away either some or all of the paint where the condensation occurs. Solvent wash is most likely to happen on parts that have paint applied extremely wet (solvent rich). Flow coating and dip coating are especially prone to solvent wash because of the high solvent content of these coatings. Solvent wash rarely occurs with atomized (spray or rotary) application of coatings.

Prevention. Whenever paint must be applied especially "wet," as, for example, how it is in flow coating and dip coating, extra flash time must be allowed for extensive solvent evaporation before the painted parts enter into the bake oven.

Chapter 17

Coating-Related Testing

Categories of Tests for Coatings

Quite a number of practical tests have been devised over the years to measure coating quality, both in the uncured bulk state and as an applied and cured coating film. Most tests are aimed at verifying the designed integrity of a particular property of the coating or finish. The less complicated tests are better suited to the quality control needs of most coating operations because they cost less and are easier to perform. Due to greater time requirements and high equipment costs, rather sophisticated test methods are more likely to be utilized by coating manufacturers, research organizations, and professional testing laboratories. Note that in addition to tests on the coating material, tests may also be performed on the substrate and on the coating delivery system to help assure satisfactory film properties. Of the literally hundreds of tests, only those that are widely useful to coating companies and coatings users are described here.

The majority of coating tests can be grouped into the following three basic categories:

1. Tests of coatings "in the bulk"
2. Tests on parts and coating application equipment
3. Tests of the applied coating film.

The test analyst should know how to select the correct evaluation method to rate test results, and then how to compare individual values from each test. Test descriptions from organizations such as the ASTM International (formerly named American Society for Testing and Materials) frequently include explanations of the preferred rating methods for tests. Pictorial standards offered by some standards groups can be particularly helpful in rating results.

Tests of Coatings "in the Bulk"

Unapplied coatings may be tested to determine:

- Weight per gallon (density) of paint
- Paint viscosity
- Weight percent solids and weight percentage volatiles in paint (and the percentage of water in waterborne coatings)
- Conductivity of paint
- Percentage of volatiles in powder coatings
- Flow rate of dry powder coating
- Weight per volume (density) of the powder

- Particle size range of the powder.

Weight per Gallon (Density) of Paint

Special metal containers holding a precise volume are used for this measurement. One such cup holds exactly 83.05 cc at 77° F (25° C). The weight of coating that fills the container, expressed in grams, is equal to the pounds per 10 U.S. gallons of the coating. Conversion to metric units can be readily done if desired.

Paint Viscosity

In technical terms, *viscosity* is the ratio of the shearing stress and the rate of shear of a Newtonian liquid. Shear (in terms of liquids) is an applied force that disturbs a liquid at rest. In a Newtonian liquid, the rate of shear is proportional to the shearing stress. In a non-Newtonian liquid, the rate of shear to the shear stress is not proportional.

To briefly explain viscosity in terms of shear and Newtonian/non-Newtonian liquids would be complex and difficult. However, more precise measures are needed than descriptions with colloquial terms such as "hardly gooey at all" or "it's as slow as molasses in January." For the purposes of this book, viscosity can be defined in simple nonscientific terms as the "flowability" of a liquid. A coating with a high viscosity has poor flowability; in common language it is said to be "thick" or "syrupy." A coating with a low viscosity has a high flowability; and may be labeled as "thin" and "watery." But for even rough accuracy in viscosity measurement, it is obvious that an instrumental method is required.

Various extremely accurate viscometer instruments, such as the Brookfield, are available for measuring precise coating viscosity. But they tend to be slow and too delicate for nearly all manufacturers who use coatings; and only researchers and coating formulators normally use them in their laboratories. A simple, fast device with timing system has been devised for coaters to measure coating flowability (viscosity). The most common device is referred to as a viscosity cup. Names of some commonly used viscosity cups are Shell, Norcross, Zahn (see **Figure 17-1**), Ford, Sears, and Fisher. A viscosity cup is a small metal or plastic cup with a precisely sized hole at the bottom. The test consists of filling the cup and timing with a stopwatch the number of seconds required for the coating to drain in an unbroken flow through the hole in the cup. As viscosity rises, the time increases for the liquid to drain through the orifice. **Figure 17-2** provides a means of converting viscosity measurements (in seconds) from

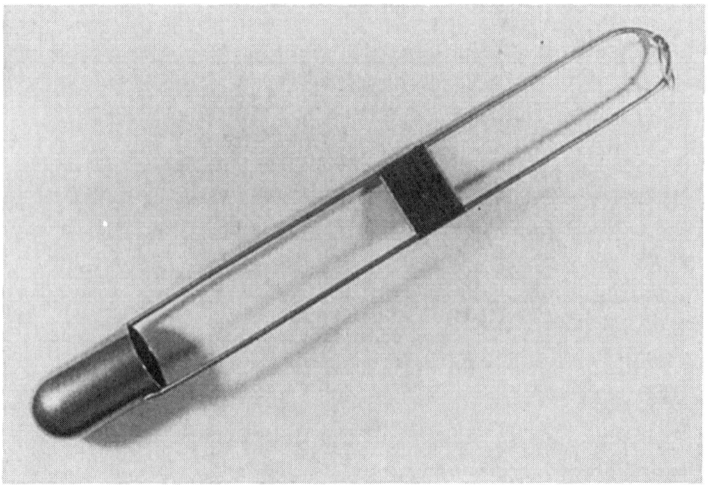

Figure 17-1. Zahn viscosity cup.

Centipoise	Fisher #1 (seconds)	Fisher #1 (seconds)	Ford Cup #3 (seconds)	Ford Cup #4 (seconds)	Zahn #1 (seconds)	Zahn #2 (seconds)	Zahn #3 (seconds)	Zahn #4 (seconds)	Zahn #5 (seconds)
10	20		5	30	16				
15	25			8	34	17			
20	30	15	12	10	37	18			
25	35	17	15	12	41	19			
30	39	18	19	14	44	20			
40	50	21	25	18	52	22			
50		24	29	22	60	24			
60		29	33	25	68	27			
70		33	36	28		30			
80		39	41	31		34			
90		44	45	32		37	10		
100		50	50	34		41	12	10	
120		62	58	41		49	14	11	
140			66	45		58	16	13	
160				50		66	18	14	
180				54		74	20	16	
200				58		82	23	17	10
220				62			25	18	11
240				65			27	20	12
260				68			30	21	13
280				70			32	22	14
300				74			34	24	15
322							36	25	16
346							39	26	17
370							41	28	18
394							43	29	19
418							46	30	20
442							48	32	21
466							50	33	22
490	(For poise value, divide centipoise value by 100)						52	34	23
514							54	36	24
538							57	37	25
585							63	40	27
657							68	44	30
777								51	35
896								58	40
1016								64	45

Figure 17-2. Viscosity conversion.

various types of cups to each other, and also into centipoise viscosity units.

Each viscosity cup maker offers cups with varying hole sizes so that drain times are reasonable depending on the viscosity range of a coating. A smaller hole is used for very low viscosity fluids so that their drain time is not too fast. The drain period can be measured over enough time (usually 15–30 seconds) so that accuracy is not compromised. Similarly, thick-bodied fluids are measured with larger-hole cups to avoid drain times that last longer than is convenient.

Coating is often purchased at usage viscosity so no additional solvent is necessary. Thus, no employee can make an error by adding improper solvents or the wrong amount of solvent. (Some plants say they do this to "idiot-proof" that part of the coating process; I'll leave it to the reader to decide if that is an insult to painters or if it makes their job easier.) The paint cost can be slightly higher this way so other plants prefer to purchase paint at a viscosity higher than their application viscosity. Solvent can then be added to reduce the coating to the correct application viscosity. Usually the final adjustment to coating viscosity is done just prior to application of the coating. Because temperature has a large effect on viscosity, coating personnel must make allowances for this variable to avoid viscosity change as the coating becomes warmer or cooler during their work shift.

Weight Percent Solids and Weight Percentage of Volatiles in Paint (and the Percentage of Water in Waterborne Coatings)

For all organic solventborne coatings, the sum of weight percent solids plus the weight percent volatiles (VOC) will be 100%, so just finding one of these values readily provides the other. The usual test involves weighing a wet sample of coating, then evaporating away the volatile components, and finally reweighing what remains, which is the solids. The test can be done simply. Fill a 5- or 10-ml syringe with coating and weigh it carefully. Into a small, preweighed beaker or disposable aluminum cup, deliver roughly 4 ml of coating. Then again carefully weigh the syringe to determine the weight of wet coating in the beaker or cup. Bake the sample for 1 hour at 235° F (113° C). Allow the sample to cool, reweigh the beaker plus the sample to find the dry-solids weight. The percent solids and percent VOC by weight can be found as follows.

Percent solids equals weight of dry coating multiplied by 100 divided by the weight of wet coating.

$$\%\ \text{solids} = \text{wt dry coating} \times 100 \div \text{wt wet coating.}$$

Percent VOC equals 100 minus the percent solids.

$$\%\text{VOC} = 100 - \%\ \text{solids.}$$

One procedure for VOC in waterborne coatings requires that the percentage of water be found using one of two procedures: by using gas-liquid chromatography, which is not readily available in most plants, or by the *Karl Fisher method.* The latter is a complex titration procedure and one virtually never attempted by coating personnel. The complete method is not detailed here for that reason. Although the Karl Fisher technique is still used, the gas chromatographic determination is much faster and easier, plus it is more accurate. For many coatings the total %VOC can be determined directly from the gas-liquid chromatographic (GLC) data; but in other cases the amounts of low molecular weight alcohols and water are hard to distinguish from each other using GLC.

Determining volume percent solids requires weighing an uncoated disposable small part both in air and in water and reweighing the part in air and in water after it has been coated and cured. It is not commonly part of plant coating testing. The details of the test can be found in ASTM Test Method D-2697-73.

Conductivity of Paint

The electrical conductivity of coating can be found by measuring its resistance with an ohmmeter adapted with wide-surface terminals to ensure test consistency. Conductivity, normally expressed in units of micromhos, is the mathematical reciprocal of resistance in megohms, so its value is calculated that way. Notice that these are reciprocal properties, therefore it is always true that the higher the resistance, the lower the conductivity (and vice versa).

Some cured coatings need to be highly conductive, such as coatings used to provide electromagnetic interference shielding or radio frequency interference shielding. In coating to be applied electrostatically it is desirable, but not absolutely necessary, that the coating have a low conductivity (high resistance) to avoid shorting out the application equipment internally through the coating itself. Paints should not have zero conductivity, however, because a small amount of conductivity is needed so that coating particles will be able to hold electrostatic charges. Otherwise, these coatings will not provide the electrostatic application benefits such as increased transfer efficiency and coating "wrap-around." Values of 0.5–5.0 megohms resistance are normal for electrostatic application. Coatings having 2–5 megohms resistance can be used with automatic application devices, since these can use higher voltages than is recommended with manual application. Manual application is most often employed for coatings having approximately 0.5–2.0 megohms resistance. Paints with too little resistivity (too much conductivity) will cause severe Faraday cage problems. Poor transfer efficiency results from coatings with very low conductivity because the coating droplets are not able to carry enough electrostatic charge.

Percentage of Volatiles in Powder Coatings

The volatile content is measured by taking the weights of a powder sample before and after baking. The loss of weight from baking divided by original sample weight multiplied by 100 gives the percentage volatiles.

Flow Rate of Dry Powder Coating

The most accurate method for evaluating flow properties uses a meter to measure powder fluidization and flow characteristics under specific conditions. A simpler, albeit less accurate, way involves timing the flow of powder samples through various sized funnels.

Weight per Volume (Density) of the Powder

Both gas displacement and liquid displacement can be used to measure specific gravity of powders; conversion to density is easily calculated from this. This author prefers liquid displacement using a weighed powder sample put into a 50cc volumetric flask that is then filled with a nonsolvent such as n-hexane. From the volume of the flask plus the weights of the flask, powder, and hexane, the powder density can be calculated directly.

Particle Size Range of the Powder

A simple method for particle sizes passes a powder sample through increasingly finer sieves to determine the amount retained by each sieve size. Suspending the powder in a liquid and then measuring the degree of laser light scattering photometrically does far more accurate size measurements. For nearly all powder applicators only the first method is practical.

Tests on Parts and Coating Application Equipment

Various informative tests can be conducted on parts to be coated, particularly tests to make sure the surface is clean and/or suitably pretreated. Besides those, other tests can be performed on the coating application equipment itself, especially if electrostatic application is used. Three important tests that can be conducted for electrostatic application are:

- Checking for complete electrical grounding of parts
- Measuring electrostatic application voltage
- Determining transfer efficiency (TE).

Checking for Complete Electrical Grounding of Parts

Positive electrical ground connections in electrostatic coating ensure maximum electrostatic attraction of coating to the product, and prevent spark-producing charges from building up on the product. The electrical pathway to ground goes successively from the product to a support such as hook or hanger, to the conveyor trolley, to the conveyor channel, to the grounded building metal framework, and finally into the earthen soil ground under the plant. A poor connection at any of these junctions will result in a marginally grounded part and reduced electrostatic coating application efficiency.

A quick test for electrical ground can be made using an ordinary ohmmeter, or a decreased-ground detector. In both cases, one terminal is attached to the product to be coated, and the other is attached to a good absolute ground, such as the conveyor I-beam. If an ohmmeter reads 1 ohm or less, the part shows little resistance to electrostatic charge flow and is therefore properly grounded. The decreased-ground detector operates similarly. It can be made to give either an audible signal or a visual alarm (or both) when electrical grounding is poor.

Poor grounding is a frequent problem and requires constant attention to prevent it. The most common cause of poor grounding is dried coating that has accumulated on hooks and hangers, and on overhead conveyor wheels and in the trolley channels. Shielding to minimize paint buildup, as well as frequent cleaning of equipment, are prudent preventative practices.

Measuring Electrostatic Application Voltage

Most electrical supply control panels will allow variable voltage and indicate the voltage output of the electrostatic power supply. The readouts are not always accurate, however. Clearly it is a good idea to occasionally double-check the actual voltage at the application device using a test meter. This will ensure that the power supply readout is correct, and that sufficient voltage for good electrostatic charging is truly reaching the applicator. Voltage testers are available for this purpose. Since electrical charges are involved, instructions should be followed carefully, and all safety precautions observed. In all honesty, however, you should know that they are not much used. While it doesn't mean no one uses them, in my 35 years of painting consulting I personally have never seen a paint shop use one of these devices.

Determining Transfer Efficiency (TE)

Transfer efficiency (TE) is a measure of the amount of coating applied to a part in comparison with the total amount of coating used. If the percent solid by weight of the coating is known, TE can be determined by using the following steps.

1. Weigh the coating container (along with the coating in it) before and after coating a number of parts. For convenience, this weight difference will be called Value A. It is the weight of paint applied.
2. Weigh the parts before they are coated, then weigh them again after coating and curing is complete. The weight difference will be labeled Value B. This is the weight of paint solids on the parts.
3. Divide Value A by 100 and multiply by the percent solids. This number, which is the actual amount of solids in the weight of paint applied, will be referred to as Value C.
4. Divide Value B by Value A and then multiply by 100 to get the percentage of paint transferred to the parts, that is, the transfer efficiency achieved.

Here is an example. Let's imagine we know our paint to be 45% weight solids. Suppose that weighing a paint pot before and after spraying parts revealed that 625 ounces (A) of wet paint had been applied, and that the weight of cured paint on parts was found to be 175 ounces (B). We can calculate the amount of solids in 625 ounces of wet paint by multiplying by 45% (0.45, or 45/100). The result is 281.25 oz of solids (C). B divided by C—here 175 oz divided by 281.25 oz—gives 0.62. Multiplying by 100 to get percentage shows our transfer efficiency was 62%.

TE for paint also can be determined by measuring the dry-film thickness (DFT) (in mils), and measuring the surface area coated (in square feet). The method assumes DFT is constant and that part surface area can be accurately determined, which is not often true. So this method will normally be used only for an approximate idea of the transfer efficiency. It is necessary to know the percent volume paint solids, either by measuring it or getting this value from the coating supplier. Measure the volume of paint that is applied, and then put the figures into the following formula to calculate the result:

$$\%\text{TE for paint} = (\text{ft}^2 \text{ coated} \times \text{mils DFT}) \div$$
$$(\% \text{ volume solids} \times \text{gallons used} \times 0.1604).$$

For example, the data below were taken from production records at a furniture company. One particular month they had 14,682 ft^2 of product to an average thickness of 1.8 mils using 165.3 gallons of 47% solids-modified epoxy paint. Therefore:

$$(14{,}682\ \text{ft}^2 \times 1.8\ \text{mils average DFT}) \div$$
$$(47\%\ \text{solids} \times 165.3\ \text{gal} \times 0.1604) = 21.2\%\ \text{TE}.$$

In this example, management was surprised to learn that true paint transfer efficiency was considerably lower than the 35-40% they had assumed.

A similar method for powder is possible using the density of the cured film, but it is not very accurate due to significant point-to-point film thickness variations inherent with powder coatings. No significant use is made of this TE determination method for powdercoating.

Tests of Applied Coating Film

Numerous tests can be performed on applied coating films. Among the most common are:

- Wet-film thickness
- Uncured powder thickness
- Dry-film thickness
- Adhesion/flexibility (bend, impact)
- Adhesion (tape)
- Hardness
- Chemical resistance
- Stain resistance
- Extent of cure
- Water immersion
- Humidity resistance
- Accelerated weathering
- Gloss
- Distinction of image (distinctness of reflected image)
- Color match.

It is important to remember that a number of coatings, especially air-dry alkyds, will continue to cure for a week, or even as long as two weeks, after the initial apparent cure. For this reason, a conditioning period after curing must be allowed before accurate testing can be done on the dry film. This is especially true for tests such as hardness, stain resistance, and corrosion resistance. A conditioning time of 24–72 hours before testing is a common practice for durability and adhesion tests. In most cases the test results will improve after the conditioning period. If tests run on parts without a full conditioning time show satisfactory performance, they will certainly also be satisfactory after the conditioning period. However, tests such as film thickness, gloss, and color match are not related to the extent of cure, so these tests can be run without a conditioning period.

Wet-film Thickness

The most commonly used wet-film thickness gauge resembles a comb with two end teeth of equal length and about 10 teeth of varying shorter lengths in between. When the gauge is placed into the wet film and then removed, the marking on the last tooth to be wetted by the paint gives readout of the wet-film thickness. Because the gauge leaves a visible imprint in the wet coating, it is a destructive test. This may limit its usefulness. It may be possible to do the test in a noncritical spot, or on a separate test panel if the comb mark would cause rework problems.

Unlike dry-film thickness, the thickness of a wet film is seldom of great interest. But at times knowing the wet-film thickness is helpful, such as when coaters are being trained, or when setting up automatic application equipment, or possibly when needed in order to satisfy specifications. Even then it is only measured in order to get an assured minimum dry-film thickness on the parts.

Uncured Powder Thickness

Some success has been achieved in using infrared photometry for uncured powder film thickness measurements. Because of the cost and relatively limited value of the readings, however, few companies have found it worthwhile to invest in these devices. It likely will find a niche, but because of economics it will never be very popular with most powder-coaters. Where the precise dry powder film thickness is critical to cost or performance, however, such as in powder coating on coil stock, photometric dry powder thickness measurements may indeed be cost advantageous.

Dry-film Thickness (DFT)

The dry-film thickness of cured coatings on magnetically responsive substrates is often measured using calibrated magnet holders inside what are called "pull-off" gauges or "roll-off" gauges. The magnetic substrates on which such instruments are utilized are primarily ferrous (containing iron) metals. (Magnetic pull-off gauges can also measure DFT on nickel-containing alloys, but coated nickel materials are only infrequently used industrially.) These thickness measurement devices have a magnet suspended by a spring inside a nonmagnetic body (usually made of plastic or aluminum). When the gauge is touched to the coated surface, the magnet and the metal are attracted to each other, but the force is weakened because it must act through the coating film. The magnetic attraction between magnet and substrate is reduced almost linearly as the coating thickness increases. The spring will be stretched whenever the magnet is pulled away perpendicularly to the surface. The thicker the coating, the weaker the magnetic attraction. Thus, the spring is stretched less when the gauge is pulled away from a thick film than when it is pulled from a thin coating layer. The person doing the test reads the length of spring extension, as indicated by a calibrated scale marked on the gauge, which provides direct readings of film thickness in units such as mils or microns. This works only for nonmagnetic coatings such as paint and powder, not for metal layers such as plating.

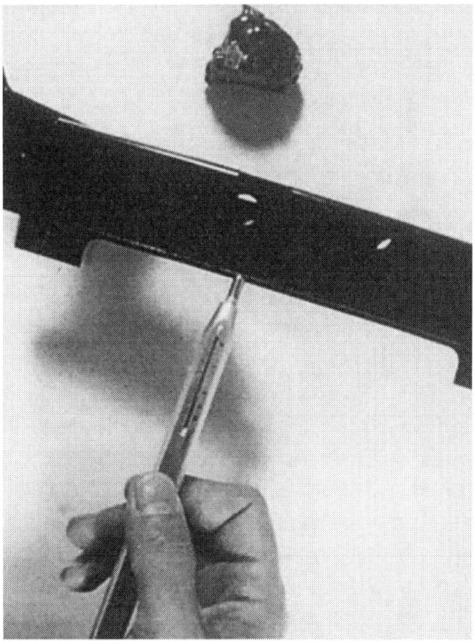

Figure 17-3. Magnetic pencil gauge.

The two main types of magnetic pull-off gauges are called "pencil" (pull-off) and "banana" gauges (roll-off) because their shape resembles those items. A pencil gauge, as shown in **Figure 17-3**, has a magnetic head attached to a spring and a pointer. Gradual manual pull-off of the gauge away from the surface gives a reading of film thickness with an average accuracy of roughly ±10 to 15%. One of the easiest-to-use pencil gauges looks and operates like a hypodermic syringe. This particular model retains the readout value and thus is easier to use than most. With many pencil gauges, however, the person using the gauge must have a quick eye, for as soon as the gauge is pulled away from the surface, the magnet retracts inside the housing and the DFT reading disappears.

The manufacturers of these instruments make clear that if the gauges are not operated vertically, their accuracy is reduced. Accuracy of nonvertical readings can be improved by plotting a correction curve for various angles away from true vertical. Since the device is not that precise, few people (at most) ever bother with this correction factor. For usage where many readings are not taken vertically, a Tinsley type 7000 gauge may be preferable. It contains a meter-like scale readout with a balanced pointer that is unaffected by its angular orientation. The Tinsley type 7000 gauge provides accuracies of ±10%.

A banana or roll-off style gauge, as shown in **Figure 17-4**, has improved accuracy over pencil gauges. Rotating the wheel slowly pulls a magnetic head away from the surface being measured. When the magnet breaks contact, both an audible "click" and a visual indicator inform the tester to cease rotating the pull-off wheel. At this point the calibrations marked on the wheel allow the film thickness to be read directly. The cost of a roll-off device is approximately double to triple that of a pencil gauge. All roll-off gauges are easier to read since they provide a read-out value that persists after the magnet retracts, unlike pull-off gauges that lose the DFT reading when the magnet head releases from the surface.

One preferred roll-off model employs a battery to rotate the wheel that pulls the magnet from the surface being tested. The roll-off gauge starts and stops rotating automatically, and therefore is faster to use. In addition, errors that can occur if the tester does not stop rotating the pull-off wheel quite quickly enough are avoided. This added feature doesn't increase the gauge price to any significant degree and is truly a valuable inclusion. (I wouldn't buy any other.)

Both pull-off and roll-off magnetic devices are portable, easy to use, and relatively low in cost. They do not damage the coating and do not require batteries unless the model just

Figure 17-4. Magnetic banana gauge.

explained is desired. However, they will not work on plastics or aluminum; they are useful primarily on iron and steel that have been coated with nonconductive coatings.

The magnet tip should be visually inspected periodically for wear and for attachment of small steel particles and other contaminants that could disturb the readings. Exposure of a magnetic gauge to a strong electrical or magnetic field can alter its strength and change the calibration. Roll-off gauges and pull-off gauges can both be recalibrated when that becomes necessary. This should be done if it does not give correct readings when checked against known calibration film-thickness standards. These standards are several small rectangles of plastic film with known thickness that come with the gauge. If lost they are available from gauge suppliers. A number of instrument manufacturers will recalibrate their gauges for a modest service fee.

Coating dry-film thickness can be measured on all metals, even on nonmagnetic metals such as zinc or aluminum using "magnetometers," more commonly called "eddy current detectors." These measure dry-film thickness based on the relative strength of eddy currents in the metal induced by a varying magnetic field in a conducting coil. Differences in the current flow (because it is alternating current) within the coil are related to the film thickness separating the coil and the metal under the coating film. These compact devices are normally accurate to ±5%. Battery-operated units are convenient since they are small and easily hand-held. Their accuracy, compact size, applicability to all metal substrates, plus their speed and ease of operation have made them popular instruments for measuring coating film thickness.

An eddy current gauge can measure DFT on any metallic substrate but it must be calibrated for each different alloy and metal substrate just prior to use. This can be done in just a few seconds using a bare sample of the substrate metal and two or three of the standard thickness plastic films always provided when the instrument is purchased. Calibration should be done at film thicknesses both slightly thinner and thicker than the film thickness to be tested.

If extreme accuracy is needed, film thickness can be measured with beta-ray back-scatter devices (50+ years ago electron beams were called beta rays and the name has stuck). They emit electrons toward the surface, and the relative amount of deflected rays are measured and correlated to film thickness. The beta-ray instruments provide extremely accurate values as long as an appreciable difference exists between the nature of the substrate and the coating. They work well measuring coating thickness on metals, wood, paper, cardboard, etc., and on many (but not all) plastics. These devices are quite costly and require careful and time-consuming calibration. Only rarely are these used for industrial coating thickness other than thin special purpose tapes. Inorganic coatings requiring extreme thickness uniformity, such as emulsion or metallic oxide coatings on photographic film or video and audio tapes, are often checked by beta-ray back-scatter instruments. Metal film thickness is also measurable with these devices, and so they are used in semiconductor manufacturing.

Measuring coating film thickness on plastic parts by beta-ray back-scatter instruments can often present difficulties because the organic coating resin and pigment is often molecularly and morphologically similar to the resin and filler of the plastic substrate. But, of course, neither pull- or roll-off gauges nor eddy current detectors are able to determine coating thickness on nonmetallic parts either. To find coating thickness on plastic parts, some alternatives to the previously discussed methods used for measuring dry-film thickness on plastic include the following.

Measure the plastic thickness before and after coating, using a micrometer or other means of measuring linear thickness.

Measure the thickness of a piece of tape before it is placed onto a plastic part, then again after the part is coated and cured. After painting, remove the tape from the part and measure the thickness of tape plus coating with a micrometer. Naturally, a tape of constant thickness is required.

Measure the thickness of a piece of steel before it is adhered onto a plastic part, then again after the part is coated and cured. Remove the steel piece from the part and measure the thickness of the coating with a micrometer. An even easier way would be to measure paint thickness on the steel piece with a magnetic or eddy current gauge.

Use a Tooke gauge. This instrument will scribe a short V-shaped groove through the coating and also into the plastic substrate. Next the magnifying glass supplied with this instrument is positioned directly over the groove. Internal scale markings projected through the viewer permit a direct thickness readout of the coating film thickness.

Gauges using ultrasonics are the newest method. These are more accurate than any of the others but are also more costly and complex to use.

Since some of these tests are destructive, plants prefer to run them either on scrap parts or in a nonappearance area on the part.

Adhesion/Flexibility (Bend, Impact)

Coating film adhesion and flexibility are interrelated properties—brittle films will lose adhesion more readily when bent than flexible films. There are several ways adhesion/flexibility can be tested. One method puts a coated panel through various types of folds or bends. Another procedure drops weighted objects from varying heights onto the front or back of a coated panel. The aim of the tests is to measure how easily coating can be made to separate from the substrate by impact or by tensile and compressive forces.

In the bend test, a panel coated on one side can be folded back with the coated surface on the inside of the fold to test for *compressive* coating adhesion/flexibility film stresses. A 90° bend is normal for this test. A panel can be folded in the opposite direction with the coated surface on the outside of the bend to test for adhesion under *tensile* (stretch) stresses. This is usually a 180° bend, and this bend is termed "0T," pronounced "zero-T." (The "T" stands for thickness.) A panel bent back onto itself with nothing between the two flat parts of the fold has a 0T bend. A panel, if bent back 180° but with another section of itself (or over another piece of panel having the same thickness) between the two folded sections, is described as having undergone a "1T" bend. The "1T" stands for a bend over one thickness. Similarly, panels bent back over two sections of itself, or over two layers of the same thickness, have a 2T bend, and so on. It follows intuitively that the tighter the bend radius, the greater the severity of the compressive and tensile stresses that strain the coating film's adhesion and flexibility.

Coated test panels may also be folded around circular rods or conical mandrels of various diameters and tapers per ASTM D-522. These two tests are known as the "cylindrical mandrel bend test" and "conical mandrel bend test," respectively. They are always performed with the paint side out, so that both check adhesion and flexibility under tensile stress. Results are usually given as the smallest diameter bend that does not result in coating adhesion loss.

Tests of coating film adhesion/flexibility by dropping weights onto coated panels are conducted using steel balls or rounded-end metal impactors of specific weight, usually 3 pounds (1.361 kg). Various impactor shapes are used, but all such tests are commonly termed "falling ball impact" tests because ASTM G-14 specifies the use of a 3-pound hemispherical impactor or "tup." The impactor may be dropped onto a test panel or an actual production part. The test weight may be dropped from various specified distances, depending upon the severity desired. Having the impactor strike the coated side of a panel is referred to as a "direct impact test;" hitting on the uncoated side is termed a "reverse impact." In some instances, tests might be specified for both sides. With direct impact tests, the concave-dented coated surface produced is examined for coating adhesion loss and cracks; while after reverse impact the convex protrusion generated is similarly examined. If a coating is overly brittle, it might crack and possibly lose adhesion in these tests. The degree of cracking and coating adhesion failure is identified by photographs or by a suitable rating method (see ASTM D-2794). As one might expect, the thicker the film, the more brittle it becomes, so thin films of a given coating tend to have better impact resistance than thick films. Additionally, the more the substrate stretches and deforms under impact, the more the coating will tend to delaminate.

A highly accurate adhesion test, but one that is virtually never performed except by companies that manufacture coatings or for meeting specifications on special coating projects, is called the "pull-off" adhesion test. To evaluate adhesion, a special metal disc to which a knob has been affixed is glued firmly to the coated surface. A calibrated hydraulic device grasps the knob and slowly lifts. The force required to remove the disc, plus the coating layer glued to it, is a measure of the film adhesion strength. Pull strength is usually given in units of PSI or kilopascals (1 PSI = 6.89 kP).

Adhesion (Tape)

Firmly applying tape to the coating film and then rapidly pulling the tape off the coating is the most common way to measure coating adhesion. Transparent plastic tape with an appropriate adhesive strength is used. A piece about 1-inch (25.4 mm) wide and roughly 3–5 inches (7.5–12.5 cm) long is pressed onto a coated area making sure it is free of any contamination, blemishes, or other surface imperfections. The tape is smoothed into place with a finger, then rubbed tightly to the surface with an eraser on the end of a pencil. A dark color under the tape indicates that the tape has been affixed firmly. One end of the tape is grasped and pulled steadily and quickly (but not jerked) back on itself as close to the plane of the tape as possible. ASTM D-3359 is considered to be the standard tape-adhesion test, although hundreds of variations of this test are used. ASTM D-3359 provides a rating scale to evaluate the results as well. The ease of the tape adhesion test is an attractive feature. As with other tests, results may vary if tape testing is performed immediately after the coating is cured, instead of waiting until after a 48–72 hour conditioning period.

The tape adhesion test may be performed directly on an unaltered coating film or on a coating film that has been scored (scratched through the coating to the underlying substrate) with an "X." The coating may also be cross-hatched with multiple perpendicular score lines in various grid patterns and then tape-tested. Further details on tape-adhesion testing are provided in ASTM D-3359.

Sometimes these tape test methods may be performed after the panel has been stressed by being subjected to varying temperature, humidity, or salt spray exposures. Occasionally, a specification will call for a tape adhesion test to be performed on a coated panel that has been immersed for an hour in deionized water. Any such testing variation is always totally permissible if agreed to by all of the buying and selling parties. Testing should be done to assure that the coatings will perform as desired; testing should never be done only to match some association or agency approved test protocol unless the affected parties agree to this. Test protocols such as those from ASTM are extremely valuable, and probably should be used unless a better method is known. But no one minds, not even ASTM, if modifications are made in the test methods when a preferred alternate is employed for sound scientific reasons. Remember, however, organizations such as ASTM are devoted to helping you make sure your tests do not produce piles of worthless data. They help testers to generate only meaningful, repeatable, and, whenever possible, correlative numbers.

Hardness

Although sophisticated coating film hardness-testing methods are available and in use by coating manufacturers, the ASTM D-3363 pencil hardness test is the simplest and most convenient test for those applying coatings. Hand-held pencils with leads of varying hardness are pushed (hardest lead first) at an angle against the coating film to determine the coating hardness. The pencils are not sharpened to a point but into a right cylinder; that is, the full round cylinder of exposed lead is squared off with a perpendicular base. The circular base edge of the lead is then pushed against the coating film at the specified 45° angle. The amount of force on the pencil must be enough either to scratch or cut into the film, or otherwise to break off the edge of the pencil lead. Check closely for marks in the coating film. Begin testing with the hardest lead, then decrease hardness until the hardest pencil that will not mar the coating is found. Some plants measure both "gouge hardness" and "scratch hardness," but most look only for "mar hardness" of any kind. (If a pencil will penetrate the coating but not produce a $^1/_8$ inch (3 mm) gash, it is "scratched" but not "gouged" per ASTM D-3363.)

Pencil hardness of a coating is rated on a scale from soft to hard as follows: 6B, 5B, 4B, 3B, 2B, B, HB, F, H, 2H, 3H, 4H, 5H, 6H. (In practice, pencil leads softer than "H" are rarely used; they are simply too soft.) The pencil hardness test is somewhat imprecise, but it's quick and adequate for most purposes. Instead of using individual pencils for this test, one can purchase a set of varied hardness pencil leads that fit into a holder much like an automatic pencil.

A hardness test is often used as a measure of the extent of cure of a coating film; undercured films will be soft, and overcured films will be excessively hard. This is allowable in the majority of cases, although overall the solvent rub test (see "Extent of Cure" section below) probably is better for this purpose. Normally a permissible range of hardness rather than just a single hardness value is written into a coating specification. Therefore, such a requirement might indicate that ". . . a 2H minimum to 4H maximum film hardness shall be required."

Chemical Resistance

In chemical-resistance tests (ASTM D-1308 and others) various chemicals are applied to coating films and allowed to remain on the surface for a matter of hours or days, or whatever time length is deemed advisable for testing the product. This chemical exposure is then followed by visual examination of the film for harmful effects. Popular chemicals for such tests include acids, alkalis, oils, and solvents. The film is then evaluated for the degree of harm, if any, that has occurred. Softening, dissolution, loss of gloss, and degradation of the film are the four most often encountered results of chemical attack on coatings.

Stain Resistance

To perform stain-resistance tests, various potential discoloring materials are applied to coating films. This is done as in ASTM D-3023 on any paint and any substrate, even though the standard itself is specific to clear films on wood. After a suitable interval, the test panel is cleaned and dried, and then the coating is examined for stains. Vinegar, catsup, mustard, iodine, bleach, and detergent are often used as agents in these tests. The appliance manufacturers, for example, are vitally interested in having a coated finish able to resist staining and discoloration from a wide variety of common foods and household products. Medical equipment makers are another illustration of the many industries concerned with stain resistance of coated surfaces.

Extent of Cure

The hardness test, as we have seen, is an approximate measure of how well a coating film is cured. A much-preferred one is the "solvent rub" test, which measures the film's resistance to dissolution in whatever solvent is deemed to be most appropriate. In this procedure, which is shown in **Figure 17-5**, the finish is tested for removal of coating (or color on the cloth) after a number of manual double rubs with a cloth or paper towel moistened (not saturated) with the specified solvent. For example, a purchaser might specify that if no color is evidenced on the cloth after 50 double finger rubs with acetone or methyl isobutyl ketone, the coating film shall be considered to be cured adequately. Only within the last several years has an ASTM test method for solvent rub testing been published, and that is why the pencil hardness test is still often used for extent of cure. The solvent rub test is quite subjective since the

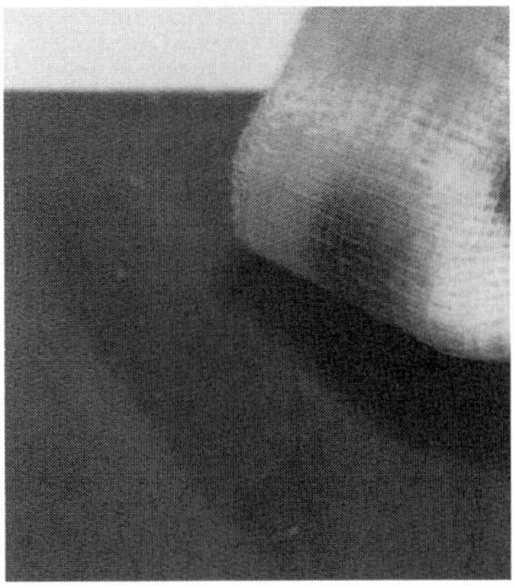

Figure 17-5. A solvent-rub test can be used to check the extent of cure. This discoloration occurred in only eight double rubs, indicating that the coating is undercured.

pressure applied during the rubbing can vary from individual to individual, so the written ASTM D-5402 procedure is a welcome aid to operators of this test.

Water Immersion

Moist and wet conditions can bring out evidence of poor adhesion and/or cause blister formation. In fact, the water immersion test checks on the tendency of coatings toward blistering. Water, especially very pure water, can be a surprisingly disruptive chemical to coating adhesion. It has the ability to penetrate most coating films readily. Once it does so, it can produce undesirable effects unless the substrate is clean, and the proper coating is correctly applied (to the necessary thickness,) and cured correctly. The tendency of a film to form blisters can be rapidly measured by simply soaking the part in deionized water. People sometimes find it hard to comprehend that distilled (or deionized or reverse osmosis) water is a more strenuous test for blister formation than soaking in salt water. This is true because the osmotic pressure that causes coating films to lift and form water blisters is greater with pure water. The osmotic pressure produced using water containing dissolved salts is always lower. (To learn more about this fascinating phenomenon so vital to all plant growth, the reader is encouraged to consult any basic science textbook.)

Humidity Resistance

In regular humidity testing, the parts are hung vertically in an atmosphere of high relative humidity at around 100°F (38° C) for a specified number of hours. Steam is used to supply the moisture. With the condensing humidity test (D-4585), coated panels are set at about a 30° angle from horizontal over a reservoir of warm water, as shown in **Figure 17-6**. In a sense the panels form a roof over the water tank. The coated sides of the panels face the water; the reverse sides of the panels are exposed to the ambient temperature

Operation at 100°F Vapor Temp.

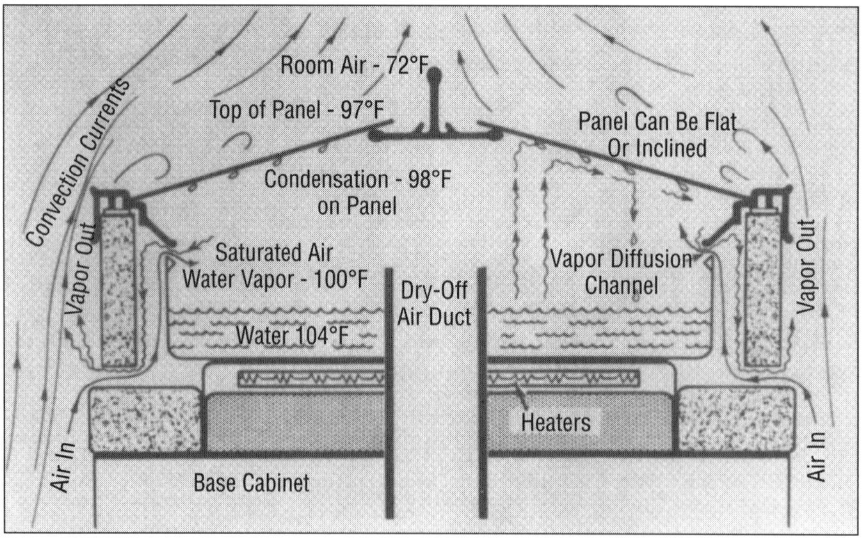

Figure 17-6. Condensing humidity test.

in the room. The hot water vapor condenses on the warm, moist inside surface of the panels and drains back down into the water reservoir. The heat of condensation and the continuous draining of the pure water (which is actually distilled during the test process) constitute a severe test of the coating's adhesion and resistance to moisture. **Figure 17-7** shows coating blisters on a film test panel produced in a condensing humidity chamber. A single condensing humidity test can be run for as long as desired, but the maximum time is generally about 24 hours. In addition to visual examination after humidity testing, the panels can be subjected to other tests, such as tape adhesion.

Accelerated Weathering

Various types of testing cabinets (see **Figure 17-8**) have been devised for accelerated weathering tests. The threads of commonality among them are ultraviolet light, heated moisture, and provisions for spraying with salt solutions. Accelerated weathering tests can have three purposes:

1. To compare the relative performance of different coatings,
2. To determine how a coating will perform in the field, and
3. To estimate how long a coating will perform in the field (a subject of intense disagreement).

To predict the relative performance of different coatings in service conditions, coatings are often tested in regard to their comparative resistance to salt spray corrosion. This may be done to determine if a coating is better than a currently used coating, or to assist in the selection of the best coating from among a number of possible candidates.

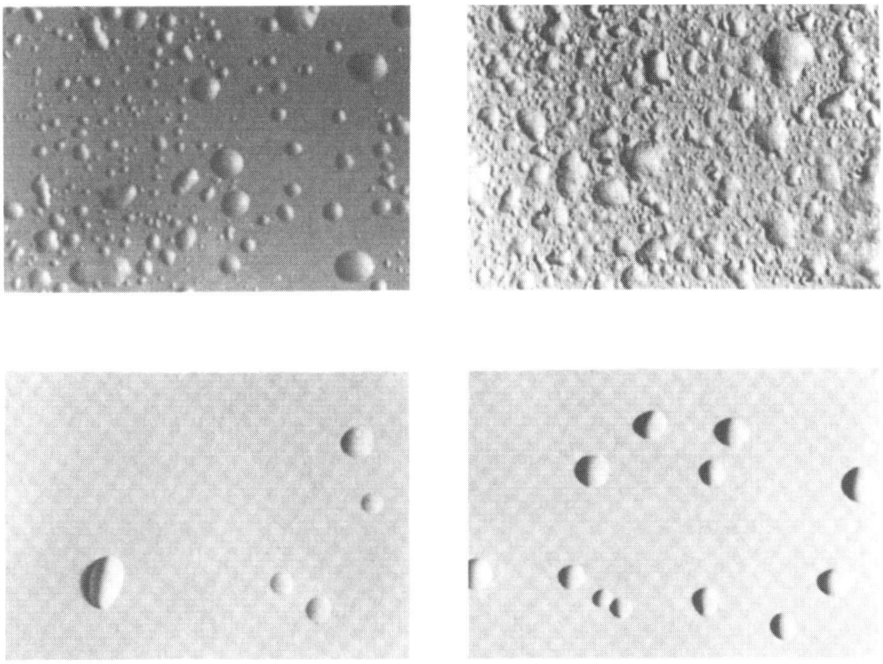

Figure 17-7. Coated panels exposed to a condensing humidity test.

Figure 17-8. Accelerated weathering test cabinet.

Numerous variations of this test are being utilized. Typically, test panels are exposed to dilute (2–10%, with 5% most often used) sodium chloride in a spray mist at 100° F (38° C) for an appropriate length of time. This duration may be as little as 24 hours or as long as 2,000–3,000 hours. Two types of failure may be noted. *Face rust* is the appearance of rust on the surface of the part or panel where the coating seems continuous. *Creepback* is the spread of rust under a coating film and subsequent loss of the overlying coating, perpendicularly away from a line scribed through the coating into the underlying metal. Tests are usually conducted according to ASTM B-117 or B-287.

However, nearly every salt spray test may be misleading, and ASTM points this out very clearly. For example, salt spray test results on coated zinc and galvanized parts often appear worse than those on coated steel parts. However, the coated zinc parts invariably provide far superior corrosion resistance in field service. Because of this anomaly, it is recommended that salt spray tests <u>not</u> be conducted to compare paint or powder coatings on zinc-coated parts versus those on steel parts.

The relative performance characteristics of coatings can probably be ascertained to a fair degree of certainty from these tests. A coating that shows failure at 200 hours of salt spray is probably not as durable as one exhibiting no failure at 200 hours. But there are more than a few recorded instances of one coating performing better than another in salt spray tests, yet holding up less well under actual field exposure conditions. This does not mean the test is not worth running, but obviously an amount of caution is necessary here.

Accelerated testing can usually determine the best performing coating among a group of potential candidates. The most durable coating in a group can be determined with fairly high certainty by testing all the candidates under the same conditions. If the test parameters were selected carefully, the coating that performs best in testing would be expected to give the highest performance in the field under normal usage conditions. This has been found to be true at least 95% of the time. Only in exceptional instances has it been found that the results of accelerated testing are reversed in actual service. If the wrong test is selected, however, the differences among the candidates may not be discovered, and all coatings may exhibit exactly the same test durability.

Unfortunately, it is not always obvious which version of salt spray testing is the correct one to use. Coated parts may need to meet a specification of 96 hours in a given accelerated test condition. Others may require 500, 1,000, or even 2,000 hours under specified test conditions.

The second purpose of accelerated weathering tests is to determine in-house how a coating will perform in the field. The warm salt spray in these cabinets severely tests the integrity of a coating film, especially on metallic substrates. It does this by speeding corrosion rates. Theoretically, the duration of a test can relate to service time in the field. Accelerated weathering tests may run from 1 hour to over 2,000 hours, and attempt to simulate a few months or many years of service exposure. The question is frequently raised, "How many days or weeks of outdoor exposure is represented by each 24 hours of salt spray testing?" No reliably accurate answer to this question is available, unfortunately. The true answer will certainly be different in the humid and salty air of Miami and Galveston than in arid cities such as Phoenix and Las Vegas.

In order to make accelerated tests reflect service life conditions as closely as possible, evaluators have run tests under an extreme variety of temperature, humidity, and salt spray concentration both with and without added corrosive agents such as copper acetate and acetic acid. In some cases, ultraviolet light is added during simulated exposure tests to induce chalking and fading. Whatever the test conditions, if they are thoughtfully selected to simulate service exposure as nearly as possible and properly evaluated, they are likely to give at least some indication of expected field performance.

Cyclic accelerated weathering tests have been developed that relate more closely to field conditions than continuous salt spray tests. Thus, 10 days in a cyclic test that tries to simulate a product's actual end-use conditions of varying thermal and chemical exposure conditions every 24 hours may be more realistic than 240 continuous hours at an unchanging temperature and humidity. One cycle will typically last 24 hours. For example, a cycle may involve this sequence:

A 15-minute soak in 5% neutral sodium chloride solution, then
1 hour at room temperature (usually 68° F, or 20° C), followed by
20 hours at 140° F (60° C) with 85% minimum relative humidity, and finally
several hours at 30° F (–1° C).

This pattern can be repeated until the specified number of cycles (days) has been completed.

In cyclic testing, the tester's imagination is the only limit on what can be devised. It is convenient if each cycle is completed in 24 hours and adapted to fit the work schedule of test personnel who will normally be at work for only one shift each day. Test specimens typically would be examined during normal Monday through Friday workdays, skipping the weekend days, unless there were special reasons for more frequent checks. Test cabinets should not be opened any more often than necessary, however.

Panels may be tested unscribed or scribed with one or more lines, or the coating may even be subjected to stone bombardment in a gravelometer before testing. The test results may be determined by visual examination measuring the amount of coating loss back from a scribe line (creepback), or tape testing may be done to determine how much adhesion will be lost after simulated field exposure tests.

Gloss

Gloss is a measure of the amount of incident light reflected from a surface, as shown in **Figure 17-9**. A standard polished surface, used as a calibration or reference that reflects all of the light, is given a gloss rating of 100. A surface that reflects no light has a gloss value of zero. The gloss number is not stated as a percent, yet the number represents closely the percent of light reflected from a surface. Some samples may reflect more light than the reference sample, so gloss values greater than 100, while not common, are nevertheless possible.

Instruments that measure gloss (gloss meters) send a beam of light (incident light) at a particular angle from the vertical directed at the surface to be measured. The instrument measures the amount of light reflected at the same angle. The angle of incident and reflected light is usually specified at 30°, 45°, or 60°. The smaller the angle (to the vertical), the more the light will be reflected, and thus also the higher the gloss reading. This explains the importance of including the angle of incident light for each gloss reading, given in statements such as, "The gloss is 88 at a 60° angle," or "The 60° gloss is 88." Smooth surfaces reflect more incident light than do rough surfaces, as would be expected (see **Figure 17-10**). But the amount of reflected light is independent of the color of the surface, contrary to what one might anticipate.

To maintain accuracy, a gloss meter must be calibrated frequently against gloss standard plaques in a range near to that being measured. Suitable calibration standards should be included when a gloss meter is purchased. The standard plaques should be kept with the instrument, and should be stored and used carefully to prevent scratches.

Some coating users have a need to describe highly specialized reflections not ordinarily significant for most coating operations. *Specular gloss* is the shininess or brilliance on highlighted areas of a part. *Sheen* refers to the specular gloss at very small angles of

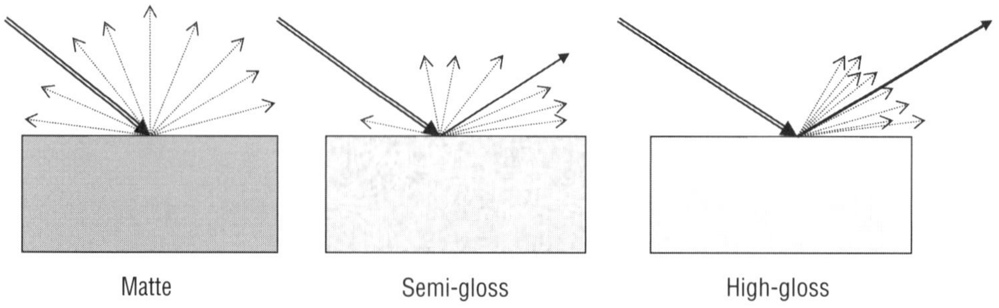

Figure 17-9. Light reflections from different surfaces.

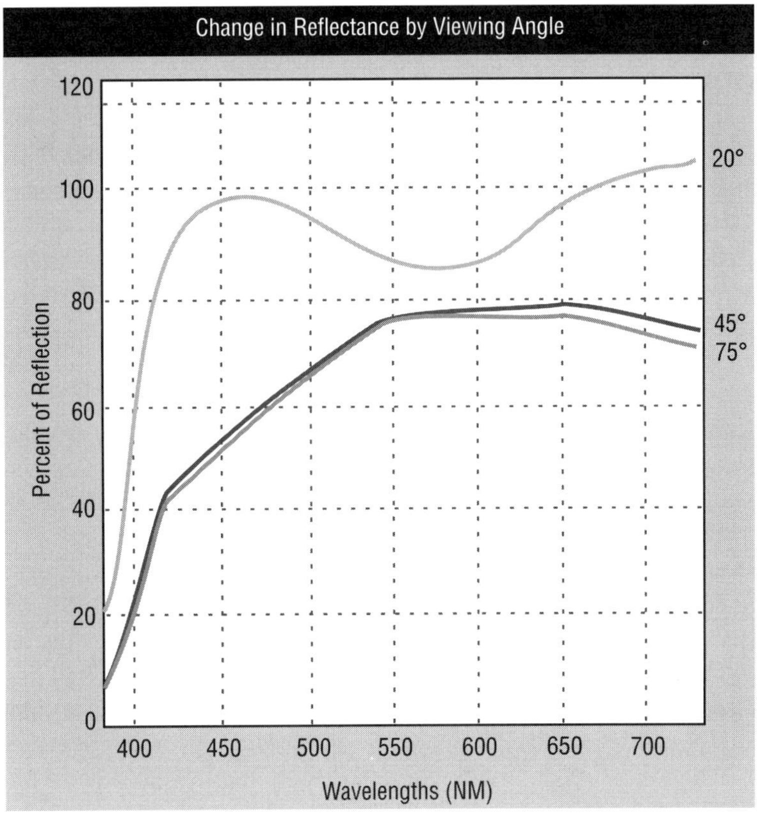

Figure 17-10. A spectral reflectance graph shows that the amount of light reflected at various wavelengths is about the same for a sample viewed at angles of 45° and 75°. However, when viewed from a 20° angle, the sample would look considerably different.

light incidence and reflection. *Brilliance* is the apparent strength of a color as perceived visually; it can perhaps be better understood if compared to the "loudness" of sound. Most paint users need not be concerned with these concepts.

Distinction of Image (Distinctness of Reflected Image)

Distinction of image (DOI) is the measure of how well a surface acts as a reflecting mirror. Distinctness of reflected image (DORI) and the abbreviation "DI" indicate the same phenomenon as DOI. The letters DOI should be used for distinction (distinctness) of image rather than the variations to avoid confusion. A polished black glass standard has a faultless DOI; it is a consummate mirror that can perfectly reflect every incident light beam without distortion.

Not all DOI measuring devices use the same comparative scale. Some DOI instruments place the highest smoothness rating at 100; at least one has a perfect rating of 20. A surface with a perfect DOI will be absolutely smooth. Therefore, in a sense, DOI is a measure of surface smoothness; it indicates how much roughness is present in a coating film. In many instances, the observed roughness is the result of orange peel or related flow irregularities in the film.

Gloss and DOI are often confused. Gloss measures the ability of a surface to reflect light. A surface, even though it is not perfectly smooth, may reflect essentially the total amount of the incident light impinging on it. However, a surface not perfectly smooth cannot perfectly reflect an image, that is, it distorts the shape of the image somewhat, even though it may reflect nearly all the light. In other words, a surface with a high DOI is always also high in gloss. But a surface with a high gloss may or may not be high in DOI. For a coated surface to be high in DOI, the substrate must be smooth, and the coating film must be free of waviness. Moderate orange peel will not detract much from gloss, but it will reduce DOI.

DOI measuring instruments project an image onto a coated surface and detect the distortion that occurs when the image is reflected from the surface. Some DOI devices rely on visual evaluations, and these have been shown to allow inconsistencies in readings. The evaluators rate a test panel based on the ability to see or not see distortions in various images projected onto the surface under test. It has been shown that values thus obtained will not only fluctuate widely among evaluators, but also that the same person evaluating panels often reports varying values if shown the same panel at different times. The Landoldt Ring (DOI light box) instruments, which project various sizes of the letter "C" onto panels, suffer from this particular drawback.

One DOI instrument, the Tension (pronounced tahn-see-own′) meter shown in **Figure 17-11**, will photograph the image reflected from the coating surface being tested with the grid pattern shown in **Figure 17-12**. The operator examines the picture of the reflected image. The DOI value assigned is the highest grid number where the operator determines that parallel lines in the picture seem not to touch. This test instrument has an additional advantage of providing a permanent photographic record of DOI. It was seldom used in North America until Chrysler Motors (now Daimler-Chrysler) bought the French

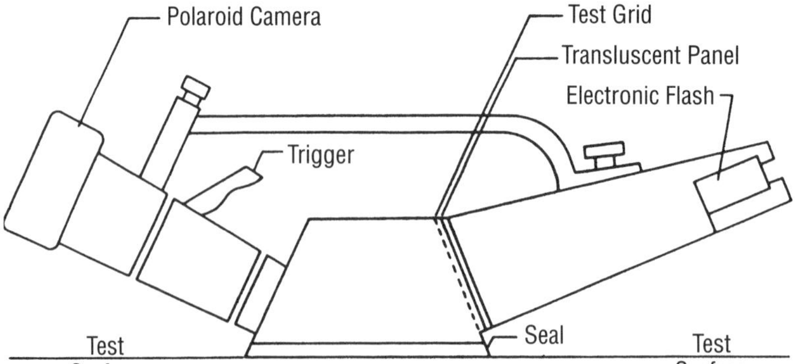

Figure 17-11. DOI tension meter.

automaker Renault who had used this test. After a brief spate of usage at Chrysler, its utilization now has nearly disappeared.

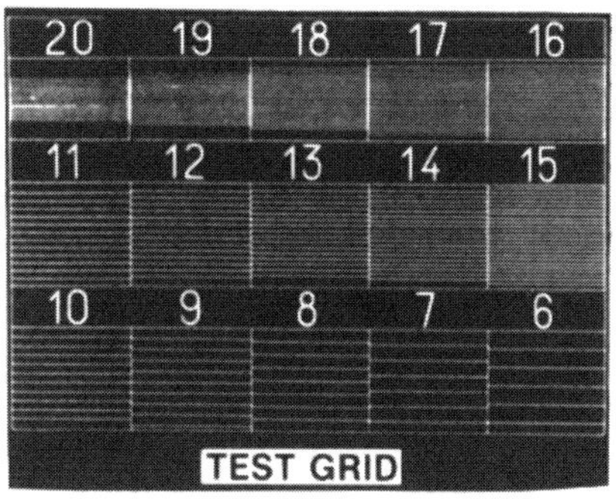

Figure 17-12. DOI test grid.

Color Match

Color does not have an actual real physical existence. Thus it cannot be handled or packaged; rather, color is a phenomenon that must be visually sensed or be determined instrumentally. In humans and animals it is a sensation arising from the interaction of electromagnetic vibrations on the eye retinas that are interpreted by the brain. Almost every sighted person has seen the various colors that are present in "white" light, perhaps when a prism separates colors from each other, or when sunlight on raindrops produces a rainbow. The colors can often be observed in a thin layer of oil floating on water. The separated color ranges are frequently classed using seven names: red, orange, yellow, green, blue, indigo, and violet. But hundreds of wavelengths are actually present in white light.

The *wavelength* of a light beam is the tiny distance between corresponding consecutive points in the electromagnetic radiation wave oscillation. Wavelengths that humans can detect visually are in a small portion of the total electromagnetic spectrum. Beyond the longest visible wavelengths are the infrared waves associated with heat, and then microwaves, radio, and television broadcast waves. Out past the opposite end of the visible range lie the short but high-energy waves such as ultraviolet light, gamma rays, and X-rays. **Figure 17-13** gives a breakdown of various types of electromagnetic radiation and their associated wavelengths. Wavelengths in the visible spectrum lie in the tiny region between about 4,000 and 7,500 angstroms (1 angstrom equals 10^{-10} meter). The longer the wavelength, the lower the energy associated with it. Thus, the 7,500-angstrom end of

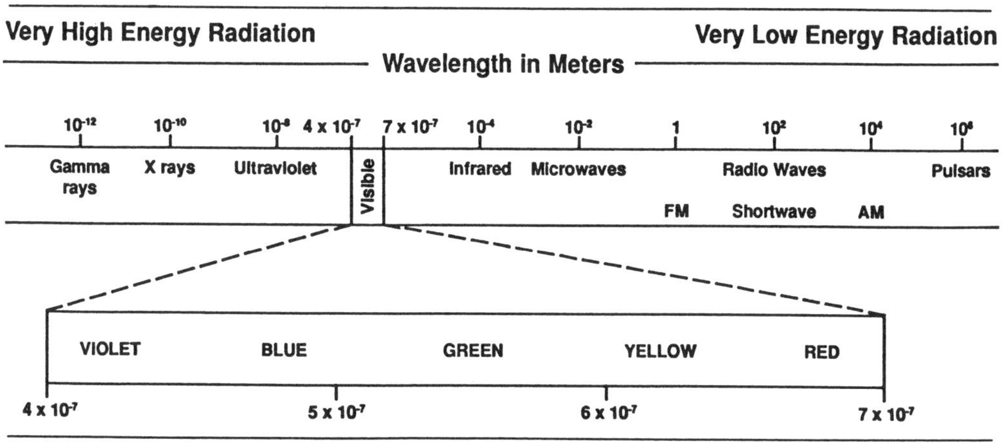

Figure 17-13. Electromagnetic radiation wavelength ranges.

the visible spectrum—the red wavelengths—is lower in energy than the 4,000-angstrom violet wavelength end.

The perceived color of an opaque object depends on the wavelength(s) of light that are reflected from it. An object we see as "white" reflects all visible wavelengths; an object we say is "black" does not reflect any of the visible wavelengths. An object with a particular color reflects those wavelength(s) required to make that color, and it absorbs all other wavelengths. In this same way, coating pigments are able to produce distinctive color shades from white to black by absorbing some or all of the individual visible wavelengths present in the light source. Thus, a red pigment reflects red light and absorbs all the others. If several pigments of different colors are blended together in a coating, the individual pigment colors are not perceived, but instead a single color will be seen that is a result of the combined pigment colors. Thus mixed white and black pigments produce a gray color; and similarly, mixed blue and yellow pigments yield a green color.

When a broad spectrum light such as daylight hits a pigment particle, in addition to absorption of visible wavelengths in major wavelength bands, some selective absorption often occurs across the entire visible spectrum. This makes the color less pure, and is said to "muddy" the color. The most desirable pigments therefore always have less muddy overtone to their colors.

Matching colors of paint can be a problem because the exact shade of a coating can depend on many factors, such as pigment particle size and shape, spray pressure, target distance, booth humidity, and film thickness. Coating colors can often be matched fairly well visually, but they can almost always be matched more precisely using a color-measuring instrument. This is not just because a significant portion of humans are partially color-blind (about 18% of men and 7% of women are affected by this visual deficiency).

Color matching, whether visually or instrumentally, should be done under several types of light source. Light sources may have different colors of light in the illumination they give off. The light sources typically have reddish color (sunset type) light or bluish color (northern sky) light. You may have observed that incandescent bulbs give off a red type light compared to fluorescent tubes that emit a more blue light. The reason several light sources are necessary for accurate color evaluation can be explained in terms of the reflected and absorbed wavelengths. The sun radiates visible colors of all wavelengths, and an object that looks blue in sunlight absorbs all the wavelengths of visible light except the blue wavelengths. A source of light that is deficient in some of the visible wavelengths, for example, fluorescent lighting, may cause that same blue object to appear a slightly different shade of blue than in sunlight. This is because sunlight and fluorescent light contain different visible wavelengths or different proportions of these wavelengths. As a result, the visible wavelengths reflected to the observer will be somewhat different in sunlight than in fluorescent light.

Two frequently used color measurement scales are the Munsell and the CIE (L*a*b*) systems. Both use a tristimulus method, meaning that three different "dimensions" are used to describe a color. The Munsell system uses the terms hue, chroma, and value to identify the particular shade of color (*hue*), the intensity of the hue (*chroma*), and the grayness (*value*) scale that varies gradually from pure white to pure black. The CIE scale uses a lightness, black-to-white L scale. In the "a" scale, positive values are increasingly red, and negative values become increasingly green. For the "b" scale, greater positive values become progressively more yellow, while more and more negative values are increasingly blue.

In the earlier XYZ system adopted in 1931, colors were also measured in terms of three components.

1. X is the color shade, or "hue."
2. Y defines how light (or how dark) the color may be, or "value."
3. Z is a measure of the vividness, or "saturation."

Color instruments read these tristimulus values that are labeled "X," "Y," and "Z." This system was standardized by the International Commission on Illumination. In 1964 standard illuminants were specified so that universal color comparisons could be made. But with the XYZ system the size of the differences between two colors doesn't at all compare to the amount of difference seen by the human eye. For this reason the L.a.b. system was developed, and later the improved L*a*b* system. It is based on the opposed-colors concept, which theorizes that all colors can be defined by three components.

1. L is the color's lightness (or darkness).
2. "a" is the degree of red to green in the color; red if it's a positive value, and green if a negative number.
3. "b" states the degree of blue to yellow; yellow for plus numbers, and blue for negative ones.

The "a" and "b" values in each case are zero for the color gray.

The L*a*b* system is more commonly used than the XYZ simply because it indicates the amounts of color differences in objects more closely akin to the degree of color difference seen by a human observer. You may want to use it instead of XYZ for color measurements.

Instrumental color matching is done with two main types of light detection devices: the simplest, and the type recommended for most manufacturing plants that apply coating, is called a 45/0° instrument. It does not use the gloss of a sample as part of the color measurement, and thus it interprets colors much like the human eye. For more precise work such as in paint formulation laboratories, the integrating sphere or D-8 color-measuring instrument is preferred. It compensates for glare so that matching among low-gloss and high-gloss samples can be done accurately. Colorimeters, as shown in **Figure 17-14**, generate red-blue-green values for samples to determine their color. Spectrophotometers are more complex and measure numerous values across the visible light wavelength range to determine a color. No matter

Figure 17-14. Checking the color of a finished golf cart in natural light with a portable colorimeter.

what instrument is used, it must be calibrated periodically to ensure accuracy. The more expensive instruments have better light filters and are thus more accurate, but for most manufacturing plants the less costly devices are normally perfectly acceptable.

Which Tests to Run?

With so many to choose from, some selection is needed. Tests selected for a coating depend on many variables, and what entity is doing the testing. Coating manufacturers will run various tests to determine if their coating measures up to the specifications set by the end user. Finishers also conduct various tests, and which tests are run is often determined by factors such as:

Plant size	Cost of parts being coated
Plant location	Color of parts
Part quantity	Application methods
Type of parts	Normal service life of parts
Indoor or outdoor service location of parts	Size of parts.

Most finishers do not test incoming coatings at all, not even viscosity. Only a very few finishers perform various tests as each batch of coating is received from the manufacturer. These incoming material inspection tests may include:

Weight per gallon	Hardness
Percent weight solids	Viscosity
Fineness of grind	Adhesion and flexibility
Color match to standard	Chemical and stain resistance.
Gloss	

Tests run by finishers for production control and quality assurance are much more common, and may include:

General overall visual appearance	Hardness
Gloss	Dry-film thickness
Color match to standard	Coating transfer efficiency.

In a majority of finishing plants, wet-coating tests are not common practice. The exception, of course, is viscosity testing that is routinely done in nearly all wet-coating operations just before applying the paint. The majority of coating tests by finishers are run on panels or parts after the coating has been applied and cured.

Before leaving this chapter, let me make an essential point on coating testing. I highly recommend the *Annual Book of ASTM Standards, Volume 06.01* (about US$150) and perhaps also *Volume 06.02* (about US$120) for anyone doing even minor amounts of coating testing. These, plus more, volumes can be acquired on CD-ROM (about US$400). These standards are revised periodically, and new standards are written as tests are developed. ASTM publishes all of its standards yearly, but existing copies tend to be suitable for at least 5–10 years, so it is not necessary to purchase these two volumes annually. It may be worthwhile to purchase this on CD rather than filling bookshelves with numerous volumes.

As long as a test is conducted identically each time, the so-called "properness" of the procedure is of lesser importance. It is informative, however, to read the test procedures. ASTM excels at stipulating how tests are to be executed. Until one has read the ASTM

TESTS ON APPLIED COATING FILMS

Film Property	**Commonly Used Test(s)**
Corrosion resistance	Salt fog; salt fog cycle; humidity
Chemical resistance; oil, acid, or alkali resistance; extent of cure	MEK rub (or similar solvent rub); Covered liquid agent(s) spot
Flexibility	Mandrel bend; "T" bends
Surface gloss	Reflected light glossmeter
Surface microprofile (orange peel)	Distinction of (reflected) image
Color match, color value	Tristimulus (ΔE, l,a,b or ΔE*, l*,a*,b*) values
Chalk and fade (weather) resistance	Xenon arc, QUV, Florida exposure
Film hardness	Pencil hardness
Impact resistance	Direct and reverse falling weight
Heat stability	Overbake yellowing
Adhesion	"X" and 90° cross-hatch tape pull
Stain resistance	Staining agent spot
Water resistance	Condensing humidity, Water immersion

Figure 17-15. Tests on applied coating films.

documents, it is hard to imagine how much detail can be specified to perform a test as straightforward as pencil hardness, for example. Following ASTM guidelines for a test helps ensure that the same procedure is performed identically for each test. In this way, the results obtained are accurate and repeatable.

ASTM's advice and instructions are also helpful in showing how to convert observed test results into meaningful numerical ratings for comparative evaluations. Without the tester having full knowledge of precisely how to run and interpret certain coating tests, the results of a lot of hard work may actually be worthless. Even worse, poorly conducted tests can convey totally misleading information.

More on: INSTRUMENTAL COLOR MEASUREMENT

Color measured by a computer employs a standard light source lamp to focus on the object under study. The light may be either a pulsed incandescent type or one that stays on. LED (light-emitting diode), and xenon flash tubes are also used. The source needs to be mathematically determinable in a normal daylight color range. The incident light that either passes through a transparent or translucent object or is reflected by an opaque surface is collected, and the color analysis unit inside the computer examines it. This is standardized black (with total light absence) and white against a certified white tile. The analysis unit converts the light input into computer signal data form.

Various instruments perform this function differently. Colorimeters have red, green, and blue filters to observe color similar to human eye receptor cones. Analytical spectrophotometers use gratings to measure light response at precise wavelength intervals across the spectrum. These give the most precise reading. Less expensive spectrophotometers utilize filters for this separation and are not as accurate. The color computer reports the results as simulated human vision by factoring in the illumination data. The output is called a tristimulus response, having three values termed X, Y, and Z. An international group, CIE, has supplied a tristimulus evaluation method using "L" as the white to black value; "a" as the red to green value; and "b" as the blue to green value. These three precisely define the color.

Final thoughts: MODIFIED IMPACT FOR ADHESION TESTING?

Can impact be used to determine adhesion strength? Unlikely as it sounds, recent reports have shown that coating adhesion can be measured by two methods — both using a form of impact. One uses microimpact, and the second method employs single-point repetitive impact on the film.

MICROIMPACT TESTING

Measurement of thin film adhesion can be done by scratching with a diamond stylus that is slowly moved across the film under a constant or increasing load until film failure occurs. Failure is established by these occurrences:

1. Observation of the wear track under a microscope.
2. Detection of an acoustic energy outburst.
3. Measuring any sudden change in frictional force between the stylus and the paint film.

Adhesion failure is dependent on interfacial adhesion, plus the mechanical characteristics of both the film and the substrate. For that reason, it is only rarely possible to calculate absolute adhesion energy values from scratch test data. In fact, sometimes the applied force can produce reattachment of the film and thereby increase adhesion.

Adhesion may be determined in some other cases by direct indentation measurements. What happens is that a sharp diamond is forced into the film and it penetrates both the coating film and the underlying substrate. Eventually, the stress generated may be enough to result in failure at the interface. This is detected either by using acoustic emission, or by monitoring displacement of the diamond stylus.

A new scanning technique to detect adhesion failure in microimpact testing impacts the wear track continuously. Effects of test probe impact immediately adjacent to previous impacts are studied to assess the factor of wider area damage from the impact. Degradation of this type does indeed occur in many typical field failure situations.

TESTING BY REPETITIVE SINGLE POINT IMPACT

Single point impact, repeatedly applied, can be utilized to measure surface fracture resistance or toughness of a film. Energy is transmitted over the test probe contact location by elastic waves generated from the reverse side of the test plaque. For a nonenergy absorbing contact between a hard probe and the sample, it is possible to produce pendulum recoil from the surface. But, after the surface fractures or delaminates, it is ordinarily no longer possible to create pendulum recoil. Therefore, it is not possible to use pendulum recoil and subsequent impact collisions to generate surface damage. Then, by constantly monitoring how the pendulum position changes, the progression of film damage can be measured.

Chapter 18

Stripping

Reasons for Removing Unwanted Paint

In industrial painting, *stripping* is the term for removing unwanted cured paint film from a surface. The unwanted paint film may be on a conveyor hook, a paint hanger, the spray booth wall or floor, a grating, or similar equipment. Paint shop equipment is stripped regularly to maintain operating efficiency. For example, paint-free hooks and hangers are needed to ensure a positive ground for maximum electrostatic paint attraction, and clean floor gratings will optimize booth airflow for reduced paint fog and greater sprayer safety.

Unwanted cured paint also may be on finished products that need to be stripped so they can be repainted. Parts and assemblies with paint defects as well as items painted in the wrong color are usually stripped and repainted only when this operation is less than the cost of a new item. Stripping is also done after paint films have undergone a life cycle and are beginning to show areas of paint loss, as on commercial airplanes (see **Figure 18-1**), ships, trains, and buses, not only for corrosion prevention but also to keep an attractive appearance for business reasons. Stripping and repainting is done on many military items for the added purpose of maintaining the full camouflage paint for tactical concealment.

Figure 18-1. Aircraft can be stripped and repainted.

Types of Stripping

Stripping of paint from industrial equipment and products is done using any of these basic methods:

- Mechanical
- Blasting with abrasive grit
- Blasting with water
- Chemical
- Solvent
- Burn-off ovens
- Molten salt baths
- Hot fluidized sand.

Mechanical

Mechanical stripping uses physical contact devices that grind, abrade, cut, or scrape. Manual techniques are labor-intensive and potentially hazardous, but power tools can help expedite the process. Hand tools and manual power tools include scrapers, rotary grinders, rotary and orbital sanders, abrasive cloth or paper, and wire brushes. These methods are slow, crude, and frequently leave a badly marred surface finish. If the paint is strongly bonded to the surface, manual methods are usually not completely effective in achieving total paint removal. But for limited amounts of stripping, manual methods are convenient.

Blasting with Abrasive Grit

In blasting with tiny grain or pellets, the grit is propelled at high speed against a cured paint film. Each individual grit media particle abrades a tiny fragment of the paint film. Repeated bombardment by thousands of particles gradually removes the paint film. Vari-

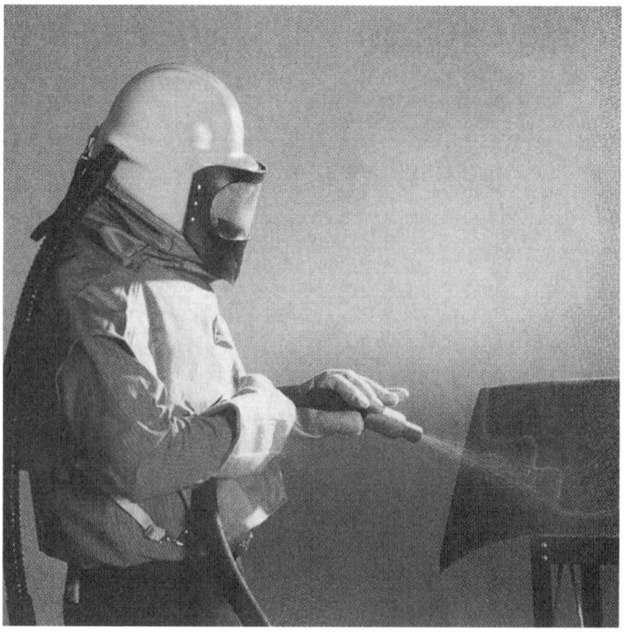

Figure 18-2. This operator is using plastic blast media to remove a powder coating.

ous methods are used to forcefully project the media onto the painted surface, such as a jet of pressurized air or a rapidly spinning wheel. The abrasive media is often not recovered, but it is reused where and when that is practical. A screening or filtering stage can separate fine dust particles of grit and stripped paint from the larger reusable grit particles.

Many different types of abrasive media are available, ranging from sand to particles of plastic. The hard types of media are used in stripping thick, tough-to-remove paint. Plastic media are used for "gentle" stripping and are often used in automotive and small aircraft repainting to strip only individual layers of paint, such as a topcoat from a primer.

In one type of media blasting (termed "cryogenic" stripping), the product and paint film are immersed in liquid nitrogen (about -320° F, or -196° C) to embrittle the paint film. The product is removed from the super cold nitrogen bath and immediately blasted with media. The cold embrittled paint film usually can be removed with blasting more readily than at room temperature. However, a large number of paint films are so durable they tend to be rather resistant to this process.

Blasting with Water

This type of stripping uses a high-pressure (3,000–10,000 PSI, or 205–690 bar) stream of water to remove paint films. It is used regularly in high-volume paint booths to strip paint from floor grates. The ultra-high-pressure water jets used for stripping operate at 35,000–55,000 PSI (2,415–3,800 bar), but require far less water. Some nozzles use multiple sapphire orifices to reduce tip wear. Extreme caution is necessary to prevent human injury from all high-pressure water streams, which can easily cut flesh. Manual operations are normally limited to pressures below 40,000 PSI (2,750 bar). Wheeled machines that can be taken into paint booths for the water stripping process are designed to shroud the high-pressure water stream to protect the operator and workers in the vicinity. Booth floor grates also can be passed through stationary water jet paint stripping machines. They are similar to belt-type parts washers. An advantage of water blast stripping is that no chemical disposal is involved, other than proper disposition of the stripped pieces of paint that are removed.

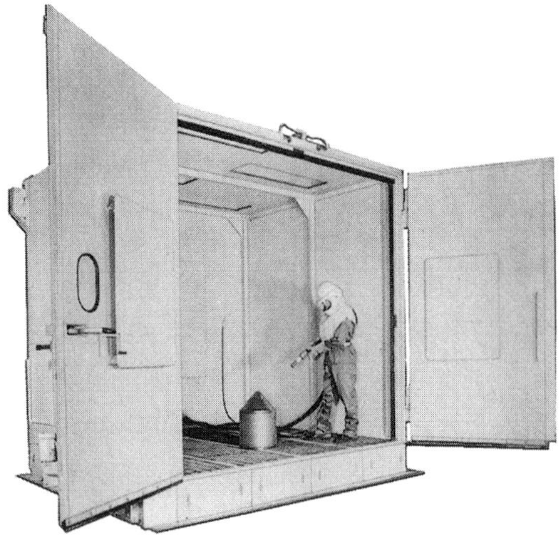

Figure 18-3. A blast room for manually stripping large parts. The doors are open for demonstration purposes; they would normally be closed during operations.

Chemical

Many chemicals are available for stripping paint films, but with any such chemicals come concerns about disposal costs as well as potential environmental pollution. This scares many organizations away from chemical strippers. The strippers are usually categorized according to whether they are heated aqueous solutions (hot strippers) or unheated organic solvents (cold strippers). Although a few hot aqueous strippers are acidic in nature, far more often they are strongly alkaline. Both types of these hot aqueous strippers destroy the bonds that hold paint resins together. This is not true of organic solvent strippers. They employ either a single solvent or a blend that will soften and/or dissolve the paint layer(s).

Alkaline strippers contain strong concentrations of potassium hydroxide or sodium hydroxide (pH = 13-14) and possibly high concentrations of trisodium phosphate. In the past some strippers have contained phenols because inexpensive phenolic compounds will act as highly effective activators for reactive strippers. But the disposal of phenolic wastes is so restricted nowadays that the use of phenolic agents in strippers is rare. Yet, in parts of the world where environmental laws are not in effect, phenolic agents are still used. Proper disposal of spent stripping baths is a costly nuisance because they contain paint sludge as well as alkali. The problem and the expense are exacerbated when the paint sludge contains toxic compounds or heavy metals.

Heat is a valuable aid in aqueous alkaline stripping because the degradative reactions on paint proceed two or three times more rapidly when hot. The boiling point of concentrated hot strippers may allow temperatures of 225–235° F (107–112° C) to be reached. Vigorous boiling action in hot tanks also provides helpful bath agitation to speed stripping. But it is also a disadvantage of hot stripping tanks that the tanks must be heated continuously, making this method difficult to use on short notice. Another disadvantage is the potential danger to personnel from fumes and potential scalding from splashes. Because water is evaporated from the strip tank, condensation of that water vapor can drip on people, the floor, and equipment to create a messy working environment.

Hot alkaline stripping solutions can rapidly degrade many paint resins to the point that the paint is completely removed, or nearly so, in the stripping tank. Some paints can be stripped in about 20 minutes; others may take an hour or more. For extremely adherent paints or durable resins, some additional physical scraping or abrading often becomes necessary on stubborn areas of the paint.

Hot caustic chemicals can be dangerous. They cannot be used on zinc or aluminum because they react vigorously, in some cases violently, with these metals. The alkali will quickly dissolve such parts. Stripping operators have been burned by hot solution when zinc parts were inadvertently lowered into caustic strip tanks. The reaction is so rapid that some of the hot solution is literally blown out of the tank.

Over time the hot stripping action will weaken. The alkaline stripping chemicals are consumed during usage and thus need periodic replenishment. At this time about half of the solution should be discarded and replaced with fresh solution makeup. During operations, water lost from evaporation and carryout on parts should be replaced regularly. Periodically the accumulated stripped paint, which forms a layer of sludge on the bottom of the tank, will also need to be removed for disposal.

Acid strippers utilize mainly nitric, sulfuric, or phosphoric acids, either alone or in combination. Acidic strippers are not used much anymore because the baths tend to be highly corrosive and release acidic vapors that cause corrosion inside the plant. Acid strippers are rarely used hot since heat would increase the emission of acid vapors that are not only corrosive, but hazardous to workers' health.

Restrictions on land burial, liquid discharge, and air emissions have placed limitations on all forms of chemical stripping. The spent stripper solutions, whether acidic or alkaline, must be disposed of properly. The stripped-off paint forms a sludge that can be removed from the bath by filtering, then the rest of the spent bath must be neutralized. All harmful materials in the spent bath must be removed. Local restrictions may not allow dilution or draining to sewer. Not surprisingly then, more and more companies have turned away from chemical stripping to thermal methods of paint removal.

Solvent

Certain solvents at ambient temperature have the capability to strip paint films, and this eliminates the need for heated solvent. Ambient temperature solvent stripping is an excellent alternative method for stripping paint from aluminum, zinc, and other metal parts that cannot tolerate stripping methods using caustic or heat. Solvent stripping is used either as a dip, spray, or brushed gel, depending on part volume and production requirements.

Solvent strippers include highly aggressive oxygenated phenols, ketones, glycols, and esters, as well as saturated and aromatic hydrocarbons. But many solvent stripping products carry human toxicity concerns. For example, phenols are highly corrosive to skin and are toxic; numerous nonphenolic solvents are toxic and flammable; halogenated hydrocarbons have toxicity, ozone depletion, and safe-disposal concerns. Nonflammable halogenated solvents such as methylene chloride, trichloroethylene, and 1,1,1-trichloroethane were used heavily before the age of toxic and environmental chemical regulations, but only a few halogenated solvents are allowed now in the modern countries of the world.

Solvent strippers remove paint either through dissolving or softening, or both. Lacquer films will totally dissolve in the correct stripping solvents. Enamel coating films, however, will often not dissolve but rather will swell, soften, and wrinkle when treated with solvent strippers. This greatly simplifies their subsequent physical removal. No hard and fast rules govern which solvent strippers work best on a coating; some trial and error is normal to identify the best one for a given paint.

Vapors are a serious problem with most solvent stripping operations. Some vapors are hazardous to workers' health, especially if workers breathe the vapors for long periods. Other solvents, while relatively safe from a health viewpoint, have odors that workers find offensive. One way of containing stripping solvent vapors is to use a wax that floats on the solvent, forming a vapor seal. Another moderately effective vapor seal is water if the solvent is not water miscible and has a high enough density to enable water to float on it. Some solvents, however, cannot tolerate water. For example, water added to halogenated hydrocarbons can form hydrochloric acid, which is highly corrosive to metals such as steel.

Paint, especially soft but undissolved enamel paint, may accumulate on the bottom of a solvent-stripping tank from which it can be separated out periodically. The remaining solvent is not always reduced in stripping effectiveness. In the case of lacquer paints, the stripper solution can become increasingly viscous as more and more resin is dissolved in it. The paint pigments, however, do not dissolve and thus will build up on the tank bottom.

Waste disposal considerations. Probably the greatest problem with using solvent strippers is how to dispose of the solvent-rich paint sludge. Although flammable solvents sometimes can be incinerated, nonflammable solvents such as halogenated solvent strippers cannot be disposed by burning. They form corrosive, irritant hydrogen halide gases when exposed to flame. To compound the difficulty, stripped paint sludge that contains either flammable or nonflammable solvent residues are rarely (if ever) permitted at ground disposal sites.

The solvent in paint-stripping sludges (and also used cleaning solvents) can be reclaimed, often for direct reuse, by distillation. Plants may do this themselves or hire a service for it. A number of solvent stills being sold for this purpose use replaceable plastic bags. The used solvents are placed in the bag and heated to distill out the volatiles. When all solvent is evaporated, the bag containing the solvent-free residue is removed for convenient and safe disposal. If no toxic substances are present, ground burial of the bag and its contents is normally acceptable at waste disposal sites.

We have observed that the use of halogenated hydrocarbons for solvent strippers has declined sharply because they cannot be burned and are no longer allowed at disposal sites. As a result, plants have no place to get rid of halogenated solvents at a reasonable cost. However, with the increased use of distillation to recover solvents of all types, there has been a slight return to using environmentally acceptable halogenated solvent paint strippers.

The cost of having a private disposal firm handle paint and solvent waste is frequently prohibitive. For small-quantity waste producers, however, contracting with an outside firm to perform waste management may be cost-advantageous. Companies of any size should look into the relative economics of such an approach to disposing of paint and solvent. A number of plants have been pleasantly surprised to find that it is less costly to use a disposal service company than to distill their own used solvent.

Figure 18-4. A wash-down booth. The operator is using high-pressure water to spray ash from hooks and racks that have been cleaned in a paint burn-off oven.

In general, however, many plants have turned completely away from any solvent stripping because of the disposal problems. Another reason for this change is that so many of the newer powder, high-solids, and waterborne coatings are resistant to chemical stripping. With many modern coatings, both chemical and solvent stripping are slow and much less than fully effective. Often, therefore, paint burn-off becomes a far more desirable method. With powdercoatings, burn-off is almost an automatic choice for stripping.

Burn-off Ovens

Burn-off ovens are batch style enclosures designed to operate at about 1,000° F (538° C). The high temperature burns away the organic portion of the paint film, leaving an inorganic ash residue. The oven design includes various methods for eliminating the emission of the smoke generated by combustion of the organic resins in coatings. Some burn-off ovens include an auxiliary unit for washing ash residue from the stripped parts. A wash unit confines the ash, and prevents it from spreading throughout the plant. If care is not taken, ash is easily spread to surrounding plant areas where it becomes a cleanliness problem or a dirt nuisance in other ways.

With all heat-stripping processes, the possible loss of favorable metallurgical properties of parts and paint hangers must be considered. Some metals lose their strength (temper) if heated excessively, or undergo softening; as a result they may deform under load. Such metals are obviously not suited for this type of stripping process. Burn-off ovens are frequently used for stripping paint from conveyor hooks and hangers, however, so the problem is certainly avoidable by proper part design and material selection.

Figure 18-5. Liquid solution strips coatings from dipped portions of the part. Note the complete removal from the areas that are submerged.

Molten Salt Baths

Molten salt baths can strip paint films almost instantaneously. The baths are composed of a mixture of various salts. Different salt mixtures are available with melting points ranging from 600–900° F (315–482° C). All salt baths strip paint by burning the film, usually in less than 10–25 seconds. As one would anticipate, the higher the bath temperature, the faster the stripping action. Large amounts of flame and smoke are immediately produced when paint-covered parts are immersed in salt baths. So much smoke forms that special emission controls are generally needed.

Hooks, hangers, and floor grates are often stripped in salt baths because they are made of heavy gauge metal and can readily withstand the high temperatures for brief periods. Production parts can be stripped as well if they can withstand the heat, but that is less often the case.

The burning of the organic resin and additives in the paint leaves behind inorganic pigment residues, which settle to the bottom as sludge. Depending on how much paint is stripped, from 3–55 gallons (11.3–208 l) of sludge may be produced a day. Each 55 gallons of paint sludge requires about 450 pounds (204 kg) of fresh salt mixture. The sludge must be separated out periodically and discarded.

Molten salt baths must be operated with extreme care and caution due to the danger from high temperatures and bath splashing. For handling items to be stripped, an overhead hoist that allows operators to stand a safe distance away is used. Conveyorized hot

Figure 18-6. This parts stripping system uses molten salt bath equipment and proprietary process chemicals to quickly remove powder coatings from both ferrous and nonferrous parts at a temperature of 600° F (315° C).

salt stripping is being done only at a limited number of locations. Conveyorized part entry and exit is ideal so that the stripping operators need not be near the hot salt during dangerous parts of the operation.

Hot Fluidized Sand

Beds of fluidized hot sand can effectively strip paint film. The sand is kept in a fluidized state by currents of air at 700–1,000° F (370–538° C). The combination of heat to burn the paint plus the abrasive action of sand provides thorough and rapid stripping of even the toughest organic coatings. But hot sand stripping is just not commonly done, although the reasons for this are likely more due to the perceived economics of installation and operation than the suitability of the technology. It does work well on items that do not entrap sand.

Selecting a Stripping Process

Various factors need to be considered before choosing any stripping process. These include:

- Cost and volume of items to be stripped
- Substrate characteristics and part configuration
- Safety
- Environmental regulations.

Cost and Volume of Items to be Stripped

Purchase and installation of stripping equipment is probably not cost effective for a majority of plants. Contract stripping companies have succeeded in marketing their services to plants unable to justify in-house stripping systems, or who for various reasons are unwilling to do their own paint stripping. If a contract stripper has a facility located reasonably close, it is possible that they can be used to do the paint stripping for a plant. However, the availability of a contract stripper is not of much benefit if items to be stripped need to be transported long distances or if they are too bulky to be easily handled; then overall costs can become prohibitive.

For many plants, stripping can turn out to be more expensive than scrapping parts. In others, hooks are not totally stripped, but instead are periodically spot cleaned by grinding or sanding at contact points in order to maintain good electrostatic grounding.

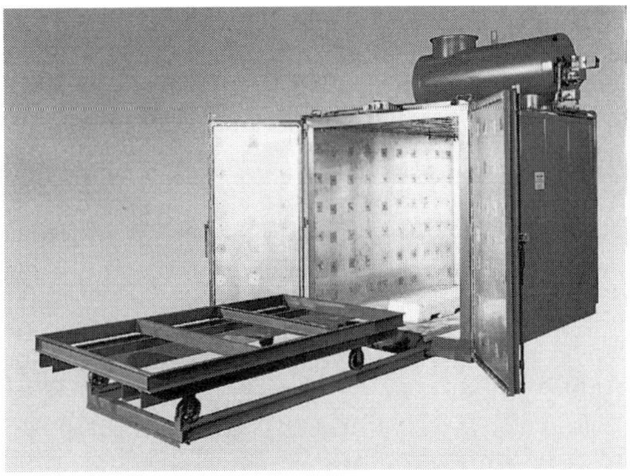

Figure 18-7. A typical controlled pyrolysis burn-off oven for stripping coating from hooks and hangers.

Substrate Characteristics and Part Configuration

The nature of the substrate must be taken into consideration before selecting a stripper. Heat can cause metals to lose their temper, and various chemicals can destroy or damage materials sensitive to stripping agents. Abrasives may degrade soft substrates. Parts with shapes that are complex may not be reachable by water jets or grit particles; others may have recesses that would carry out too much sand or hot salt.

Safety

Personnel must be required to wear the prescribed protective equipment and clothing for all types of paint stripping. In addition, plant areas surrounding the stripping need to be checked thoroughly for damage susceptibility from the stripping. Proper ventilation must be available if stripping is to be done inside the plant, as is normally the case. Occasionally, stripping is conducted outside in locations where the ordinary weather conditions allow this, such as the less affluent, warm weather countries of the world. Where this is possible, it is still prudent that wind, temperature, and precipitation conditions be monitored to prevent mishap.

Environmental Regulations

Environmental and waste disposal limitations must be considered when selecting a stripping method. It is not unusual for a plant to discover that if they were to use chemical stripping, disposal cost for used chemicals would be so high that it would be cheaper to throw parts away rather than strip and repaint them.

The drawbacks presented here should not leave readers with the impression that stripping is always an excessively expensive operation, but it is important to realize how economically vital it is to determine the most suitable stripping method.

Chapter 19

Special Considerations for Painting Plastic Parts

Understanding the Chemistry of Plastics

The word *plastic*, when used as an adjective to describe materials, is defined in the dictionary this way, "applies to substances soft enough to be molded yet capable of hardening into the desired fixed form." *Plastic* in this chapter context is the generic name given to a broad category of materials produced either from petroleum-based or agricultural plant-derived organic polymeric resins. Plastics may contain various additives to produce certain desired physical and chemical properties. Pigments may also be added to plastics to yield various shades of color. Plastic products have been molded into practically every conceivable size and shape with an array of textures and properties to facilitate an almost infinite assortment of end uses. Natural plastics have been with us since time began, but not until they were made synthetically in great quantities did we see an expansion into nearly limitless use of plastics. In that sense they are truly the wonder products of the last century, and remain so in the 21st century as well.

Plastic resins are characterized by complex organic chemical structures, made from the chemical bonding or linking together of varying numbers of individual molecules. Each molecule involves mainly carbon, oxygen, and hydrogen atoms held together primarily by single and double chemical bonds. Resin molecules differ greatly in their individual size, chemical makeup, and degree to which they undergo polymerization. They may be made up of any number of smaller molecules that have linked together chemically. That number can range anywhere from approximately 7 to 700,000 molecules; there is really no theoretical limit. In most cases at least many thousands of molecules join together to form the coating resins.

More importantly, all resins are categorized into either one or the other of two possible behavioral types: thermoplastic resins or thermoset resins. *Thermoplastic* resins are characterized by having various lengths of polymer chains that are held together by relatively weak, mutually attractive forces among the individual molecules. Thermoplastic resins can always be reformed into a new shape with the application of sufficient heat. *Thermosetting* resins, by contrast, consist of multiple polymer chains that are bonded together by comparatively strong cross-linking chemical bonds. Thus, they become to some degree permanently rigid and tend to resist softening when heated. Thermosetting resins cannot be reformed readily into new shapes by the application of heat. In actual practice, there are materials that behave much as thermoplastic resins but which technically are thermoset resins. Thermosetting resins can have varying degrees and types of cross-linking, so some degree of heat softening is normal in paint resins before the resin is cured. "Curing" for thermosetting paint resins is synonymous with "forming chemical bonds" and "cross-linking."

Plastic resin chemical types include acrylic, polystyrene, polypropylene, acrylonitrile-butadiene-styrene (ABS), polyester, polycarbonate, nylon, polyphenylene oxide (PPO), acetal, polyvinyl, urethane, cellulose acetate butyrate, phenolic thermoplastic olefin (TPO), and ethylene propylene dienemonomer thermoplastic rubber (EPDM). Most of these can be manufactured to be either thermoplastic resins or thermoset resins, so the chemical name alone does not indicate which of the two types it is. There are countless other chemical resins as well, but the ones named above are the most important finishing resins in use today (see **Figure 19-1**).

Some resins have acquired generic or trade names that often relate to how they are used. This can be deduced from a few examples listed below.

- Sheet molding compound (SMC), a thermosetting polyester resin with 25–32% fiberglass reinforcement, is made into sheets that are placed into molds.
- Reaction injection molded urethane (RIM).
- Reinforced reaction injection molding (RRIM) is RIM with 8–32% fiberglass reinforcement added.
- IMR is RIM or RRIM resin containing internal mold release agent(s).
- Plexiglas and Lucite are trade names for clear polymethyl methacrylate, which is commonly used as a glazing substitute in aircraft, building, and home windows.
- Styrofoam is polymeric stryene that has been blown into a foam.

Plastic formulations can include single resins or blends of various resins. They can contain a host of different additives, such as fillers and plasticizers. One additive (chopped fiberglass) is used in a category of plastic called fiberglass-reinforced plastics (FRP). In addition to the glass fibers, these plastics may consist of various resin blends, and possibly other additives as well.

	Chemical Name	Common Name
ABS	Acrylonitrile Butadiene Styrene	ABS
EPDM	Ethylene Propylene Diene Monomer	EPDM
PA	Polyamide	Nylon
PC	Polycarbonate	Lexon
PPO	Polyphenylene Oxide	Noryl
PE	Polyethylene	Dylan
PP	Polypropylene	Profax
PS	Polystyrene	Lustrex
PVC	Polyvinyl Chloride	Geon
PUR	Polyurethane	Bayflex
EP	Epoxy	Epon
VP	Vinyl Polyester	SMC

Figure 19-1. Commonly used plastics.

Why Bother to Paint Plastic?

Pigments can be added to produce plastic formulations having just about any desired shade of color. Just one example is the wide array of ballpoint pen colors. Pens of all shapes and sizes are available in hundreds of molded-in colors. So are items such as garden furniture and children's playground equipment. Why, then, does anyone bother to paint plastic items? Well, sometimes there are strong reasons for doing so. But whether to mold-in the color or to paint a plastic item (or both) is a decision that needs to be made considering the requirements and economics of each individual case. The governing parameters are always cost, appearance quality, and product performance properties. Although it is simpler just to add colorants to the plastic itself, the reasons for painting plastics are manifold. Some of the important reasons include the need for:

- Highlighting
- Texturing
- Protection of the plastic
- Appearance uniformity
- Hiding defects
- Functionality
- Improving conductivity
- Coating of the second surface.

We will examine each of these in more detail to clarify the importance of each.

Highlighting

Only some selected areas of plastic products may be painted to highlight structural details. Areas may perhaps be painted in contrasting colors for decorative effects, or painted with identifying markings such as company names, symbols, and logos. Examples are the plastic wind deflectors mounted atop truck cabs, which commonly contain company names, identifying patterns, or advertising. Plastic hulls of boats frequently have glamour stripes painted in horizontal arrays. Each ball in a croquet set will have a differing color in multiple stripes around it. Plastic containers, both the reusable and single use types, usually have brand names painted on for identification and to increase sales appeal.

Texturing

Coatings on plastic can give desirable texture patterns or a tactile character to an entire part, or just a portion thereof to enhance the product's look or "feel." Computer keyboards and various electronic housing with their spotty color-on-color texture are an example of this. "Soft-feel" and "leather-feel" paints are used on decorative plastic items. Automotive interiors, especially steering wheels, are examples of where these tactile coatings are utilized.

Protection of the Plastic

Numerous plastics are sensitive to solvent etching, thermal crazing, and mechanical abrasion. In such cases, protective coatings can be applied to enhance the durability of the products. For example, stone chipping and solvent damage to polycarbonate automotive headlight lenses is nearly eliminated by painting the lenses with a clear polyurethane coating.

Appearance Uniformity (and Appearance Modification)

Gloss and textural differences from one area of a plastic part to another are seen on many unpainted plastic parts. Painting can be done to produce a uniform gloss across the product. The paint gloss can either be higher or lower than that on most of the original part. In general, molded plastic parts tend to have glossy surfaces—sometimes too glossy. Plastic furniture parts, if painted, usually receive a low-gloss paint to eliminate the somewhat "cheap" appearance of glossy plastics. But toys are often painted with high-gloss colors to "uniformize" the color and gloss on them.

Wood-graining paints, used in still another appearance-altering process, can make plastic look like wood. It can provide a matching pattern to items constructed of both wood and plastics so the entire product seems to be made out of wood. By selecting the paint pattern and graining effect to be applied, finishers can duplicate just about any kind of wood (oak and birch are especially popular) and any color of stain or paint.

Plastic parts are sometimes metallized to make them look as if they were chrome-plated metal. The plastic parts are clearcoated to make the surface very smooth before metallizing. Then after metallizing they are again clearcoated, this time for abrasion protection.

Hiding Defects

Swirls and related visual defects are impossible to avoid when molding reinforced plastics such as glass-filled resins. These appearance defects can be completely hidden by painting. Painting will also conceal other markings on plastics that detract from part appearance quality. When ejection pin marks, parting lines, or scratches on plastic parts are sanded down to give a smoother surface, tiny scratches or a dull-looking sanded area always shows. Coatings can be used to "uniformize" the appearance and hide the fact that sanding was done on the part.

Functionality

Many protective clear-plastic cover plates or lenses, such as the covers over the instrument panel in an automobile dashboard or in an airplane cockpit, are painted. However, the painting is frequently done only in selected areas. Interior sections may receive a reflective paint to increase light output; while selected exterior areas may be painted black to stop light from escaping from other locations of the lenses. Other coating may stop reflection of light. Optical device interiors are normally painted black to prevent stray reflected light from reducing the readability or sensitivity of these instruments. The inside of every camera is a low gloss black so stray light doesn't diminish the sharpness of the image to be preserved.

Improving Conductivity

The low electrical conductivity of most plastics may require special conductive coatings on plastic parts to meet requirements for electromagnetic interference (EMI) shielding and radio frequency interference (RFI) shielding. Electrical safety grounding requirements may also require special conductive coatings. Nickel- and copper-containing paints are often used for these purposes if high conductivity is mandated, otherwise coatings containing carbon may conduct electrical charges satisfactorily. Conductive paints are also used to prevent the accumulation of static charges on plastic components.

Coating of the Second Surface

Second-surface finishing (a better name might be "reverse-side finishing") is for obvious reasons unique to transparent plastic parts. For example, motorcycle helmets of clear plastic are often painted on the inside of the shell with various decorative colors, such as metallic reds, golds, and blues. Vending machines for drinks and snacks often have numerous second-surface painted plastic sections for product displays and pushbars. The paint is visible through the clear plastic shell, but it is protected from scratches and abrasion by the clear shell. The decorative "reverse-side" paint inside the shell is normally applied and cured, then coated with an opaque paint to prevent light leaks and also to protect the back of the decorative coating from abrasion.

Cleaning Plastic Before Painting

Just as with all other surfaces, plastics should be thoroughly clean before painting. If the plant performing the molding also does the painting, it may be possible to rush parts from the molding machines to the paint line without the need for intermediate cleaning. But when mold releases are used this is normally not a good option, even if so-called *paintable mold releases* are utilized. Most plastics can be cleaned with aqueous detergent solutions; aqueous cleaners will also neutralize any static charges on the parts. A few types of plastic such as nylon may absorb too much water to be cleaned this way. The absorbed water is not fully driven out in the dry-off oven, and so in the paint oven it causes blisters in the paint film.

In practice, most plastic contamination problems develop from three principal causes:

- Fingerprints
- Dust and lint
- Mold release.

Fingerprints

Fingerprints (salt and skin oils) can usually be completely removed with detergent cleaning. Dirt from fingers and hands can be avoided altogether by requiring all handlers from molding to painting to wear lint-free gloves. The gloves need to be changed regularly so they do not become a source of contamination.

Dust and Lint

Because nearly all plastics are nonconductive, they readily build up static charges that hold lint and dust tenaciously. Painting over these contaminants can result in visible defects that show through the applied paint. Wiping plastic parts with a tack cloth in many instances will not remove all of the contaminants. In fact, wiping the plastic may generate additional static charges and actually make the problem more severe. It can be shown that a person just touching a plastic part for one second with a latex or nitrile rubber-gloved finger or hand will create a negative charge of 3–4 kilovolts at that point. Touching with cotton gloves does not generate much charge, but cotton can be a rich source of lint fibers.

One of the preferred methods of removing statically attracted lint and dirt is to use a destaticizing air blow-off (see **Figure 19-2**). One such system generates some airborne ions that are positively charged and others that are negatively charged. The ion source is a weak radioactive source positioned inside a blower housing. When this "destaticizing air" is blown across the part, the ions in the air neutralize all static charges on the parts, and the air curtain gently blows away dust and dirt particles. Vacuum removal of dirt at this stage is strongly advised to prevent soil redeposition. The destaticizing step should immediately precede the painting operation, otherwise the parts can once again accumulate charges and become recontaminated.

General plant dust and lint is well controlled by conducting plastic painting operations inside a "clean room." Such a room contains carefully filtered air and is constructed of nonlint-producing materials. Only authorized personnel who are wearing proper attire (lint-free clothing, hairnets, shoe covers, etc.) should be allowed to enter the area. Be certain that only lint-free rags, packaging, mops, etc., are allowed inside the clean room.

Mold Release

Mold-release agents are often used to facilitate separation of plastic products from their molds. It may be nearly impossible to remove some molded parts from the mold cavity without first spraying on mold release, so mold release cannot always be omitted. Nevertheless, a large number of these mold-release agents interfere greatly with paint adhesion.

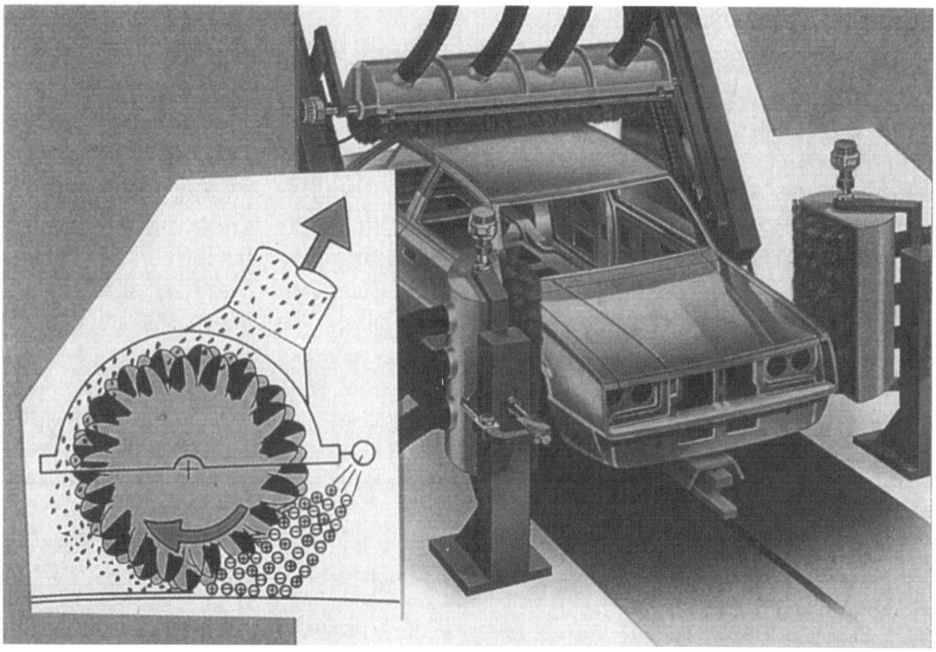

Figure 19-2. Feather duster using destaticizing air and vacuum to clean body just prior to painting.

Probably the most notorious mold release agents, as far as painting is concerned, are those containing silicone compounds. Silicone materials have a way of spreading throughout a plant, causing untold problems with paint adhesion to plastic products. They can be spread by a plant's air-conditioning system, causing widespread contamination of parts stacked just about anywhere in the plant. Wax-type mold releases can sometimes be removed by solvent cleaning, but waxes are not recommended if parts are to be painted. Solvents may attack the plastic substrate so caution is needed. Water-soluble mold releases that can be removed readily with aqueous detergent solutions are preferred.

Sometimes release materials, termed "internal release agents," are blended directly into the plastic formulation. During and after molding these can migrate to the part surface and cause paint delamination months or years after a part has been painted. For painting, internal mold releases must be avoided. Plastic raw material stocks held in bins or hoppers are sometimes accidentally contaminated with plastic containing internal mold release. This can happen in the mold shop when regrind material containing internal mold release is mistakenly dumped into the container of mold release-free resin pellets.

Certain types of plasticizers used in plastic formulations can also cause problems with paint adhesion similar to those from mold releases. Plasticizers are added to some molding resins to increase their impact resistance. The plasticizer can slowly migrate to the surface and weaken the interface between the plastic and the paint film, causing adhesion loss. Delamination of the paint film may occur weeks or months after the paint has been applied, even though initial adhesion tests showed excellent adhesion.

Tracking down the source of plastic surface contamination can require some disciplined detective work. Plants are not always willing to pursue the problem with that much diligence, for it can be a frustrating and time-consuming task. In such an investigation every possible dirt source must be relentlessly examined in hopes that the contamination can be identified and eliminated once and for all.

Preparing Plastic Surfaces for Painting

Even if perfectly clean, some types of plastics are inherently low in surface profile and tend to give poor paint adhesion unless the surface is roughened by chemical or mechanical means. The most common way of overcoming surface smoothness is to micro-dissolve (or micro-etch) the surface with solvent to create a microscopic roughness that provides mechanical anchoring sites for the paint. Paints are chosen so that solvents in the paint being applied are able to perform the etching. This solvent must be selected carefully because different solvents etch plastics at varying rates. Insufficient etching will not provide proper adhesion; while excessive etching can damage the plastic. Too much etch may also expose filler particles, and possibly even create sites where materials in the plastic formulation can bleed into the coating. Solvent selection can be critical for other reasons as well. Some plastics—polycarbonate and polystyrene, for example—will crack or become totally surface-crazed from attack by selected solvents to which they are susceptible. This can totally ruin parts.

When molding some part shapes there are areas where plastic injection creates friction from rapid plastic flow. These areas get substantially hotter than other part areas. The parts at those "hot spots" in the mold will often form highly cross-linked or glazed skin areas that tend to resist solvent micro-etching. Unless additional steps are taken, paint adhesion will be poor in these areas. Tumbling in mildly abrasive media or blasting with light grit can deglaze these areas enough to allow satisfactory paint adhesion. Treatment in hot solvents or in solvent vapors can also be effective.

When deglazing is not desirable and solvent etching is ineffective, as in the case with extremely nonpolar plastics, it may be necessary to use a chemical reaction to create polar oxidized groups on the surface. Examples of plastics treated this way are thermoplastic olefins such as polyethylene and polypropylene, and EPDM (ethylene-propylene-dicyclopentadiene monomer). These and similar plastics that have both low polarity and low surface energy may be quickly passed close to an ordinary gas flame. This causes no visible melting or charring but oxidizes the surface enough to form sufficient polar chemical groups on the surface to promote good paint adhesion. Passing parts through a strong electrical arc (corona discharge) has also been done; the ozone in the arc causes similar surface oxidation reactions.

Plastic parts with complex shapes frequently have areas that are internally stressed as a result of the molding process. Stress-relief action by some solvents can actually produce visible cracks in these stressed parts. Many plastics can benefit from stress-relieving, which is done by heating parts to around 150° F (65° C) for 20–45 minutes prior to painting. Overheating can distort thermally sensitive parts, however, and parts with high molded-in stress are likely to warp unless held in a fixture during heating. The effect of heat on plastic is discussed further in this chapter in the section on plastic properties that affect painting.

Low-polarity plastics are frequently treated with photosensitizers and then exposed to ultraviolet light. The UV decomposes the photosensitive compounds to form free radicals that combine with oxygen in air to generate polar groups on the plastic surface. This method of making plastics paintable is by far the most commonly used one. Sometimes UV treatment only is sufficient and photosensitizers aren't even needed. Some bulk molding compound (BMC) made from unsaturated polyester plus an internal mold release agent such as zinc stearate can be impossible to get paint to stick on. But when BMC is exposed to the light (and heat) of UV lamps, surface volatiles will evaporate and disperse the mold release agent enough to provide good paint adhesion provided the parts are painted immediately afterwards.

Cold gas-plasma technology can also pretreat plastics and dramatically improve surface properties for paint adhesion. Gas-plasma has been described as a "fourth state of matter." If any gas is given enough energy, it becomes ionized (a plasma). People are often at a loss trying to understand what a "gas plasma" is—convinced it is too complex and theoretical for them. But examples include things as simple as fluorescent lights and arc welding. The glow in each case is caused by excited ions in gas plasma falling to their lowest energy state. Plasma processes usually do not change surface appearance of parts so any material can be plasma-treated without causing discoloration. Plasma conditioning not only allows plastics to be painted with excellent adhesion, but this excellent adhesion is achieved with the same paints that are used on metals. This is important to some manufacturers since assemblies that contain both plastic and metal can be painted with the identical paint, making color match very simple compared to having to use two separate coatings.

The plasma reactor typically includes a sizable vacuum vessel fitted with an access door for loading and unloading parts. A radio-frequency generator provides excitation of the gas inside the vacuum chamber. When the reactor is started, a glow discharge can be observed. This process can use any gas or mixture of gases within the safety limitations of the system. Gases such as oxygen, nitrogen, air, helium, argon, and ammonia are commonly used. Gas-plasma surface treatment micro-etches and activates the surface. A brief treatment will make the surface polar and give it a high surface energy so that coatings will "wet" parts completely and achieve strong adhesion.

About 20 years ago chemical oxidizing agents were occasionally added to paints to improve their adhesion, although this is no longer common practice. These agents mildly oxidized the surface to achieve some polarity—at least enough to provide good paint adhesion. Reflectance infrared spectroscopy has verified that all the treatments listed above do indeed produce oxygenated (hydroxyl, carbonyl, and carboxylic acid) groups on the plastic surface. The polarity of these groups increases the surface energies of plastic parts and hence enhances paint adhesion.

Plastic Physical and Surface Properties Affecting Painting

A number of plastics have inherent properties that tend to cause difficulties in painting. In addition to the solvent sensitivity already described, these include:

- Water absorption
- Heat sensitivity
- Porosity formation from sanding
- Lack of electrical conductivity
- Color matching with metal
- Gloss variations compared with metal
- Spot-to-spot gloss variations on a part
- Wicking.

Water Absorption

Aqueous detergent solutions can do an excellent job of cleaning plastics for painting. Normally, water will neither corrode nor etch plastic. However, it was mentioned earlier that some plastic parts will absorb water during aqueous cleaning, and that this will lead to bubbles and blisters when the paint is heat cured. Highly polar plastics may also be susceptible to water absorption during storage and handling, especially in humid conditions. Absorbed water can be removed by drying the plastic resin pellets at a moderate heat for an extended period. It is better to dry the resin before molding to avoid part defects from moisture and then, if possible, to paint the parts promptly. In this way parts get painted before they can absorb much water. As an added benefit, if the freshly molded parts can be painted before they become contaminated, they won't even require cleaning.

Heat Sensitivity

Heat used for dry-off or baking may cause structural shrinkage, warping, twisting, or related distortion of plastic parts. The extent of the heat distortion will depend on temperature, part configuration, nature of the resin, and especially molding parameters. Placing the part in a fixture for rigidity, and then stress-relieving the part at only a slightly raised temperature, may sometimes correct minor distortion. This slow process would only be economically feasible for small numbers of parts, however. **Figure 19-3** lists various plastics and their heat tolerance temperatures, but this is for stress-free parts. If molding stresses are present, the distortion temperatures are much lower.

The thermal stability of many plastics is already quite low. For example, the maximum temperature that styrene parts should endure is only about 140° F (60° C). Acrylic and ABS parts can withstand no more than about 180° F (82° C). Not all plastics will distort at these low temperatures; polycarbonates and acetals do not deform below about 275° F (135° C). When plastic heat stability problems are encountered, instead of using a shorter, hotter bake, the applied paint can often be baked at a lower temperature for greater lengths of time and still achieve a full cure. However, longer bake times increase manufacturing costs because it requires either slower line speeds, or extra long ovens.

Porosity Formation from Sanding

Air or other gases are deliberately introduced into molten plastics to create structural foam parts. This will economize on the amount of plastic that is used in the molding. In many cases, depending on their size and shape of the gas pockets, they actually provide strength equivalent to or even greater than a plastic without a cellular structure. That fact seems totally amazing to many people.

Abrasion from operations such as cutting, trimming, or sanding to remove defects such as mold parting lines may open up the hollow cells in structural foam plastics. Porosity from sanding can later lead to blisters and popping problems during paint baking. Even if no cells are opened, the sanded area may be rougher than surrounding areas. This sometimes results in a dull-looking paint area that may require prepaint priming of that area. **Figure 19-4** lists various problems, probable causes, and remedies for finishing structural foam parts for business machines.

Lack of Electrical Conductivity

The minimal electrical conductivity of plastics precludes painting them by electrodeposition. Perhaps a heat stable conductive plastic could be ecoated, but to my knowledge

Heat Tolerance in °F / °C			
ABS	170 / 77	Polystyrene	130-145 / 55-63
Acetal	220 / 104	Polysulfone	350 / 177
Acrylic	180 / 82	EPDM	225 & up / 107 & up
Alkyd	300 / 149	PRIM	250 / 121
Cellulosic	220 / 105	PVC	170-225 / 77-107
Nylon	300-425 / 149-218	PVDC	170-210 / 77-99
Phenolic	225-500 / 107-260	RIM	250 / 121
Phenylene Oxide	180 & up / 83 & up	Structural Foams	140-250 / 60-121
Polycarbonate	250 / 121	TPO	230 & up / 110 & up
Polyester SMC	400 & up / 204 & up	TPR	250 / 121
Polyethylene	175-250 / 80-121	TPU	250 / 121
Polypropylene	250 / 121	Xenoy	210-225 / 99-107

Figure 19-3. Plastics and their heat tolerance temperatures.

Problem	Probable Cause	Remedy
Swirls	Compound or mold surface too hot.	Adhere to recommended melt temperatures. Control mold temperatures carefully. Use high-volume solids primer filler and texture coating.
Pinholes/Craters	Solvent attack with accompanying release of gases.	Use only mild solvent or waterborne primers.
Blisters	Absorption of solvent with subsequent release of gases.	Air-dry, prebake or store foam part at room temperature for 72 hours before coating.
Bubbles in Substrate	Inadequate filling of mold, Inadequate foaming of part.	Fill mold as fast as possible. Keep careful control of material-fill levels.
Poor Adhesion	Too much mold release.	Use only minimal amounts of mold-release agents.
Wicking	Mold too hot, material too Viscous, solvent too active.	Adhere strictly to recommended melt temperatures. Keep careful check of mold temperature through careful placement of gates. Use minimal amounts of mold release.

Figure 19-4. Chart for finishing structural foam.

it is not being done. Plastics in general are also more difficult to paint electrostatically than metals because of their low conductivity. Conductive precoats can be applied to enable plastics to be electrostatically coated, although this does occasion an extra step in the coating process. Conductive precoat is applied by various methods, including dipping and spraying. Fast-evaporating alcohol solutions of conductive organic salts have been used as a prep coat to plastics electrostatic conductivity, but the alcohol must be counted as VOC. Aqueous solutions of calcium chloride and lithium chloride are more likely to be used because they have no VOC components. Neither type of conductive coating interferes perceptibly with paint adhesion if the prep coat is properly applied.

Color Matching with Metal

Color matching between mated painted metal and painted plastic can at times be a vexing problem, especially if lower bake temperatures are used with plastic to avoid heat distortion. Unfortunately, differences between metal and plastic surface textures, heat capacities, and densities can cause an apparent color variation even when they painted as an assembly with the same paint and cure schedule. The different heat-up rates (heat capacities) of materials and different degrees of light reflection from their surfaces can produce apparent color mismatches that can be extremely difficult or impossible to fully overcome.

Gloss Variations Compared with Metal

Gloss variations between different plastics and between plastics and metals can occur, usually for the same reasons that cause color differences. By far the most frequent cause, however, is surface texture differences between metal and plastic.

Spot-to-Spot Gloss Variations on a Part

Spots that are dull are sometimes called *dive-in* or *sink*. They are the result of different plastic densities from one region to the next within a single plastic part. The density changes are caused by variations in the mold pressures and temperatures across the part during molding. Priming these parts before topcoating becomes necessary if a uniform gloss across the entire surface is required.

Wicking

Air entrapment or solvent absorption at fiberglass-to-plastic interfaces can occur with reinforced plastics. This *wicking* of the solvent deeper into the plastic results from the greater thermal expansion and contraction coefficients of plastic resins compared with the thermal coefficient of glass fibers. When the hot plastic in the mold cools, it contracts to a greater degree than the fiberglass. So the plastic may shrink away from the fiberglass, leaving micro-gaps that act as capillaries. When paint is applied, the paint solvent can be drawn by capillary action from the wet paint into the body of the part. Solvent travels inward rapidly along the open space at the fiberglass/plastic interface. Because the solvent has been pulled deeper into the part, it heats up more slowly than surface paint. The solvents may thus not evaporate until after the paint has started to cure. By the time the internal portion of the plastic part is finally warm enough to evaporate the "wicked-in" solvent, the paint film has already skinned-over to some degree. The solvent in the capillary is forced up under the paint skin, forming a "paint pop" or blister.

Sometimes low-molecular-weight polymer molecules in the plastic substrate will cause an identical problem when volatilized by oven heat. Molding resins free of volatile materials should be used if the parts are to be painted.

Plastic Paintability Information Resources

The paintability of various plastics can be sharply different. It is unlikely that a truly helpful chart could be prepared giving specific instructions for painting all types of plastics. It is not sufficient to specify that the plastic is a material such as phenolic, nylon, or acetal. Generic names are too imprecise. Even a trade-named plastic such as Noryl (General Electric's polyphenylene oxide) has over 100 different grades available. And unfortunately, some of the grades have different paintability than other grades. No plastic painting guide is given here because the chart must be so general it would have little value to painters. If you see such a chart, remember that it is only a general help and not an exact specification.

In each case, the specific plastic blend must be known to accurately predict paintability. Even then, some trial-and-error experimentation will be needed to determine which coatings are suitable for a particular plastic or a specific part. If you are not able to test on your own, paint companies are almost always willing to test their coatings if the plant will send them with a sample of the plastic part(s) to be painted. Be sure to take advantage of this service.

Plastic versus Metal Physical Properties

(General properties of these classes are compared. Individual material values may vary.)

Property	Metal Value	Plastic Value
Heat Stability	Very High	Low – Medium
Impact Resistance	High	Low – Medium
Thermal Expansion Coefficient	Low	High
Density (Water = 1.0)	1.6 – 9.0	0.8 – 2.4
Water Wettability	High	Poor – Fair
Solvent Resistance	Impervious	Poor – Good
Tensile Strength	High	Low
Softening or Deflection Point	High	Low – Medium

Chapter 20

Conveyors for Painting

The Need for Conveyors

Industrial painting is done by batch processing if the number of parts is modest. However, if the parts volume is large it is more appropriate to use a manual or powered conveyor to move parts through the various pretreatment and coating steps. In the typical batch-processing method, relatively small quantities of products to be painted are picked up and moved manually from station to station. If parts are small they are picked up and repositioned by hand. Heavy items require special lifting equipment such as a hoist, crane, or forklift. Picking up freshly painted parts by hand to get them out of the spray booth would often mar the soft paint. But waiting until parts are touch-dry ties up the booth for too long. Some plants solve this dilemma by using a monorail conveyor with manually moved parts. This works well for low volumes, especially if large, heavy items are painted. When large quantities of parts of any size are to be painted, the work is most likely to be moved automatically along the paint line by a power conveyor.

Types of Conveyors

The objective of all paint line conveyors is to move parts at an appropriate pace through the scheduled processes on the paint line. In addition to size and load bearing differences, conveyors are constructed according to the following general types, each of which has distinctive features that make it especially suited to certain kinds of parts:

- Flat belt, roller, or track conveyors
- Chain-on-edge conveyors
- Chain-in-floor conveyors
- Overhead chain and trolley conveyors
- Inverted conveyors.

Any of the conveyor types listed above may be constructed in variable width and length, and have varying load-supporting capacities suited to the overall dimensions and weight of products being handled. Conveyor design variances are common. The overhead types, for example, may support the conveyor chain or cable with the trolleys riding an open I-bar, or with trolley wheels held within an enclosed U-shaped track. The enclosed U-track may be installed with the opening up, down, or to one side. Up- and side-openings with enclosed style tracks help prevent some of the dirt from falling down on parts. Additional conveyor classifications are made, based on special features.

Flat Belt, Roller, or Track Conveyors

A motor-driven flat continuous horizontal conveyor will carry parts set on it, allowing various finishing steps to be performed as they travel. The flat continuous belt conveyor is probably the earliest general type of paint line conveyor, and includes many different style belt constructions such as polyurethane, wire mesh, dual chain with cradles, cross slat, pin, perlon filament, and flatbed. The main advantage of all flat belt type conveyors is their simplicity. Roller conveyors, either motor-driven or manual, also carry parts horizontally and so are variants of this conveyor style. Powered flat conveyors are especially chosen for finishing single sides of coiled paper and metal, and for painting flat stock such as partitions, shelving, and wood paneling. Spray, roll coating, or curtain coating are the application methods most often used to paint horizontally transported parts.

Of course, the parts need not be finished while they are actually resting on the conveyor. They can be removed (by hand or with lifting equipment) from the conveyor, taken into an adjacent paint application booth, and then returned to the conveyor. Although some plants do this, in most cases handling freshly coated parts is either inconvenient or impossible to do without marring the paint.

Chain-on-edge Conveyors

Chain-on-edge conveyors consist of a light-duty conveyor chain equipped with vertically mounted spindles to carry parts to be painted, as shown in **Figure 20-1**. These conveyors are used in disc, bell, and spray-painting booths and then to transport the coated parts through the flash zone and cure oven. Unless extreme design modifications are made, chain-on-edge conveyors are not at all suitable for carrying parts through aqueous cleaning and conversion coating operations. The cleaning solutions and acidic conversion coating solutions strip lubricants from the chain and leave salt deposits that foul the chain. Protecting the chain with flexible "boots" has been tried, but boots usually wear rapidly and become ineffective. A few times I've seen the chain held inside a down-facing U-shaped support with the part fixtures curved around and up. This prevents solution

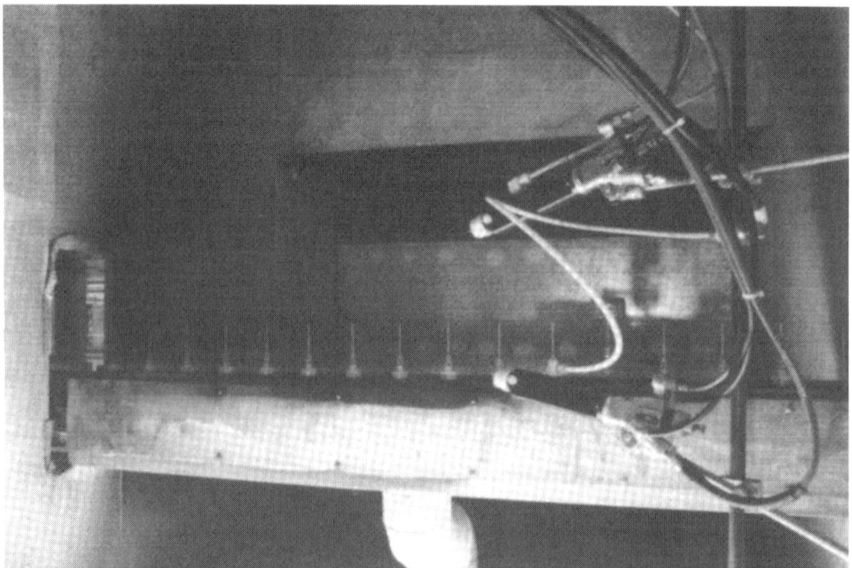

Figure 20-1. Chain-on-edge conveyor.

from contacting the chain, but fixtures in this curved configuration tend to wobble rather severely.

Chain-on-edge conveyors can be arranged in either a horizontal circuit configuration or in an over-and-under (vertical loop) arrangement. The over-and-under configuration is similar to a belt conveyor with spindles. Both styles are employed regularly for coating parts. In the horizontally arranged loop, the spindles always remain in the upright position, so parts can rest on the spindles at all times. In the vertical over-and-under configuration, half of the time the spindles are inverted as they travel the lower section of the vertical loop. While this does enable an automatic unload feature, parts obviously can only be held on the conveyor half of the time.

A chain-on-edge conveyor is not suitable for heavy objects; so, of necessity, parts to be coated tend to be lightweight and small to modest in size. The loaded spindles in most instances are rotated while they pass in front of the application device, so parts should be somewhat circular or squareish with consistent rectangular or ovate cross-section. This enables all areas to be painted uniformly as parts are spun while they are being coated in the paint booth. Each spindle is equipped with a sprocket (pinion) that engages a rack near the spray guns, which forces the spindle to rotate. The main chain-on-edge advantages are quick load and unload, design simplicity, and high efficiency. Their chief limitation is that they are useful primarily for the painting and baking operations. They are not very suitable, as we have seen, for use during cleaning or pretreatment operations.

Chain-in-floor Conveyors

A chain-in-floor conveyor consists of a driven chain embedded in a channel slightly above or just below the level of the plant floor. The latter design is preferred so that wheeled vehicles such as forklift trucks can easily cross the conveyor line. Either a horizontal or over-and-under style is easily workable. The chain-in-floor conveyor's function is to pull large heavy wheeled products (or skids and trolleys bearing the weighty parts) through coating stages. The items are manually hooked on and off of the conveyor chain as needed to carry parts to successive finishing steps. Typically the chain is large-gauge, designed to handle heavy loads, although any gauge can be used if lighter items are to be moved. This type of conveyor is most often used to pull heavy tracked or wheeled agricultural machinery, earth moving equipment, and large pieces of military equipment through cleaning, pretreatment, sanding, spray-painting stations, etc. In addition to its strength, the conveyor's main advantage is simplicity. Its basic disadvantage, if it can truly be called that, is its restriction to use with wheeled equipment or items that can be placed on special wheeled carriers.

Overhead Chain and Trolley Conveyors

The overhead power conveyor has a motor-driven chain or steel cable to which trolley wheels and hooks are attached at regular intervals. The chain or cable is supported by the trolleys that ride on a continuous supporting track—normally an I-beam or a "C" or "U" style channel. **Figure 20-2** shows examples of the most common designs. The overhead chain-and-trolley power conveyor is the workhorse of the finishing industry. It can carry large volumes of parts through all finishing steps: sanding, pretreating, drying, painting,

Figure 20-2. Three examples of overhead conveyors.

and baking. An overhead conveyor can also be manually powered, a desirable low cost alternative to a motor-driven conveyor if only comparatively low numbers of bulky or heavy parts need to be moved through the finishing stages.

The overhead trolley support track may be open or enclosed. The enclosed type has a box enclosure around it, except for a slot opening that runs its length to permit hanger passage. Again, the opening may be on the bottom, top, or sides. About the only disadvantage of an overhead conveyor is that it tends to collect dust and dirt, which can fall onto parts being finished. Less debris can fall down onto painted parts with the slot located on the side or top of a "U" channel. Dirt-catching continuous trays (called sanitary trays) can be installed under the conveyor track to serve the same purpose. On the chain, C-shaped hangers that curve around the sanitary tray are used to position the hook of the hanger under the sanitary tray. **Figure 20-4** shows car bodies being moved through a spray booth by an overhead conveyor equipped with a sanitary pan and coverings to prevent conveyor dirt from falling onto the painted bodies. Note the "C" portion of the hanger.

Overhead conveyors are driven by electric motors with gearboxes and sprockets. A take-up station allows tightening the slack from time to time as the cable or chain stretches with use and the components wear. Conveyors are usually equipped with automatic oilers that periodically spray lubricant on the trolley wheel bearings. (Over the years I've seen many of the oilers either spitting too much oil or not delivering any oil whatsoever. Why this should be I'm not sure. Perhaps dust and fibers accumulate in the constantly oily tip and foul the delivery tube orifice.)

Many conveyors have an automatic brush-type cleaning device that runs continuously to keep the chain and trolleys clean and free of oversprayed paint. Passive brushes are also utilized but they quickly wear down and lose their effectiveness unless they are repositioned frequently. It is vitally important in electrostatic painting to have a clean I-beam surface (on which the trolley wheels ride) in the paint booth, and clean trolley wheels to complete the electrical circuit to ground. Without continuous cleaning, the I-beam riding surface and wheels tend to collect dust and dirt, and consequently lose a portion of their electrical grounding ability.

Overhead conveyors should be protected from solutions and coatings materials in spray washers and spray booths, and from excessive oven heat. A conveyor running unprotected inside a spray washer will be saturated continuously with cleaner, rinse water, phosphate solution, and sealer rinse, all of which tend to wash away lubricant and bring quick failure to wheel bearings. With improved designs the overhead conveyor runs above a washer. A continuous slot in the roof of the washer allows hangers to pass. Plastic brushes close off the gap to minimize solution and moisture escape onto the chain and trolleys, as shown in **Figure 20-3**. An unprotected conveyor inside a spray booth can collect an inordinate amount of overspray. This not only fouls the trolleys and chain, it reduces and eventually eliminates electrical grounding. In the oven, high heat can drive out chain lubricating oil or grease components. Heat may also, over time, cause some lubricants to become gummy.

One type of conveyor protection—when the chain is run inside the washer or spray booth—consists of a simple slotted metal or plastic shroud around the chain. A steady flow of air pumped into the shroud maintains a positive air pressure, preventing pretreat solutions or paint overspray from reaching the conveyor chain. However, despite its simplicity, I've rarely seen this style protection actually being used on painting lines. Shield panels are far more common.

Most conveyors can be operated at a range of speeds. A line speed of 1 ft/min (0.305 m/min) is slow; while 40 ft/min (12.2 m/min) is fast. A good estimate is that 90% of power conveyor paint lines operate in a range of 5–15 ft/min (1.5–4.5 m/min). The top running speed is necessarily limited by the slowest operation along the paint line. For example, if a part must

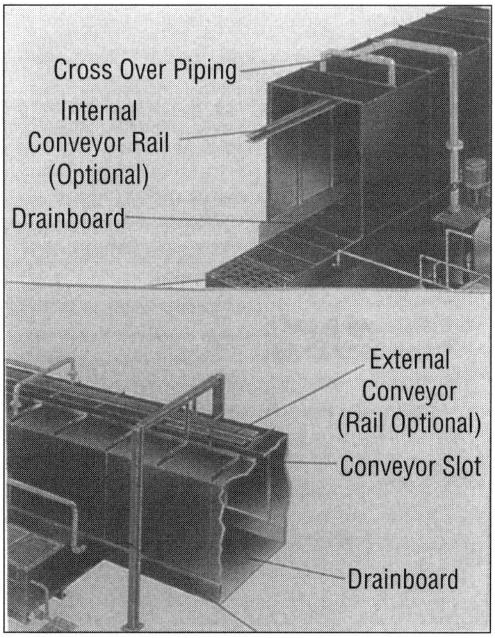

Figure 20-3. Nylon brush seals in an external overhead conveyor keep cleaners and conversion coatings away from trolley wheels and chain lubricants.

Figure 20-4. Car bodies moving through spray booth using covered overhead conveyor.

be in a 24 foot (7.3 m) spray booth for 3 minutes to allow it to be completely painted, the top conveyor speed would be 24 feet divided by 3 minutes, or 8 ft/min (2.44 m/min).

Overhead power conveyors can be simply power conveyors or power-and-free conveyors. The power-and-free feature allows one or more trolleys (with or without parts on their hangers) to be stopped at any point on the line, even though the conveyor chain continues to move. Two vertically paired conveyor tracks are used: a free lower track supporting the parts trolleys, plus a power track directly above it in which a chain with regularly spaced "catch arms" moves continuously. The catch arms snag "dogs" at the end of each trolley and push the trolley along the lower conveyor track. The dogs are fixtures that are made to extend and catch the moving chain catch arms, and thus enable the chain to move the trolleys along. The dogs are made to retract or "undog" when their path is blocked. If the trolley path is not clear, it undogs and cannot be pushed along, but instead the trolley stays where it is until released. Power-and-free conveyors allow individual hooks to be disengaged (undogged) from the conveyor line for as long as one wishes. Not all parts on the conveyor line are forced to move at the same speed as on a regular power conveyor. This allows for accumulating parts, such as in an oven for longer dwell times, or for routing any parts onto separate sidetracks. Parts requiring different processing, or varying processing times, are easily segregated this way. With a power conveyor, stopping one part automatically makes all parts stop moving; but with power-and-free much greater flexibility is allowed. It becomes possible to stop one part and yet allow the other parts to keep moving. This convenience is not inexpensive. A power-and-free conveyor will cost roughly four or five times more than an equivalent power conveyor.

Power-and-free conveyor systems can also be linked with transfer stations that allow one conveyor loop to pass some parts or all parts to another conveyor loop. The conveyor can also be designed with elevators to lower and raise parts for various processes such as pretreatment baths, electrocoating solutions, and batch-type ovens. So it can be seen that although power-and-free conveyors cost much more than straight power conveyors, they may quickly pay for themselves in handling ease, flexibility, and convenience.

Inverted Conveyors

Inverted conveyors usually run either at floor level or a few feet above floor level, and are equipped with regularly spaced parts mounting fixtures. They are in some ways "flipped-over" versions of overhead conveyors, and in other ways they resemble chain-on-edge conveyors. They are often used to transport parts for which dirt-free finishes are critical, especially those that will be topcoated and then clearcoated. Inverted type conveyance is used for clearcoated items such as automobile bodies and exterior rearview mirrors, or outboard motor covers to move them through spray booths and ovens. A major advantage of this type of conveyor is that the conveyor mechanism—which tends to collect and distribute dirt—is below the part being finished, thus dirt cannot fall onto parts. But a big drawback with inverted conveyors is that they restrict access inside the plant. Personnel can usually walk under an overhead conveyor line, but an inverted conveyor mounted above the floor is difficult to cross. Unless a bridge is constructed, personnel may be forced to walk around rather than over the inverted conveyor line. Inverted conveyor loops are used alone and in concert with overhead conveyor loops. Both power and power-and-free styles can be utilized for inverted conveyor systems.

Figure 20-5. This tow conveyor provides the flexibility needed for a variety of finishing tasks.

The Importance of Good Conveyor Design

Modern finishing systems feature advanced automation, which is made possible by innovative conveyor design and computerized control. Each paint line conveyor system is custom-designed for a particular product or blend of products. Demands on conveyor systems can be diverse and complex. The conveyor system may be required to move products of widely different shapes and sizes through the same finishing stages. A plant might require the accumulation of temporary banks of products at various locations. In addition, the conveyor needs to be at the proper height above the floor for convenient part loading and unloading. If loading or unloading cannot be done expeditiously, the line will suffer from lower production capacities—a grave business error—but one that this author has seen surprisingly often.

Considerable thought and planning should be devoted to new conveyor system selection and line layout. The impact it has on manufacturing is felt during every minute of operation, and changes are sometimes difficult and costly once it is in place. Poor conveyor design, inadequate conveyor cleaning, and improper lubrication can virtually guarantee a short trouble-plagued conveyor life, plus repeated major maintenance problems. Fortunately, with the proper conveyor design augmented by continuous brush cleaners, correctly operating automatic lubrication systems, and regular maintenance, conveyors will operate for many years with few problems beyond normal maintenance requirements.

Chapter 21

Finishing Robots

Programming Industrial Robots

Painting robots have been mounted to the floor, wall, or ceiling of paint booths, and are built and programmed to move about various axes to perform specified functions. Most robots have an arm that mechanically resembles a human arm with a wrist, and can perform hand-like functions such as aiming or gripping. Robots use a control center and a teachable electronic memory so that they can be instructed to perform any selection of preprogrammed instructions. Almost any degree of operation complexity, including machine vision, can be incorporated into a robot. But obviously greater sophistication costs more, and cost/performance criteria must be weighed.

The arm in a painting robot can be designed to have movement along three axes that can be thought of as the familiar "*x*-*y*-*z*" axes in mathematics:

Translation—an axis along a horizontal base

Elevation—a vertical axis perpendicular to the base

Reach—a horizontal axis perpendicular both to the base and to the vertical axis.

The wrist part of the arm may have movement along (or about) three other axes:

Yaw—angular right and left

Pitch—angular up and down

Roll—angular around.

Three additional axes of motion can be achieved by moving the entire robot as follows:

Parallel to conveyor travel (to track a moving target)

Perpendicular to conveyor travel (to move toward and away from the conveyor)

Up and down vertically.

A robot with full arm, wrist, and base motions would therefore have nine axes of motion.

The electronics to control each of these robot motions can be programmed *point-to-point* or *continuous-path*. Point-to-point programming moves a robot in a series of straight lines that connect a series of points relatively widely separated in space; this requires only a limited amount of computer memory. Point-to-point programming is intended for regularly shaped parts and not for programming intricate movement or extensive control. Continuous-path programming is in a way similar to point-to-point, but with a major difference in the number of points identified. In continuous-path programming, the movements are between points that are very close together. Constant feedback allows the motion to follow the series of points to simulate moving along a continuous irregular curve. Continuous-path programming requires extensive memory and computing capacity. This type of programming is useful where intricate contoured paths need to be followed, such as in spraying parts with complex shapes.

Robotic motion is usually achieved with servo controls for each axis. A *servo* is an electrical/mechanical device that provides an output motion according to a given input signal. When equipped with electrical feedback circuits, a servo's output motion can be continuously monitored and adjusted (corrected).

Robots are powered either directly through an electric motor or indirectly through a hydraulic or pneumatic motor. Characteristics of these types of robots include the following.

Electric robots tend to be highly accurate and carry light-to-medium payloads.

Hydraulic robots are characterized by an ability to "give" when under stress and typically carry heavy-to-medium payloads.

Pneumatic robots are often used in fast-velocity applications with light payloads.

Industrial robots are categorized according to their design function. The categories include assembly, material handling, welding, and spray painting. Assembly robots are programmed to perform one or more tasks on an assembly line as their names imply. Material-handling robots are programmed to move various products from one location to another. Welding robots are programmed for use in automatic welding applications. Painting robots (see **Figure 21-1**) will apply wet or powder coatings. They are almost always programmed to manipulate a spray gun or rotary bell through a series of programmed motions. Applying paint by robots fitted with a rotary bell is being done on an increasing basis but is not as common as robotic paint spraying. It is possible to program a robot to coat products either while the target is stationary, or while parts are moving along a conveyor line. For practical reasons, however, most coating is done on stationary parts

Figure 21-1. Parts rotating on a table are sprayed by a stationary robotic arm.

because it is easier (and therefore less expensive) to coat a part when it is not moving.

The four categories of robots are very similar in appearance and design. With minor modification, most robots could operate in any category. About 35% of industrial robots are used for assembly, 35% for material handling, 20% for welding, and 10% for spray painting. Material-handling robots vary in size according to the weight of the product being handled. A material-handling robot in a laboratory would likely be bench-mounted and be very small, perhaps with an arm about 1 foot (0.305 m) long. A material-handling robot operating in a foundry to move heavy castings would be very large and be designed with a large, sturdy arm capable of lifting the heavy parts. Gantry robots are capable of lifting heavy items also. A robot mounted on a gantry rather than on a very lengthy arm will better handle long reaches such as those involved in painting airplanes.

Rotary Bell and Spray Painting Robots

Robots can be used for painting operations that are uncomfortable or of marginal safety. For example, robots are being used with absolute safety to apply many two-component polyurethane paints that have a toxic isocyanate component. Robots are suitable to apply dirty, malodorous materials and to work in hot, humid, or other unpleasant environments. They can also perform physically demanding contortions and mechanical manipulations that would prove unendurable to a human operator if done for any length of time. Long reaches that require an extension or pole-gun for a manual sprayer can be accomplished easily with a robot.

Electrical robots, like anything that operates in paint spray booths, must not generate sparks or, if they do, the sparks must be enclosed in a fail-safe system to prevent the possibility of the spark igniting flammable materials such as solvents. Early model painting robots were hydraulic-powered and spark-free, averting the possibility of igniting a flammable atmosphere. However, more accurate electric-powered spray painting robots were put into service after a fail-safe system was devised to prevent the possibility of sparks, especially in the electric motors. All potential spark-generating equipment is housed in an enclosed compartment that is filled with a nonflammable gas at a positive pressure, which prevents entry of flammable solvent fumes. The fail-safe feature is that if the compartment should lose its positive pressure, the electric power goes off.

Electric spray painting robots became popular because they eliminated messy hydraulic fluid leaks and because the electric robot is a modular package. A hydraulic robot needs a separate source of hydraulic power. Also, a hydraulic-powered robot's ability to handle heavy payloads is not particularly useful in painting, where a spray gun payload may weigh only several pounds and a rotary paint or powder bell less than 20–25 lb (9.0–11.3 kg). However, the increased accuracy capability with electric-powered robots isn't much of an advantage in most painting operations, which frequently don't require critical accuracy. Where high precision and accuracy are needed, however, only an electric robot can deliver the performance desired.

The number of axes required in a painting robot depends on the product being painted. A simple nonconveyorized product such as a flat panel could perhaps be painted with a robot having only three axes. Painting a highly complex shape such as a moving automobile body could require a robot with all nine axes. The robot package itself might possibly get by with just six motion axes if ancillary equipment were available. A platform axis to track the moving car body would be the seventh axis; while the ability to position the entire robot by moving it closer and away from the car body would be the eighth axis. Finally, the platform would need the capability to raise and lower the entire robot and this would be the ninth axis. **Figures 21-2** and **21-3** show relatively simple multiaxis robots in use.

Robots are most often used with liquid paint spray guns or liquid paint bells, but robots have been shown to be able to successfully paint with powder guns or bells as well. Few if any plants have large-scale employment of robots that are spraying with a powder bell. There is no reason why powder bells cannot be used this way, so this situation may change.

Programming Robots for Painting

Robots applying powder or liquid paint with guns or bells are usually programmed in combinations of point-to-point and continuous-path, depending on the shape and size complexity of the product being painted. The usual way is to program the robots *off-line* (see **Figure 21-4**). In this technique, an operator programs the robot in a special station equipped with a computer and the appropriate software.

Off-line programming can be used in combination with conventional in-booth teaching systems where an operator programs the robot. The robot is taught new programs by

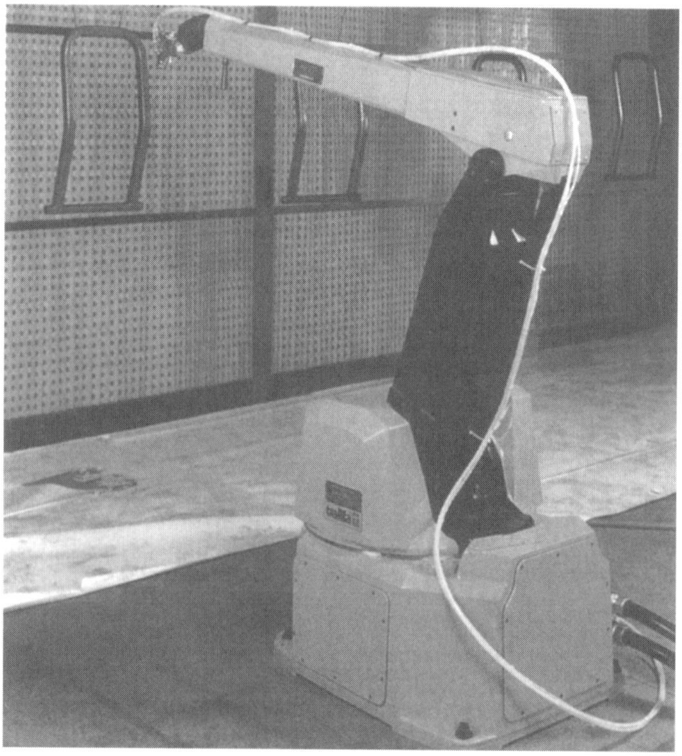

Figure 21-2. Robot sprayer has auxiliary arm for painting interiors.

manipulating the teaching pendant controls through the desired paint applications motions, as shown in **Figures 21-5** and **21-6**.

All robotic painting programs need to be synchronized with conveyor loading and conveyor speeds. The programs also need to include start-stop functions and color-change operations. Because a robot does not "see" the paint line, parts need to be hung on a conveyor the same way every time. No variation in the alignment of the hanging parts is permitted. Crooked paint hooks and hangers, or parts that sway or swing as they move past the robot, may momentarily shift outside the programmed robot path. This would result in poorly painted products and possibly leave some areas with no coating. Two-point parts hanging is often done when painting by robots to reduce the likelihood of sway or misalignment of parts from periodic jerky movements of the conveyor line. This is shown in **Figure 21-7**.

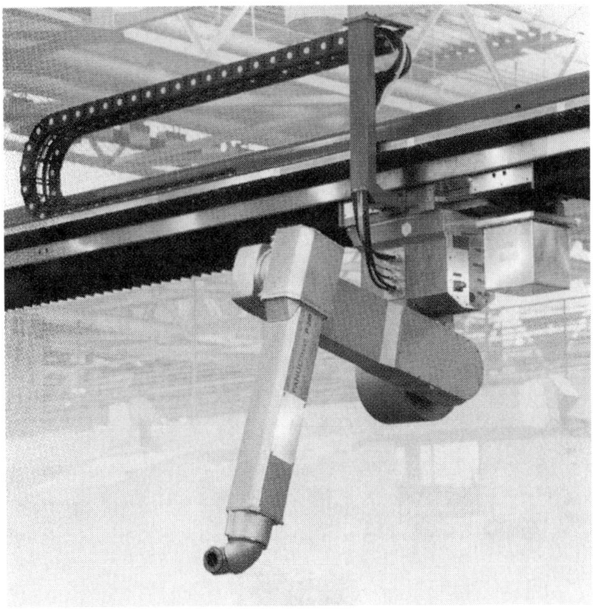

Figure 21-3. An overhead, rail mounted, six-axis paint robot that allows large parts to be coated with a single robot.

Machine vision could be used with painting robots to solve this problem. Futuristic machine vision systems for painting aim a video camera at a specific area on a product to be coated. If the camera position-detection system observes that the part is positioned slightly askew, it can alert the robot control center, which can shift the robot's program to compensate for the off-positioned part. This type of system is already being used in automotive sealer applications where a 3/8-inch-wide line of sealant is applied by robots to cover corrosion-prone seam welds. However, current painting machine vision applications tend to be relatively simple in scope, being limited to shape recognition and location misalignment. Full-bloom vision systems that could replicate human eyesight and brain function are still a long way down the road. In the meantime, perhaps approximately equal results will be obtained with more sophisticated photocell detector arrays. Simple photocell setups have been triggering spray guns (and bells, etc.) on and off for years.

Requirements for Painting Robot Installation and Operation

Before installing a painting robot, a great deal of planning must be done. The space requirements for a painting robot tend to be more demanding than for a human. Most painting robots are larger than humans and have a much bigger work envelope. To avoid crashes, the robot's physical reach must not exceed the constraints of the booth. The robot will have to be housed in a booth to capture overspray and exhaust the VOC vapors. A spray painting robot also requires considerable peripheral equipment. An off-line teaching booth in close proximity to the paint line is useful for the preparation of new paint application programs. Having an off-line booth will eliminate the need to interrupt production runs to perform robot-teaching operations.

The weight and size of the robot is considerable. An overhead lift will be needed to position the robot for installation, and as required to relocate the robot or move it for repairs. Good access must be provided for servicing electrical and mechanical components.

For ease in maintaining a painting robot, the equipment should be designed to include:

1. A computer with self-diagnostics
2. Interface electronics with only one printed circuit board per axis of motion
3. Precalibrated board backups for each drive axis
4. Easy calibration procedures.

To protect people working nearby from being struck by the robot, restraining guardrails, gates, fences, or other protective devices must be installed. A number of other important safety measures are needed to protect people and equipment in the area.

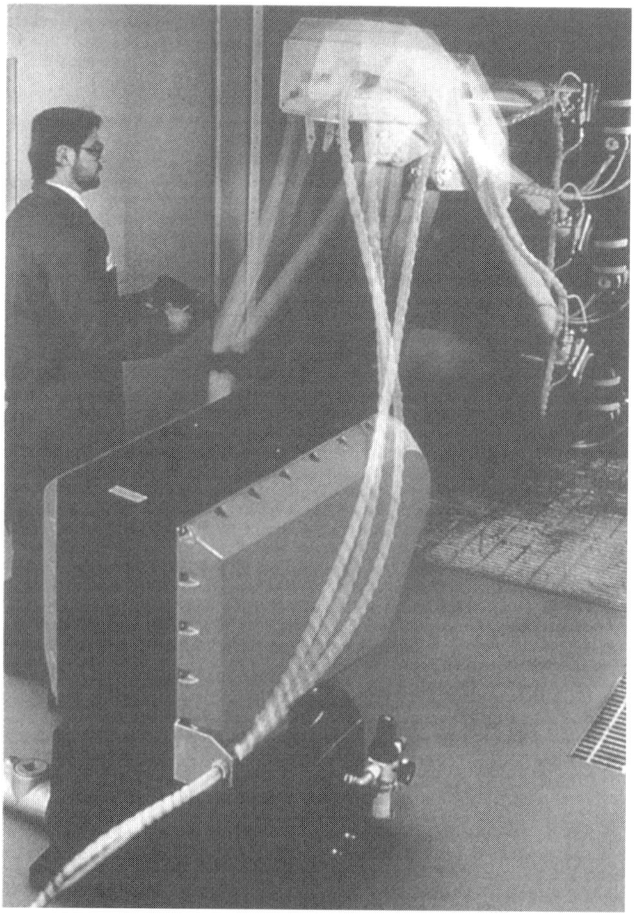

Figure 21-4. Robot being programmed off-line.

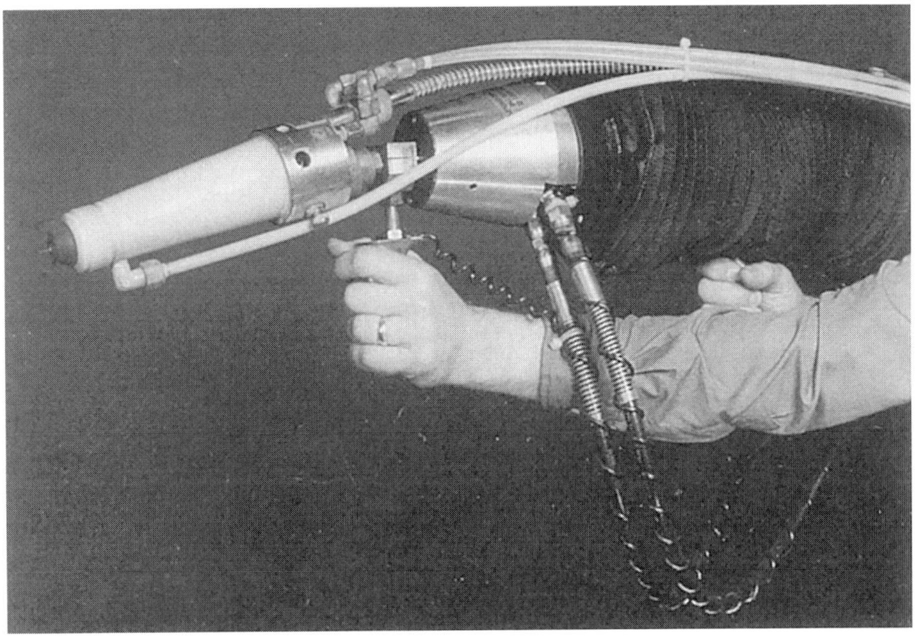

Figure 21-5. Operator teaching robot desired spray motions.

Even with optical devices that inactivate the robot if a person or an object approaches too closely, auxiliary protection is required in case of malfunction. Runaway robot protection is needed so that an out-of-control robot does not cause injury. A warning bell or light can alert employees to this condition. Alarms for runaway robot conditions should be loud enough to be easily heard above plant noise. Fail-safe brakes are necessary on robot arms to prevent wild movement in the event of a failure. Emergency shut-off switches must be located in a highly visible spot to which there is quick and easy access.

Although hydraulic paint robots are now rare, it should be said that pressure-relief valves are needed in such systems to prevent overload of the pump and motor. A velocity "fuse" in each hydraulic supply line can prevent high flow rates. Software limits to ensure safe stoppage of the robot arm in all major axes should be backed up with electrical limit switches and mechanical shock absorbers. Robot arms can be nylon-covered to preclude any sparking in the event of a collision with a metallic object.

Perhaps most importantly, a bell or spray painting robot's cost will vary considerably, depending on the size and performance capability desired. In U.S. dollars, a low-requirement unit may cost in the mid "five figures." A top-performance painting robot will cost in the mid "six figures." There are too many variables to give more precise dollar figures.

Added costs include robot installation and training for personnel to learn how to operate and maintain the device properly. In addition, all paint shop personnel must be trained in general safety precautions associated with the robot. The average robot requires an annual maintenance expenditure of roughly US$5,000–$10,000, including end-of-shift cleanup and periodic preventive maintenance.

A preliminary study should be undertaken to determine the feasibility of having a robot for paint application. Payback time is the critical economic point to consider. If the payback time meets the company's financial guidelines, authority to purchase the robot can be sought. A formula to determine a robot payback can be expressed as follows:

$Y = Rc \div (Ws + Ps + Rd - [Mc + Sc])$

where Y = years payback, Rc = total robot purchase and installation costs, Ws = replaced worker salary for a year, Ps = production savings for a year (paint savings, energy savings, etc.), Rd = robot depreciation for a year, Mc = maintenance costs for a year, and Sc = staffing costs for robot for a year.

In a hypothetical case, suppose that:

Rc = \$500,000; Ws = \$30,000; Ps = \$25,000; Rd = \$85,000; Mc = \$5,000; Sc = \$60,000.

Then, Y = \$500,000 ÷ (\$30,000 + \$25,000 + \$85,000 – [\$5,000 + \$60,000])

Y = \$500,000 ÷ \$75,000 = 6 years and 8 months payback time.

Spray Painting Robot Advantages

The advantages of spray painting robots can be categorized into two areas: cost reductions and quality improvements. The most obvious cost reduction involves labor. A typical robot installation will displace one to three painters due to its speed capability. If the robot is used on more than one shift a day, the savings are multiplied. The robot arm can simply move faster than a human arm—and without tiring.

Figure 21-6. Robot being programmed for two-part urethane application by operator wearing special uniform and air-supply headgear.

Another cost reduction that can be substantial is the savings from the reduced amount of paint used when a properly programmed robot applies it. The replacement of an average semiautomatic application system by a robot will reduce paint consumption by approximately 15–20%. This savings accrues from precisely accurate repetitive painting motions, finely tuned gun and bell triggering, and multiaxis movement that can closely follow part contours, even those with highly complex geometry. The actual material savings, it is realized, will depend on the volume of production, the shapes of the parts being coated, and the type of paint applied. A fringe benefit associated with reduced paint usage from robot painting is that lower paint use translates directly into reduced VOC and HAP emissions. Paint that is not applied, naturally, will not emit VOC. Since lowered emissions are being required almost universally, any finishing system that reduces HAP and/or VOC emission totals will help a company obtain (and retain) their operating permits from jurisdictional regulatory agencies.

Another important saving is energy savings. One of the most significant can be the energy required to heat booth makeup air. Because no human operators need be present in a robot-equipped paint booth, OSHA regulations allow the amount of air exhausted from the booth to be reduced. This can result in enormous energy savings, especially in cold climates during winter months when frigid outside makeup air needs to be heated to around 75° F (24° C). Ventilation cost is reduced nearly 50% for robotic spraying compared with manual operations.

As for quality, another type of saving results since robots provide absolute part-to-part uniformity of coating. It is not unusual for plants that have switched to robot spraying to report a 75% reduction in paint rejects. Robotic application is so consistent that it improves quality uniformity; simultaneously it reduces (and in some cases practically eliminates) paint rework. Fewer quality variances will cut down sharply on field repainting due to paint problems. Depending on where the product is shipped, it has to be determined whether it is less expensive to repair the paint at the site or return it to the plant. Often it's cheaper to freight it back to the plant, repaint it, and then ship it back to the customer—cheaper perhaps, but it's still very expensive. Any necessary recalls damage the manufacturing company's image; it may also make the customer wonder if they should buy from another source.

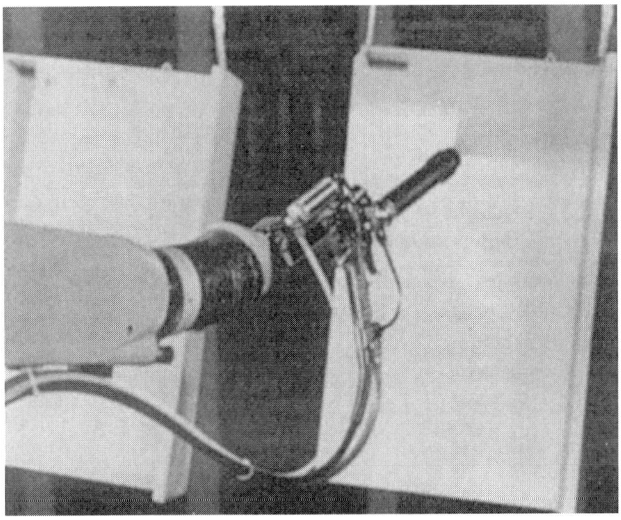

Figure 21-7. Robot spraying desk parts utilizing double hanging hooks to minimize swaying.

Case History

The following actual case history describes how one plant saved an appreciable amount of money by switching to spray painting robots. The plant was manually spraying plastic auto parts on two separate conveyor lines. To meet production, it had been necessary to paint reworked parts on overtime. One line was altered so that four robots and four stationary electrostatic guns did all the painting. The second line was left unchanged and used just for spraying repair parts; it operated on straight time only. The switch to robots on that single line allowed eight sprayers per shift (a total of 16 for both shifts) to be reassigned. Three gun technicians were added. The net reduction of 13 positions resulted in an annual saving of over $500,000. The savings of one hour of overtime per day per operator for each shift gave additional yearly savings of over $330,000. Although no exact dollar value can be assigned to quality and VOC/HAP reductions, both VOC and HAP emissions were reduced. Paint finish quality was improved noticeably by robotic application. The total implementation cost was high—slightly over $1 million—but the annual saving was close to $1.25 million. Certainly this is an unusually rapid payback, and the wages for a robot technician/mechanic were not factored into this number. Nevertheless, payback periods of less than four years are common.

Future Developments

Spray painting robotic research and development through the years up to 2015 will likely focus on the following areas, but gains in all except the last two are going to be small:

- Sensing devices
- Painting speed
- Precision of operation
- Repeatability
- Minicomputer controls
- Voice command recognition
- Artificial intelligence.

Improved sensing devices could increase the efficiency of color changes, improve product identification, detect and correct misaligned parts, and identify/select parts from mixed products. Precise machine vision is available now but costs are prohibitively high. Increased painting speed could reduce robot costs and thus lower payback time, but overly rapid gun motion with both wet and powder paints would distort spray patterns. Improved precision of operation and repeatability would increase quality. Minicomputer controls, voice command recognition, and artificial intelligence are all steps toward giving the spray painting robot some of the capability of human brain functions.

Final thoughts: WHEN TO USE ROBOTS

Robotic painting may be appropriate if these criteria are met:

- Coating cost is high.
- Overspray disposal is costly.
- Paint repair is difficult or costly.
- Labor cost is high.
- Coating appearance is critical.
- Coating components are toxic.
- Parts are complex or large.
- EPA or OSHA compliance is difficult.
- Skilled painters are hard to locate.
- Production volume is high.

Chapter 22

Spray Booths for Liquid Painting

Spray Booth Basics

Spray booths for applying powder finishes are described in an earlier chapter; this chapter will deal with spray booths for liquid paint application. While similar in concept and in function to an enclosure for applying powder coatings, spray booths for liquid painting are ordinarily not equipped to capture coating material for reuse. Incidentally, they are all called spray booths, even though paint may be applied by other means, such as by rotary atomization.

A spray booth for liquid painting functions as an enclosure equipped with a means of safely moving overspray paint away from the part and the sprayer, capturing overspray paint, diluting and exhausting solvent vapors, and replacing the exhausted air with (clean, it is hoped) makeup air. In a fully enclosed booth, makeup air may be supplied to the booth by fans and ducts; this air is filtered before it is blown into the booth. The enclosed booths are a bit more costly to install and operate; but they are able to provide a cleaner atmosphere for painting, especially when plant air has dirt, dust, pollen, welding fumes, etc. In a semienclosed booth, ambient plant air is allowed to replace the air that is exhausted out of booth. Some plants have booths in which plant air replaces exhausted air, but the back wall of the open booth is made up of two large doors fitted with air filters. The doors have several cutout openings into which the replaceable air filters are placed, filtering dirty plant air somewhat before it enters the spray booth. The air filters in the doors hold out most of the larger contaminant particles that otherwise might be drawn into the spray booth. Fine particles such as welding smoke and oil mist will pass right through these filters, however, so this style booth is not as clean as one that uses a separate makeup air supply.

Most spray booths have openings on both sidewalls to allow entry and exit of the conveyed products to be painted. For low production (batch spraying), a spray booth without such openings can be used. Access for product entry and exit is through the opening at the back of the spray booth.

The need for forced air makeup and forced air exhaust in an enclosed spray booth presents a potential air balance problem. The air passing through the booth must first of all lower the concentrations of paint solids and solvents enough to satisfy fire insurance underwriter regulations. OSHA dictates higher air throughput requirements when people are present in the booth than if it is unoccupied, as might be the case when automatic application is used. Air makeup must accomplish two things besides dilution of solvent vapors: 1) it must replace the exhausted air and 2) it should supply enough additional air to maintain a slight positive pressure inside the booth. A positive booth pressure will prevent entrance of plant air that might contain oil mist, dust, lint, or other particulate matter. The extra amount of air makeup (the positive booth pressure) should not be too large, or excessive amounts of overspray will be forced out the conveyor openings and into the plant, instead of out through the exhaust system.

Makeup air is drawn into enclosed booths from outside the plant to avoid disturbing the air balance in the rest of the plant. If significant volumes of air were taken from inside the plant for booth makeup, a negative pressure would be created inside the plant. Plants normally prefer to maintain a slight positive pressure to avoid drawing unfiltered and unheated air into the plant. In northern climates, the air makeup is not only filtered but also heated and sometimes humidified during cold weather. During summer temperature extremes, the makeup air in some plant locations may be air-conditioned or passed through water spray humidifiers to cool the air, even though this also increases the relative humidity. High humidity levels are not much of a problem for most solventborne paints; but excessively high temperatures are. Adjusting the temperature of booth makeup air downwards by mechanical refrigeration requires a great deal of energy and so is rarely done except by water spray humidification.

The amount of water vapor in the air has a significant effect on the rate at which paint solvents (and water) evaporate from wet paint films. If the air is too dry, the solvents evaporate too rapidly and as a result the finish is less smooth and lower in gloss. In ad-

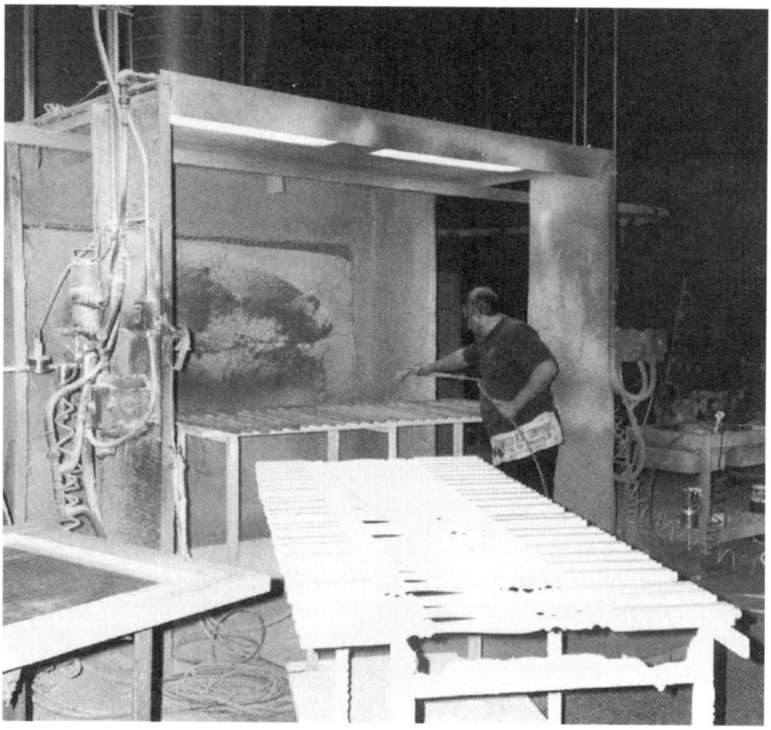

Figure 22-1. An open spray booth for coating nonconveyorized parts. This operation, which has limitations but is very low in cost, is typical of thousands of paint lines worldwide.

dition to using water to cool makeup air, sometimes moisture is introduced into dry makeup air to prevent excessively fast evaporation of volatiles from the applied paint film. This extra moisture may be especially needed in frigid weather because the extremely cold outside air, drawn in and heated for use as makeup air in the spray booth, is able to hold just a tiny amount of moisture. Automotive assembly plants in the northern tier of states, for example, routinely humidify their paint booth makeup air during cold periods.

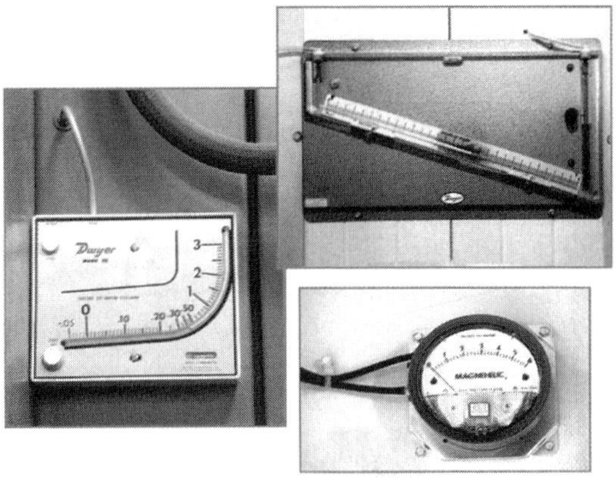

Figure 22-2. A conventional manometer (left), slanting manometer (top right), or a Magnehelic gauge (lower right), is attached to the outside of the spray booth to measure pressure differential across the booth. Spray booths usually operate at about 0.7 inch of water (1.74 dyne/cm^2) column pressure.

If the primary movement of air makeup and exhaust is from the booth ceiling to the floor, the booth is classified as a "downdraft" type. **Figures 22-3** and **22-4** show a downdraft booth. Fresh makeup air blows down through filters in the ceiling, and as it flows through the booth it carries overspray and solvent vapors with it. The contaminated air is then exhausted through the floor grating to a filtering system. If the air movement is mainly horizontal, the booth is labeled a "side-draft" booth. In both types, the air movement permits a spray operator to work in a safe breathing envi-

Figure 22-3. Downdraft spray booth.

ronment without needing a separate fresh-air supply. However, when coatings containing toxic substances are sprayed, a separate breathing air supply needs to be provided for operators. Fresh air is usually delivered to a facemask or shroud that completely encases the operator's head. OSHA specifies the maximum allowed concentrations in the breathing air for each solvent to which workers may be exposed. If any of these levels is exceeded, worker protection is mandated. OSHA *approves* separate breathing air supplied hoods for most polyurethane spraying, although it may *allow* some cartridge filter face masks also. The hoods are recommended because they are safer for the sprayer; facemasks have no fail-safe indicator to show if they are working correctly or not. In automated paint booths where people do not enter during operation, airflow rates may legally be reduced. Electrostatic application reduces paint usage (at least in theory), and thus also reduces solvent emissions. For this reason, booths with electrostatic application equipment are permitted by OSHA to have lower air flush rates (measured by booth air face-velocity) than booths with nonelectrostatic application devices. For example, if a booth contains an airspray gun, it may be required to have a makeup air face-velocity of 100–150 ft^3/min. However, if the airspray gun is converted to an electrostatic airspray gun, OSHA would demand only 75 ft^3/min face-velocity for the makeup air.

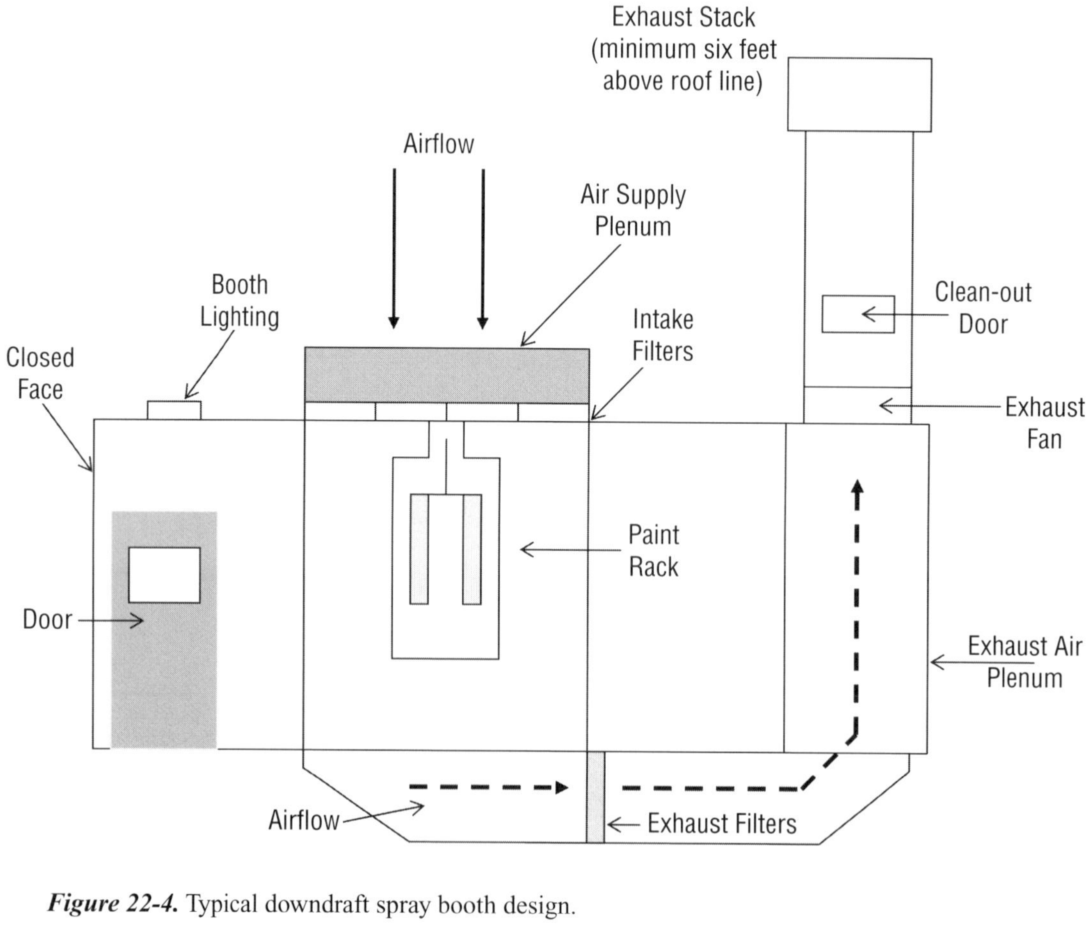

Figure 22-4. Typical downdraft spray booth design.

Paint overspray cannot be exhausted directly from a spray booth outside the plant because paint particulate has the potential to foul air ducts. It is also possible that exhausted overspray paint might drift onto nearby structures and vehicles. If unfiltered spray booth exhaust deposits overspray paint into the exhaust ducts, it would create a fire hazard. Overspray buildup on blades can slow the exhaust fan and disturb booth efficiency. Overspray paint accumulations can unbalance fan blades, causing damaging vibration, inordinate bearing wear, and eventually damage to the exhaust fan housing. Overspray paint being exhausted directly outside plants and accumulating on nearby vehicles or structures are hardly rare occurrences. A sizable number of insurance claims are filed annually in North America for the "accidental painting" of cars in the vicinity of a plant with ineffectual filtration and capture of spray booth exhaust.

Since the air inside a spray booth is filtered to catch overspray particulate before it is blown outside, the method used to filter the overspray provides another way of categorizing spray booths. If the overspray is removed by replaceable air filters that are similar to large furnace filters, the booth is termed a "dry-filter booth." If the overspray is removed by a turbulent air/water mixing system, it is called a "water-wash booth" or simply a "wet booth." A few booths will use both on hard-to-capture paints: an initial dry filter to stop the majority of larger overspray particles, plus a water-wash to capture the rest of the hard-to-stop small particles.

Atomized paint particles exiting a spray gun consist of various sized droplets in flight toward the vicinity of the part being painted. As the droplets move through the air, solvent evaporates from them continuously, sending solvent vapors into the booth air. The evaporated solvent, both from atomized particles and from the freshly applied paint film, is comprised of individual, extremely tiny molecules. Solvent vapor molecules are so small they totally escape capture by dry filters and are exhausted outside in the vented booth air stream. Overspray paint droplets, which are relatively large, are made up of resin, pigments, additive components, and usually at least some residual solvent. These particles are also carried along in the booth air exhaust stream; but because they are much larger in size than solvent molecules, the exhaust air filtering system captures all but the tiniest particles. In this way they are kept within the booth and prevented from being blown outside the plant.

Dry-Filter Booth

In a dry-filter booth, the evaporated solvent molecules and atomized paint particles created by the paint application process are borne by the air circulation to a bank of air filters. The air stream then flows into the exhaust system. Nearly all dry-filter booths are side-draft types, although other variations are possible. **Figure 22-5** shows a side-draft dry-filter booth. Note that in this design air enters the booth behind the spray guns or bells and exits the other end through filters into the exhaust plenum. Clean, filtered makeup air goes into the booth through ceiling plenum filters and descends behind the spray guns and parts to be painted. It then travels through the booth, permeates filters on the back wall, and flows out the exhaust stack. It is much preferred, however, that airflow into and out of booths be totally laminar (linear). Straight-in and straight-out airflow results in a more efficient and noticeably cleaner booth than turbulent airflow. Some older designs directed

Figure 22-5. Side-draft dry-filter booth.

incoming air down through ceiling filters in the back of the booth and then out through the filter wall at the front of the booth. This curved air path is not as desirable as laminar flow. To this end, the entire inside front wall of the booth should hold air filters; no solid sections should be used in this wall construction.

Dry-filter booths employ easily replaceable disposable filters that vary in size, composition, and particulate capture efficiency. The filters may be of the strainer type or the baffle type. The baffle type can take heavy loading levels, but are not highly efficient at removing all of the overspray particulate. The strainer types are effective at removing most particles but tend to "face-load" or "blind" (clog) quickly. This is a nuisance because it requires frequent line stops to allow filter replacement. To avoid this, manufacturers have for some time been making variable-density filters or dual-material strainer/baffle "sandwich" filters. A baffle-type filter is placed just ahead of a strainer-type filter to obtain the advantages of both styles. These combinations avoid the rapid face-loading of the strainer type but still achieve high overall capture efficiencies.

When filters load up with paint to the point that the flow of air is significantly impaired, the used filters must be replaced and disposed. In some locations plants are required by law to bake their used booth filters before they are allowed to discard them as landfill. The reason is that baking expels residual solvents from the trapped overspray of the filters. Landfill disposal of the spent filters is usually straightforward when no solvents or toxic substances are present in the entrapped paint residues. The exceptions are paints that can ignite spontaneously when the used filters are stacked together. Stacking does not permit the escape of heat generated by oxygen reacting with the finely divided organic materials on the dirty filters, thus with some paints spontaneous combustion can result. When this problem is encountered, the filter makers recommend that filters be plunged into a drum of water and the filled drums tightly sealed. This is effective in preventing combustion, but most plants hate having to do this because it adds considerably to the overall weight and to disposal costs.

Getting rid of filters that contain toxic paint is expensive. Depending on the applicable regulations, used filters laden with toxic paint components, such as lead or chromium compounds, may need to be disposed in facilities licensed to handle toxic wastes. Somewhat surprisingly, in at least three states, if water leach tests show that only low amounts of toxins are released, the filters do not require handling and disposal as toxic waste. Fortunately, most states do not allow any toxic materials to be discarded this casually.

Dry-Filter Booth Advantages and Disadvantages

Among the advantages of dry-filter booths is that there is no water to pump and treat with chemicals, and no wet gunky paint sludge to handle. More importantly for many companies, the initial capital investment for a dry-filter booth is modest compared to a wet booth. Installation is straightforward and easy; floor loading is much less than for a wet booth. Modularized booth kits are available in several sizes so that plants can purchase and install spray booths themselves for added savings.

However, dry-filter booths all suffer from diminishing airflow. Due to paint buildup on filters with usage, the booth airflow velocity and volume gradually decrease. To maintain good airflow, the filters must be replaced periodically. If filter replacement is needed more than once or twice per shift, production tends to become slowed down. The expense of replacing the filters includes cost of the new filters, the labor to switch out the old used filters for the new ones, and the cost to dispose of used filters—all add up to a significant operating expense. Plus, when the replacement needs to be done during manufacturing hours, it cuts into production output. Storage space is needed for used and replacement filters and this also has a cost. We have seen that spontaneous combustion of used filters can pose a safety concern, as well.

Water-Wash Booths

In water-wash booths, the overspray is exhausted through one or more curtains of water or through overlapping pressurized sprays of water that are designed to entrap and remove the atomized paint overspray. Water curtains or pressurized sprays are usually located on the side of the booth behind the wall adjacent to parts being painted. Some booths have the water-impingement system located below the floor grating. Effective water-wash booths can be of either side-draft or downdraft design. **Figure 22-6** illustrates a side-draft water-wash booth; and **Figure 22-7** shows a downdraft water-wash booth.

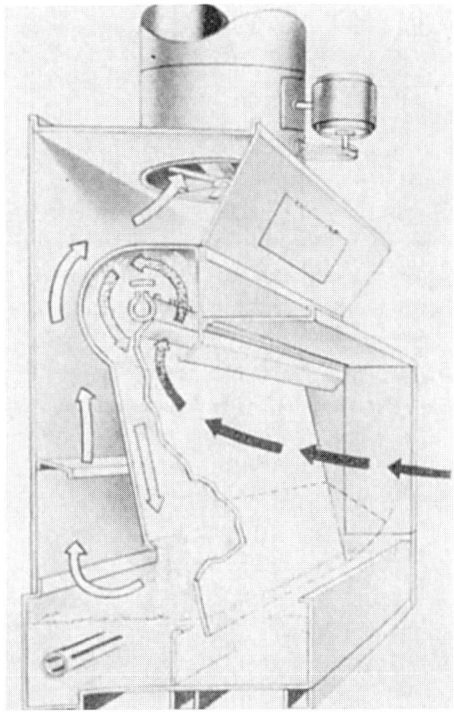

Figure 22-6. Side-draft water-wash booth.

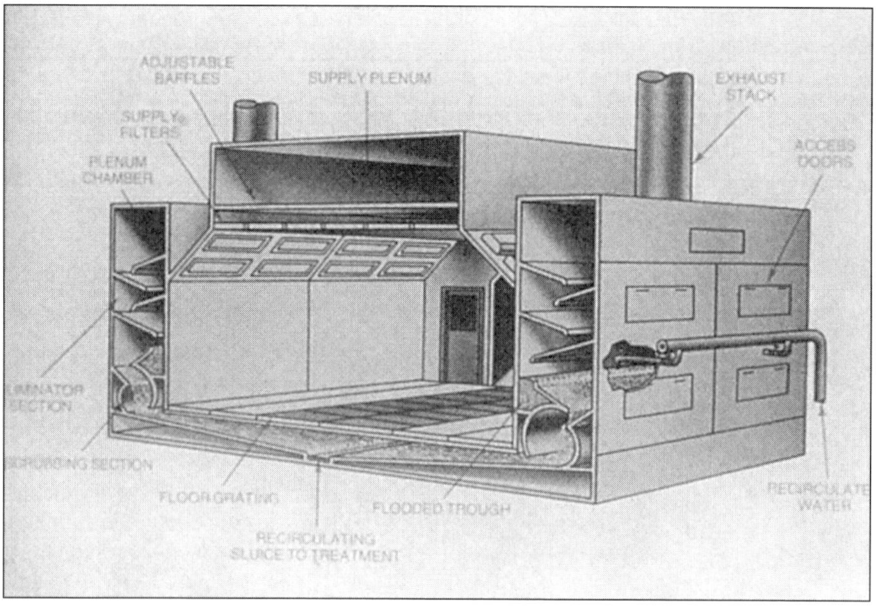

Figure 22-7. Downdraft water-wash booth.

The water curtain or spray is pumped from a reservoir tank and returned to the tank. The volume of the reservoir tank depends on the size of the spray booth and may range from several hundred gallons for a small booth to thousands of gallons for a very large booth (see **Figure 22-7**). A number of booths may use a common reservoir of recirculating booth water; frequently this is a large "pit" tank located below the first floor level of the plant.

If the overspray paint particles collected in the circulating water were not chemically treated, they would clump together or stick onto everything in the system reservoir. The sticky paint would plug up piping, pumps, headers, and spray nozzles. Chemicals added to prevent this, called *detackifying agents*, make the paint particles "unsticky" by several mechanisms. **Figure 22-10** illustrates the detackification process; this is commonly referred to as "killing" the paint. Chemicals can be used either to float the detackified paint for skimming, or sink it to the bottom of the reservoir for separation from the supernatant water. Other chemical systems are used to form a suspension of the paint sludge. The suspended "killed" material then can either be filtered out or removed by centrifuging the water.

To accomplish paint kill, various detackification agents use one or a combination of the following methods.

- Absorbent materials such as clay.
- Alkaline materials that chemically degrade many paint resins (high alkalinity can "sink" the sludge; lesser amounts of alkali float the killed paint, which permits removal by skimming).
- Polyelectrolyte materials that chemically surround and encapsulate the individual paint particles, rendering them unable to stick to anything.

Detackifier formulations are available in solid form or as concentrated liquids. Best results in separating the killed paint from water are obtained when paint particles are agglomerated by coagulants into large particles, a result that facilitates both floating and settling. Some paints are easy to kill, others are not. Many waterbornes and most high-solids paints tend to be quite difficult to detackify. Once the paint particles are killed, they must be separated from the booth water and collected.

Detackifying agents may generate foam, so antifoamers or defoamers may be added to prevent it. Foam can overflow onto the booth floor, cause pump cavitation (pumping of air instead of water), and produce rapid erosion of the pump's impeller. Foam also interferes with the paint/water separation, this in turn delays killed paint particle settling and produces excessively wet sludge. If paint sludge is too wet, it makes disposal more costly because of the extra weight, increasing both handling and transportation expenses. Additional environmental restrictions may also apply to sludge having a high water content. Many disposal sites charge considerably more for wet waste material than for dry.

Figure 22-8. Conveyorized water-wash booth.

A common problem from wet booths that are not properly maintained is the vile odor given off by the water. Water that contains organic material such as paint particles can be an attractive food source for fungi and bacteria. Antimicrobial agents are used to prevent slime formation, unpleasant odor, and corrosion pitting from the acids they produce. To further minimize these problems, the water can be kept mildly alkaline, i.e., at a pH of about 7.5–9.0.

Sludge separation and collection is generally done by one of four main ways. Gathering killed paint is accomplished:

- By skimming if it can be made to float
- By centrifuging the mixture of water and killed paint particles
- By flat bed gravity filtering or pressure differential filtering
- By allowing the solids to settle to the bottom of the reservoir. Periodically the water above the sludge is pumped off. The sunken sludge remaining on the bottom of the holding tank is next scraped up, most often manually using buckets and shovels—not an enjoyable job.

With all these four methods, any toxic materials and solvents remaining in the water wash after sludge separation must be eliminated by water treatment before the water may legally be released to drain. You should be aware that when solventborne paints are applied in a wet booth, the residual sludge normally still contains about 1–3% solvent. Disposal of wet paint sludge has become an increasingly expensive problem, especially if toxic components or excess amounts of solvents are present. In light of so many health hazards being traceable to chemicals that were improperly disposed of, most areas are

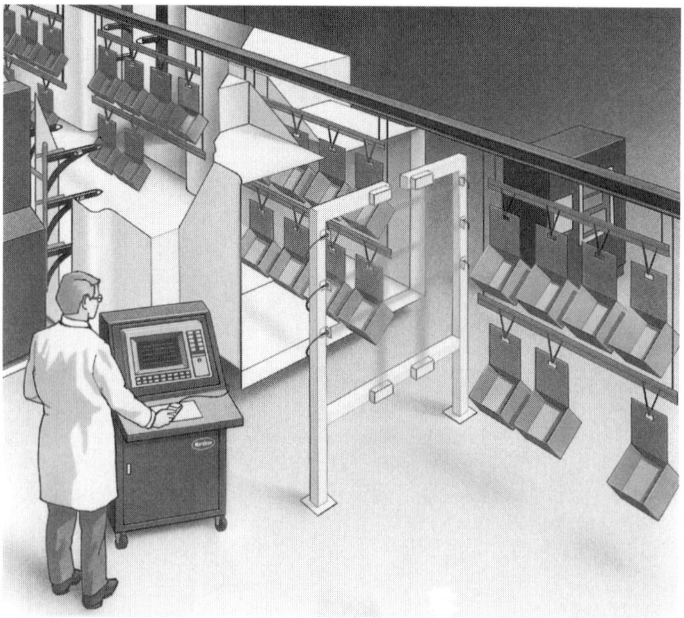

Figure 22-9. Light curtains can be used to trigger guns or bells on and off as needed. The increased transfer efficiency and greater booth cleanliness make them a wise investment for many plants.

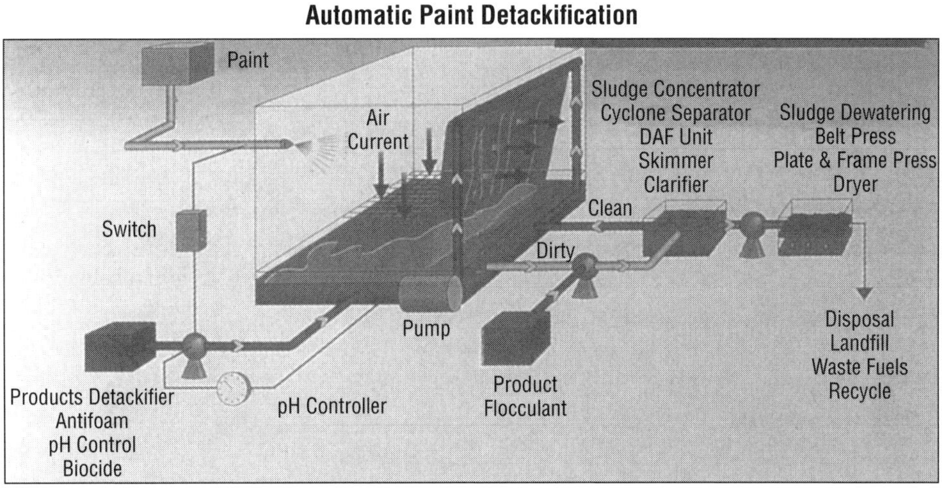

Figure 22-10. Automatic paint detackification.

exceedingly reluctant to allow transport of any manufacturing wastes, much less burial or incineration, in their communities. Many regions either have enacted or are considering laws to ban all hazardous chemical disposal except on the original plant usage site, and then only by approved and certified methods. Think about this: how would your plant handle disposal of all wastes right on your own company property?

Some efforts have been made to recycle paint sludge, but it is usually less costly for small- to moderate-sized paint users simply to dispose of it and buy new paint. If no toxic substances are present, sludge can be pressed into briquettes and burned for its heat value. Pressing to remove water is necessary because wet sludge burns poorly and it may require almost as much heat to evaporate the water as heat generated by combustion. At least one company makes equipment to heat and dry paint sludge and then crush it into a free-flowing powder. The powder is then put into landfill because the question of where or how to reuse this powder remains difficult to answer. Even after all these years, no one seems to have come up with a good method of paint sludge reuse.

Water-Wash Capture Booth Advantages and Disadvantages

Booth airflow velocity is constant in all water-wash booths, so no down time is required to change filters, and no costs for the labor and materials in changing used paint filters are incurred. On the down side, the equipment and installation are more costly and complex. Water used in the process becomes contaminated, so water treatment is needed. "Kill" chemical and paint sludge disposal costs are incurred. Finally, floor loading may be considerably higher and this will limit the places where the water reservoir can be located.

Dry-Filter or Water-Wash Booths?

Whether your plant should install a dry-filter or a water-wash booth depends largely on the amount of painting to be done. Equipment and operational costs of a dry-filter booth tend to be lower than for a water-wash booth until at least 75–100 gallons (284–378 l) of paint are applied each day. If having to stop painting for 10–20 minutes to replace booth filters causes dirt or impedes production, or if filter changes are necessary more frequently than 2–3 times per shift, these impediments can be avoided by employing a wet-type spray booth. The particular paint being sprayed has an effect on the relative costs as well. Remember too that how readily the paint is chemically detackified, the cost for disposal of wet paint sludge, and the method by which paint is applied are all important factors when choosing between wet or dry booths. Most paint shops are on the first level where floor weight is not an important concern. But if the booth is to be installed in a restricted-load location of your building, the floor weight of the water reservoir may be a deciding factor against a wet booth.

High-Solids Paint Overspray Recovery

The recovery and recycling of high-solids paint overspray by capturing it before it reaches the dry filters or the water-wash has been a goal of a few dedicated people, but it is still of little interest to all but a small handful of manufacturers. Yet for several large manufacturers who use a lot of paint, high-solids overspray recovery and reuse is producing meaningful cost reductions in a number of ways. Most obvious is that paint purchases are reduced. But capturing overspray also prevents it from reaching the filters, thus it sharply lowers filter replacement costs in dry-filter booths. Similarly in wet booths, both water-treatment costs and sludge-disposal expenses are lessened. With most high-solids paint formulations, overspray paint remains tacky and sticky no matter how long it stays on booth surfaces. The reason is twofold: to prepare high-solids coatings, the paint makers need to utilize lower molecular weight resins; and secondly, the low molecular weight resins are unable to undergo cross-linking at room temperatures. The tackiness may also be exacerbated by their relatively low content of fast evaporating solvents. These characteristics can, in some instances, make paint recovery and recycling fairly convenient, but the overspray viscosity must not be too high if this process is to work well. The elevated viscosity of high-solids overspray renders most of these paints unsuitable candidates for recycling except for polyesters. Low-solids solventborne paint overspray cannot usually be recycled because it has a lot of fast evaporation solvents and thus dries too quickly. The difficulty in recovering nearly all waterborne overspray arises for this same reason: few, if any, waterborne coatings stay wet enough (low in viscosity enough) to permit economical recovery.

Most recovery and recycling methods utilize vertical baffles stacked behind parts being painted. This works only in a side-draft booth, not a downdraft type. A large portion of the overspray is drawn toward the flat overlapping or spiral-channeled baffles and is deposited on them. More and more paint collects on the baffles, and if the viscosity is low enough gradually flows downward into a trough. The trough slopes into a collection container where much of the overspray paint that would have been caught either by dry filters or by a water-wash is recovered. Augers are used in some designs to move the collected overspray out of the trough because the highly viscous material itself flows far too slowly.

Once the overspray is collected, it needs to be adjusted to a lower viscosity by solvent addition and filtered before it can be reused. Color correction by pigment addition is required in all cases where more than one color is collected; single color paints may also need adjustment to restore the precise color.

VOC/HAP Removal

Because solvent vapors readily pass through dry filters and water-wash booths without being captured, VOC and HAP emission regulations for industrial painting are structured to require low-VOC coatings to be used. For example, a category of finishing (automotive, appliance, etc.) may require that coatings contain no more than 2.8 lb VOC/gal (335 g/l). As long as this requirement is met, the VOC given off during coating operations is within compliance. However, EPA regulations require that if coatings that exceed the VOC maximum are used, "add-on" VOC abatement methods, such as incineration or carbon adsorption, must be used.

Newer and smaller booth designs attempt to reduce exhaust air volumes, but only a maximum of 17–18% reduction in volatile emissions has been achieved thus far. Nevertheless, smaller booths have the added benefit of requiring less treated booth air makeup. Also, spray booths with automatic application equipment can be operated legally at lower exhaust levels than permitted for occupied zones, as long as no people are permitted in them when the application equipment is running.

Worker-free zones enable two air reduction techniques: air recirculation and air cascading. Partial air recycling of booth air is being utilized in some plants for VOC emission reduction and for cost effectiveness in these cases. A portion of the exhaust air from an unoccupied booth can be filtered free of particulate (including overspray paint) and then recirculated back through the same booth as long as the overall painting operations are not adversely affected and fire hazards are not increased. Both bag houses and electrostatic precipitators are methods used to remove particulate. If necessary, this filtered air can be humidified and heated before reintroduction back into the booth. In air cascades, the exhaust from an occupied booth might be filtered and then passed on into a booth where paint is being applied with an automatic device such as a bell, reciprocator, or robot. Air may also be cascaded from spray booths and then used as makeup air for dry-off or paint curing ovens.

In the incineration of VOC, the fumes are burned at about 1,250–1,400° F (677–760° C) to convert the VOC to carbon dioxide and water vapor. The incineration process in a VOC fume oxidizer unit (such as a recuperative oxidizer unit, **Figure 22-11**) is inherently energy-inefficient unless the energy of the hot exhaust can be redirected. It might be used on the finishing line or elsewhere in the plant. For example, it may be enlisted to pre-

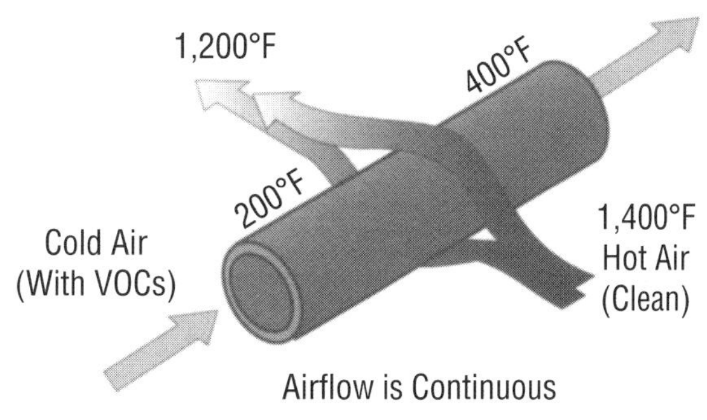

Figure 22-11. Recuperative thermal oxidation.

heat air makeup for ovens and other burners, or to provide heat for dryoff ovens. **Figure 22-12** shows a regenerative thermal oxidizer designed to reuse energy and control VOCs. Often the exhaust gases are passed through heat exchangers of varying designs. A "shell and tube" heat exchanger will allow over 50% primary recovery of the heat generated by a VOC fume oxidizer unit. Frequently the heat exchanger is used to raise the temperature of the gas stream to a level just under combustion. Attempts are usually made to reduce the incinerator exhaust temperature to as low as possible before it leaves the plant.

Fume incinerators are used extensively on coil coating lines, where low-solids coatings are often used. As a result, VOC emissions from the coatings application stations and bake ovens are high. The VOC concentration in the exhaust air often is great enough to serve as fuel for the incinerator, allowing a cutback in natural gas use. Incinerator use on coil coating lines has proved to be cost-effective.

Although both thermal and catalytic incinerators use gas burners to create the heat necessary to oxidize organic fumes, the thermal type is more frequently used than the catalytic type. Catalytic incinerators allow burner turndown once operating temperatures are achieved, but they are expensive and also susceptible to "poisoning" (fouling) by various agents. Catalyst contamination may occur in the presence of heavy metals or organic compounds containing either halogens or silicon. However, thermal incinerators operate at rather elevated temperatures, which form a lot of nitrogen oxides (NO_X, a hazardous air pollutant), far more than catalytic oxidizers that run at about half this temperature. Another alternative is a regenerative oxidizer that uses ceramic heat sinks to reach efficiencies above 90% in some cases. Although NO_X production is low, the large size and high capital and maintenance costs are major disadvantages of this type of oxidizer.

As a general rule, incinerator use is economically limited to destroying VOC from bake ovens where solvent fume concentrations are fairly high, and where the VOC temperature (F) is already roughly 300° F (95° C) above ambient. This obviously will require much less heat input to raise gases to the incinerating temperature (approximately 1,200° F,

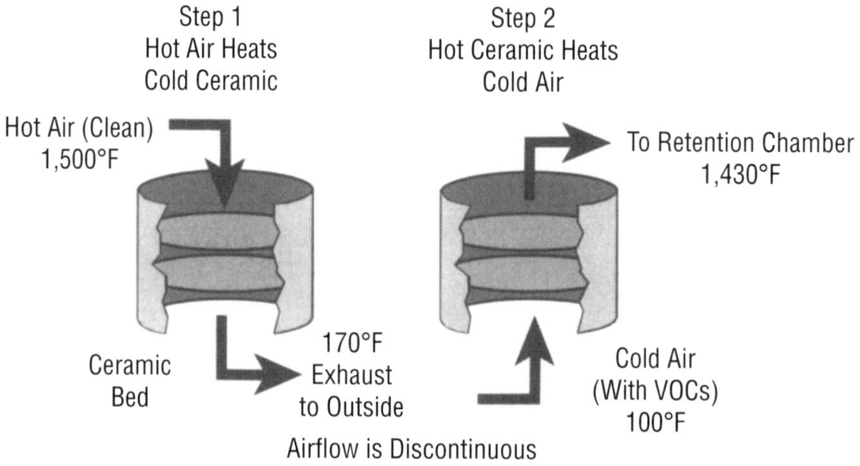

Figure 22-12. Regenerative thermal oxidizer.

or 650° C). As a means of destroying VOC or HAP from a spray booth, incineration is economically unattractive except in a few special instances. This is because the high air volume throughput requirements for spray booths results in an excessive amount of air to handle. The air is not warm but only at room temperature, and it only contains a very low concentration of combustible organic vapors. Nevertheless, large plants that use high volumes of paint do use incineration on paint booth exhaust air to remain in VOC compliance. Finishers who apply only modest amounts of coatings should comply with the applicable VOC per gallon limits and not even consider VOC incineration. Incineration as a means to achieve EPA compliance is completely impractical for plants whose total VOC emissions are low and when alternative compliance coatings are available.

Solvent Adsorption Systems

Activated carbon-bed adsorption units (see **Figure 22-13**), while totally effective in cleaning solvent vapors from booth exhaust air, also tend to be prohibitively expensive except for large painting operations. Huge beds of carbon are needed plus facilities for periodically stripping solvent-saturated carbon beds with a flushing material. In most cases, live steam is used for regenerating the carbon adsorbing medium, although other fluids such as nitrogen gas can be used for this purpose. Equipment to separate both soluble and insoluble solvents from the condensed mixture of steam and organic vapors adds to the high cost of adsorption systems. Several newer designs continuously strip and concentrate the booth air/solvents stream ahead of incineration. A rotating carbon-filter wheel (see **Figure 22-14**) cleans the exhaust air of all solvents. Part of the hot incineration exhaust gas back-flushes the adsorbed solvents off the carbon wheel and concentrates them into

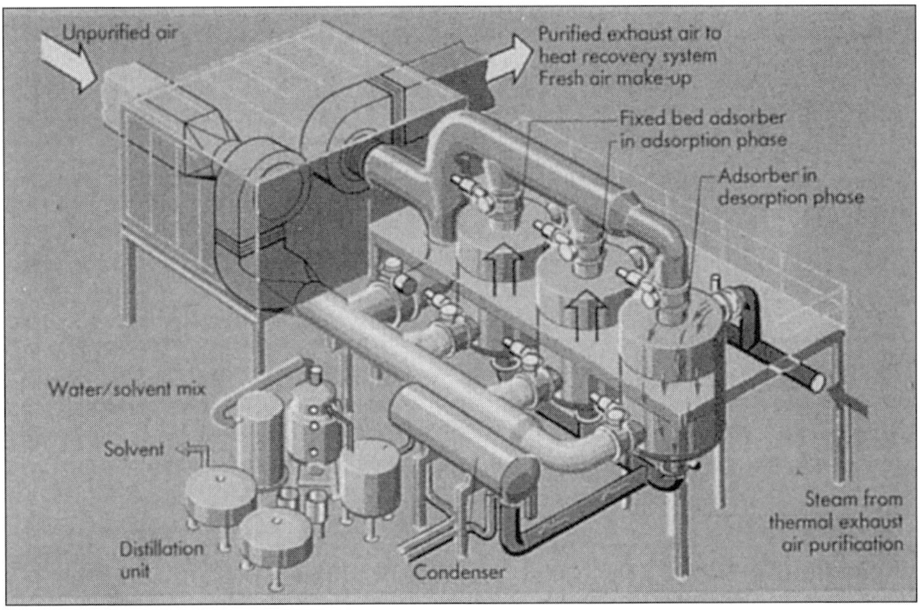

Figure 22-13. Carbon-bed solvent adsorption.

Figure 22-14. Rotating carbon filter VOC concentrating incinerator (interior and exterior).

as little as one-twentieth of their original volume. The reduced volume of more concentrated solvent-rich air can be incinerated with less thermal energy from the burner. It also uses smaller ductwork and fans with lower horsepower motors. These reliable and easily operated concentrating systems are very efficient (up to 99% is claimed) and they form almost no NO_x. But it is absolutely essential that all paint particulate be removed from the exhaust flow before it reaches the concentrators or the system will be fouled. Two-stage filtration using multiple stage bag filters on the booth air exhaust stream, located upstream of the concentrator, is normally recommended.

Some forms of VOC are amenable to capture in aqueous alkaline scrubbers. Ultraviolet rays can then activate or even begin to decompose the VOCs in solution before they are transported to an aqueous chemical oxidation chamber. A variety of oxidants are used: ozone, chlorates, hydrogen peroxide, hypochlorites, and additional oxidizing agents have been used in this process. The systems are called ultraviolet plus oxidation (UV/OX) emission reduction systems. VOCs that are not destroyed are adsorbed on carbon or zeolite beds and destroyed by oxidant treatment or by combustion after desorption. UV/OX is utilized most effectively for water soluble VOCs at modest concentrations, but operating costs tend to be quite high.

Attempts to devise filters or additives in booth water wash that will catch or somehow destroy organic volatiles have had little commercial success. Some manufacturers have shown interest in biofilter systems, in which exhaust air is passed through soil materials, such as compost or peat moss. These soils contain bacteriological microorganisms that can consume solvents by converting them to water and carbon dioxide. This method forms no NO_x, but it cannot handle hot exhausts that would kill bacteria. Space requirements and installation costs are large, but operational expenses are small. A similar scheme using huge booth water circulation volumes involving a pool of water containing

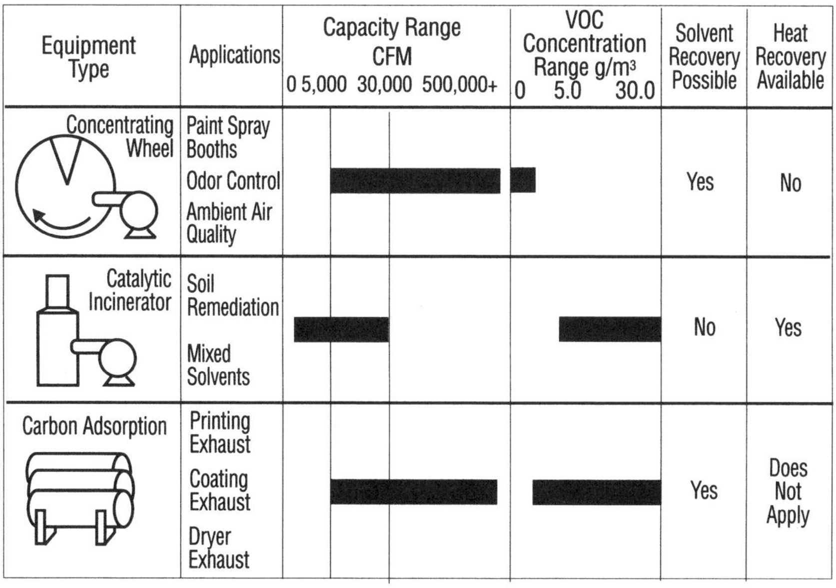

Figure 22-15. A comparison of VOC controls.

solvent-destroying bacteria was tried many years ago. It worked well in pilot plant scale tests but, alas, under production conditions it did not fare well. The major culprit was heavy amounts of cleanup solvents dumped into the water. It was suspected that workers assigned to clean the paint application equipment at night after the line was shut down were doing this. The dumps of solvent killed the bacteria, which then had to be replenished at considerable cost. This occurred so frequently—despite efforts to curtail it—that the project was abandoned as impractical.

Glossary

Adhesion Loss. The premature separation of a paint film from its substrate.

Air Bearings. A stream of air that supports a spinning shaft (as opposed to roller, ball, or similar mechanical bearings). Air bearings have limited load carrying capacity but require no lubricants.

Air Dryers. Used to remove moisture from compressed air. Dryers have three basic styles of operation: 1) deliquescent types have disposable drying agents and tend to be marginally effective for painting; 2) refrigerated dryers cool the air to condense out the water (most paint systems use this type); 3) desiccant types have a double bed dryer and are able to achieve the lowest dew point air. The beds are alternately on-stream and back-flushed to regenerate their moisture-absorbing qualities. Some plants with critical finish requirements use this style of dryer to reach dew points of -40° F (-40° C).

Air Knife. A slotted jet of compressed air will act as an effective air knife to quickly blow superfluous water from parts, often before they enter a dry-off oven.

Air Turbine. 1) Electrical motor-driven fans that create volumes of relatively low-pressure atomizing air for spraying. Their output is referred to as turbine air. 2) An air-driven precision fan that is used to spin a paint atomizing disc or bell head.

Airless Spray. A method of paint application in which the paint's fluid pressure is used to atomize the paint into particles, and a spray gun is used to direct the particles onto the surface being painted.

Alkali. A substance that neutralizes acids, synonymous with caustic. Alkalis are helpful in aqueous cleaning by speeding soil removal and suspension.

Amino Resins. Nitrogen-containing compounds such as urea, melamine, or diethylene triamine are too rigid to be effective paint resins. The amino resins are used to cross-link polyester, epoxy, acrylic, alkyd, etc., resins to enhance their durability.

Ampere (Amp). An electrodynamic unit of measure for the quantity of current in a steady electric flow.

Anode. The electrode at which chemical oxidation takes place. In electrodeposition (ecoating), the anode is indicated on diagrams by the positive (+) marking.

Anodizing. A type of conversion coating in which an aluminum oxide coating forms on the surface from contact with an anode in an electrochemical cell.

Anolyte. The water used to flush solubilizer molecules that form inside an electrocoating anode box. If used to flush a cathode box, it is termed catholyte.

Aqueous Cleaning. A system in which water plus detergent, and frequently also small amounts of acid or alkali, are used to clean surfaces prior to painting without emitting solvents.

Arab Oil Embargo. In 1973 a cartel of producers in Arabian countries withheld oil from the international market for a number of months in an attempt to raise prices for crude. Their efforts were devastatingly effective on all products that use petroleum, including paint products.

ASTM. Acronym for the American Society for Testing and Materials International. The Society publishes extensive standards used widely in manufacturing. Their address is 100 Barr Harbor Drive, West Conshohocken, PA 19428-2959. Phone 610-832-9585; FAX 610-832-9555.

Atom. Molecules are made up of combinations of these tiny bits of matter found in nature. There are 98 types of atoms.

Autodeposition (Autophoretic). A precipitation reaction of an organic resin occurs by the action of an acid that etches a metallic substrate. The ions of the oxidized metal co-deposit with the vinyl emulsion resin in the autodeposition coating process.

Azeotrope. A mixture of two miscible liquids that co-distill at a specific composition at a fixed temperature lower than that of either single pure liquid.

Back Ionization. A term that has been used to explain powder film defects and diminished transfer that can occur during electrostatic powder application. Voltages set too high and positioning the gun too close are major causes. A more accurate term for the phenomenon is "electrostatic charge repulsion." The problems are associated almost exclusively with corona charging.

Barytes. A colorless crystalline solid; a form of barium sulfate (also called "barite") used as a paint pigment.

Bell. A rotating head that is shaped to deliver paint and powdercoats forward in a circular pattern. The bell may be directed at any angle and be moved on robots or reciprocators, just as spray guns are. Limited manual bell application is performed in repainting operations.

Benard Cells. Minor circulation patterns that appear when metallic paints are applied extra wet and solvent evaporation cools at the surface.

Bleeding or Staining. The discoloration seen when solvent migrates from the topcoat into the primer, leaching out the pigment and discoloring the topcoat.

Blister. A small dome or raised area in a paint film. The blister contains, or once contained, moisture (water, water vapor, or both). Tiny blisters are called boils or (incorrectly) pinholes.

Blocked Isocyanates (Blocking Agent). Isocyanates, normally extremely reactive with water, can only be used in waterborne coatings if they can be prevented from reacting before the water is baked out of the paint film. Capping or blocking the isocyanate group with a thermally decomposable chemical does this. In the bake oven the water evaporates, the chemical cap decomposes, and the isocyanate cross-links the paint. Blocked isocyanates are often used for ecoat curing.

Blocking. When freshly painted parts are stacked, they sometimes stick together if the paint is not fully cured. Antiblock paint additives can reduce sticking or blocking.

Blushing. A defect that occurs whenever so much evaporative cooling takes place that the surface temperature of the paint film drops below the dew point.

Bounce-off, Bounceback. Paint droplets from air-atomized application have a tendency to rebound or bounce away from the surface due to the blasting effect of the air. These particles are called bounce-off or bounceback, as is the phenomenon itself.

Brilliance. The apparent strength of a color as perceived visually.

Bubble. A small domelike area in a paint film that contains, or once contained, solvent vapor. A tiny bubble is called a solvent pop.

Bulk Coating. The painting of large masses of small unhangable parts by a variety of possible techniques such as dip-spin and dipping.

Burn-off Ovens. Paint stripping can be accomplished by combustion of the coating in gas-fired, burn-off ovens in which upper temperatures are controlled by injection of water spray into the oven.

Capacitance. The ability of ungrounded objects to retain electrical charges.

Carcinogen, Carcinogenic. Cancer-producing agent.

Cathode. The electrode at which chemical reduction takes place. In electrodeposition (ecoating), the cathode is indicated on diagrams by the negative (-) marking.

Caustic. Substances that neutralize acids, synonymous with alkali. They are helpful in aqueous cleaning by speeding soil removal and suspension.

Cellosolve. Originally a trade name for the solvent family of mono-alkyl ethers of ethylene glycol, but the term is used generically now. A much-used solvent is butyl cellosolve, for example, which chemically is ethylene glycol monobutyl ether.

Centrifugal Coater. Same as Dip-Spin Coater.

Chemical Agent Resistant Coatings (CARC). Coatings used on military equipment that might become contaminated by nuclear, biological, or similar chemical substances.

Chlorofluorocarbon (CFC). A general name for a multitude of halogenated hydrocarbon compounds that have either chlorine or fluorine atoms, or both, substituting for hydrogen atoms in the hydrocarbon molecule. Since many, if not all, cause ozone depletion in the earth's upper atmosphere, their manufacture and use are restricted or banned in many countries.

Chroma. In color determinations, it is the intensity of a hue.

Coil Coating. A process wherein rolls of material such as metal, paper, cloth, or plastic are unrolled and painted, then after the paint is cured are recoiled in a continuous flow process.

Cold or Dead Entry. In ecoating, it is when the parts are immersed in the tank while the voltage is turned off. With current on the immersion is "hot" or "live."

Color Mismatch. A deviation between items that are intended to be the same shade.

Commission International de Eclairage (CIE). a.k.a. International Commission on Illumination; in 1931 this group established the L,a,b color scale. Updated versions of this scale are still widely used today.

Continuous Coater. An enclosed automatic spray booth that recovers and reuses oversprayed paint. A continuous coater is suitable for coating large volumes of similarly sized parts.

Conventional Air Spray. One of the most common types of paint applications in which an air-atomizing spray gun is used to break down a thin stream of paint into tiny particles and direct them toward the surface to be coated.

Conversion Coating. Substrates may be chemically treated prior to painting to change the chemical composition of their surface. This is most often done to enhance substrate corrosion resistance and the adhesion of applied coatings.

Cosolvents. Waterborne paints frequently require water-miscible organic solvents (cosolvents) in addition to water for easier manufacture and improved application properties.

Creepback. The spread of rust under a paint film and subsequent loss of the overlying paint.

Critical Pigment Volume Concentration, CPVC. The maximum volume of pigment that can be introduced into a paint formulation without causing a diminution of desirable film properties.

Cross-linking. In enamel curing, the resin molecules react to form an extensive network polymer system. The chemical reactions whereby the separate molecules unite by chemical bonds into a single macromolecule are called cross-linking.

Cup Gun. A spray gun used with a siphon cup attached below the gun or a gravity-feed cup mounted above the gun.

Current Density. A measure of the total electrical flow across a given surface area, frequently expressed in units of amps/square foot (A/ft^2) or amps per square meter (A/m^2).

Curtain Coating. A coating system in which paint flows from a reservoir through a slot, creating a curtain- or waterfall-like flow of paint onto the surface to be coated. The part being coated moves through the paint curtain during this process.

Cyclone Separator. A device that moves a particle-laden stream of air inside a funnel-bottomed enclosure rapidly in a circular path. The relatively high mass of powder coating particles causes them to be thrown to the sides of the enclosure. There they slide down through the funnel into a collection hopper for reuse.

Deionized Water. Water that has all contaminant ions removed by a double-bed ion exchanger that switches H+ (hydrogen) ions for positive impurity ions and OH- (hydroxide) ions for negative impurity ions. The hydrogen ions and hydroxide ions then combine to form HOH (H_2O). Deionized water is equivalent in purity to distilled water but is much less costly to produce.

Detackifying Agents. Chemicals added to the water wash of a wet booth to prevent the collected paint particles from sticking to the sides of the reservoir and clogging the piping, pumps, headers, and spray nozzles; also called kill agents.

Diluent. While true solvents can be added in unlimited amounts to lower paint viscosity, it may be more economical to lower viscosity with less costly diluent solvents. Diluents alone cannot dissolve resins, but when added to a prepared paint they will lower the viscosity just as effectively as a true solvent. However, if too much diluent is added, the resin will separate out of solution and the paint becomes unusable.

Dip-spin Coater. Bulk painting of small and unhangable parts can be accomplished by dipping a mesh basket of parts, followed by rapid rotation of the basket to remove excess paint. Parts from the dip-spin coater are dumped onto a belt for curing.

Dirt. In relation to paint, this refers to any and all contaminants including lint, dust, small clusters or improperly mixed pigment, tiny particles of dried or overspray paint debris, and oil mist.

Discs. Discs have rotating heads that deliver paint horizontally 360° around the rotating head and use an omega loop conveyor line. The disc is mounted horizontally, usually on a vertical reciprocator.

Dispersion Paint. If the paint resin is not fully dissolved but is uniformly spread as tiny globules throughout the formulation as a stable mixture by stirring, the paint is termed a dispersion coating.

Distinction (or Distinctiveness) of Image (DOI). The measure of how well a surface acts as a reflecting mirror.

Distinction of Reflected Image (DORI). Same quality as DOI above.

Dive-in or Sink. Gloss variations and dull spots in the paint film on a plastic part.

Ecoating, Electrodeposition. In a method closely paralleling electroplating, paint is deposited using direct electrical current. The electrochemical reactions that occur cause water-soluble resins to become insolubilized onto parts that are made to be electrodes in the ecoating paint tank. Subsequent resin curing is required.

Eductor. Eductors (venturi nozzles) located along ecoat return headers spaced laterally at intervals across the tank help to agitate the paint and prevent settling of pigments, resulting in cleaner deposited films.

Electrocoating or Ecoat. The electrical deposition of paint film from a waterborne organic solution onto a part.

Electrons. Subatomic particles of negative charge contained in all matter. Current flow in electric circuits can be thought of as a stream of electrons traveling through conducting wires.

Electrostatic Atomization. Electrostatic charge repulsion that pushes apart or atomizes paint into fine particles.

ELPO. This term is unique to General Motors where it is used to mean electrodeposition coating (ecoat). It is a word derived from Electrodeposition of Polymers.

Emulsion Paint. A paint in which the resin is present as dispersed globules coated by an agent (emulsifier) that prevents their agglomeration.

Enamel. All powder or liquid paints that form cross-linking chemical bonds during curing. The majority (perhaps 98%) of industrial finishes fall into this category.

Face Rust. The appearance of rust on a continuous film of coating as opposed to rust at a paint scrape, chip, or gouge mark.

Faraday Cage Effect. Electrostatic application causes paint particles to be attracted to the nearest grounded object. The attraction force is often strong enough to pull paint particles out of their intended flight direction. Recessed areas on parts, since they require a slightly longer path for paint particles, often receive no paint or insufficient paint coverage. As a result, Faraday cage areas may need touch-up painting with nonelectrostatic spray.

Fatty Edge. An excess bead of paint that forms on the bottom edges of parts when they receive excess paint. It is common following dipping or flow coating.

Fisheye. A paint film defect that resembles a small depression (the eye) with a mound or dome of raised coating in the center (the pupil).

Flat Spotting or Striking In. The appearance of gloss-deficient patches in a paint film.

Fluidized Bed. A powdercoating hopper in which finely divided powders can be made into a fluid-like state by passing air through the porous plate hopper bottom. This permits hot parts to be dipped into the fluidized bed of powder particles to coat them. Fluidization also allowed powders to be pumped in a manner similar to liquids.

Fluidizer. Also known as a solvent, a fluidizer is a liquid that lowers paint viscosity enough to allow proper application.

Flushable Electrode. The anode in cathodic ecoating is often placed inside a semipermeable membrane enclosure so that excess solubilizer generated at the anode can be continuously removed (flushed) by overflowing water pumped into the bottom of the enclosure. Flushable electrodes can also be used (but rarely are needed) for the cathode in anodic ecoating.

Free Radical Polymerization. Certain organic compounds will form highly reactive electron configurations by the action of UV light (or other activation sources). These reactive species are called free radicals because, to an extent, "free" electrons are available for bonding. In free radical polymerization, the reactive electrons form chemical bonds to join together adjacent molecules and thereby produce a cured paint film.

Freeboard. In a vapor degreasing tank, the freeboard is the space between the condensing coils and the top of the tank. Freeboard of 1–2 feet (30–60 cm) helps contain vapors in that zone.

Gas-liquid Chromatography. Both qualitative and quantitative analyses of gaseous and liquid mixtures are accomplished by separating the components carried in an inert gas stream through liquid-lined chromatographic columns; sometimes this is simply termed gas chromatography.

Gloss. A measure of the ability of a surface to reflect light.

Grain Refiners. Weight for weight, smaller zinc phosphate crystals provide superior corrosion resistance and paint adhesion compared to large crystals. Grain refining agents or "grain refiners" are used in water rinses prior to zinc phosphating, or in the zinc phosphating bath itself, to produce smaller crystals.

Ground (Electrical Ground). An object so massive that it can lose or gain overwhelmingly large numbers of electrons without becoming perceptibly charged, in an electrical sense. Our planet Earth is the best electrical ground possible.

Halogenated Hydrocarbons, Chlorofluorocarbons (CFCs) (Halogenated Solvents). Halogens—including mainly chlorine, bromine, and fluorine—can be substituted into hydrocarbon molecules to change both the physical and chemical natures of hydrocarbon compounds. These compounds are collectively called the halogenated hydrocarbons. Their nonflammability is essential for operations such as solvent degreasing, but their lack of biodegradability makes them have limited desirability.

Hazardous Air Pollutant (HAP). The emission of HAPs is subject to stringent air quality regulations and restrictions because they can cause major harm to living organisms and our environment.

Heat of Vaporization. The amount of energy required to convert a substance from liquid to vapor without a temperature change. The heat of condensation has the same value and is released when vapor condenses into liquid form.

High Bake Paint. Paints that require a cure temperature of 195° F (90° C) or greater.

High-solids. Solventborne coatings that are approximately 50% or greater in volume solids.

Hot or Live Entry. In ecoating, when the parts are immersed in the tank while the voltage is turned on.

Hot Water Curing. A method of curing autodeposition coatings by immersing parts in 180° F (82° C) water. Hot water curing is faster than oven curing for parts that act as a large heat sink, but the process is normally not used since it gives reduced corrosion resistance.

Hue. The particular shade of a color.

Hydrophilic. Water loving, or attracted to water.

Hydrophobic. Not soluble in or not attracted to water; literally water hating or water fearing.

Hydroxide. The chemical opposite of an acid. Hydroxides (e.g., sodium hydroxide, potassium hydroxide) and acids will neutralize each other when mixed. Hydroxides are also known as "caustics" and "alkalis."

Hygroscopic. Describes a substance, solid or liquid, which absorbs water strongly. One example of such a compound is calcium chloride, which is sometimes spread on dirt roads to absorb water and thus keep down dust.

Ionized Air Cloud. Around the tip of an operating electrostatic spray gun is an invisible small cloud of air molecules that have picked up excess electrons. The electrons from the power pack flow off the end of the needle electrode at the gun tip. When paint droplets pass through the ionized air cloud, they accumulate electrons that enable electrostatic attraction of the droplets to parts being coated.

Isocyanate. A somewhat toxic chemical compound or chemical group in a compound that is used to cross-link paint resins. It is a common component in 2-part urethane (polyurethane) coatings.

Isolated System. An electrostatic painting system in which the paint supply and hoses are isolated from ground in order to prevent the highly conductive paint from grounding out the entire system.

Karl Fisher Method. A chemical titration procedure used to determine the water content of materials.

Lacquer. Any powder or liquid paint that does not cure by forming chemical cross-links but forms a continuous film by polar attractions among molecules of the resin.

Latex. An emulsion of any organic resin in water. Waterborne paints that contain emulsified resins are called latex paints.

Low Bake Paint. EPA definition of a paint that can be "forced" to cure at temperatures of 194° F (90° C) or lower.

Low-solids. Coatings characterized as having a low percentage of solids (resins and pigments) both by weight and volume percent.

Machine Vision. An optical device used to locate, identify, and determine the orientation of parts in a continuous operational flow. Machine vision may also be used for automated paint surface inspection.

Megohm. An electrical conductance unit equal to one million ohms (10^6 ohms).

Metamerism. The appearance of a color shift in a paint film caused by a change in the nature of the incident light.

Mho. A standard unit measurement of electrical conductivity, the reciprocal of an ohm, a siemens.

Micromho. One-millionth of a mho.

Micron. Metric unit of length equal to one-millionth of a meter (a micrometer).

Mil. Unit equal to one-thousandth of an inch. One mil roughly equals 25 microns.

Mold Release. Material used to facilitate the removal of molded objects, usually plastic, from the mold in which they are formed.

Molecules. Compounds exist in identical groups of submicroscopic chemically bonded elements. Each individual group is called a molecule of that compound. Water, for example, consists of molecules having 2 hydrogen atoms and 1 oxygen atom. The chemical formula, H_2O, indicates this molecular composition.

Molten Salt Bath. Mixtures of inorganic salts will melt at 650–900° F (343–482° C). Painted items immersed in these paint stripping mixtures are rapidly rid of coating by combustion of the paint.

Monomer. The completely unpolymerized molecular form of a compound that is able to undergo chemical cross-linking. See also *Oligomer* and *Polymer*.

Mottle. A defect that occurs when metallic paint is applied excessively wet, and color pigments separate from metallic flakes to create visible dark rings that range in size from 0.039–0.118 inch (1-3 mm).

Multicoat. Two or more coatings applied to a surface, each of which has a unique function to perform.

Nonrinsed or React-in-place Conversion Coating. Several pretreatments for steel and aluminum that can be used when corrosion resistance requirements are not extreme. They react fully enough to avoid the need for being flushed off the surface with water.

Nubbing. Localized overetching of zinc alloys.

Off-line Programming. With robotic paint application, the programming of a robot at an off-line location.

Ohm. A standard unit of resistance to electrical flow equal to a conductor carrying one ampere of current from a one volt potential.

Ohmmeter. A device used to measure resistance in an electrical circuit.

Oligomer. Slightly polymerized form of a substance that is able to be polymerized to a greater degree. Also known in some instances as *Prepolymer*.

Omega Loop. The conveyor for rotating disc paint applicators shaped to produce a circular path around the vertically reciprocating disc that delivers paint from all 360° of its circumference. The "omega" term is used because the shape of the conveyor loop resembles the capitalized form of the Greek letter omega.

Orange Peel. A microroughness in a paint film surface caused by poor paint flowout that is similar in appearance to the skin of an orange.

Paint. Any organic resin containing materials used as decorative, protective, or functional coatings. Four materials typically found in paint are resins, pigments, fluidizers, and additives.

Permeate. The resin- and pigment-free output of electrocoating bath ultrafiltration used to rinse freshly coated parts; also called flux or ultrafiltrate.

pH. The relative acidity or basicity (alkalinity) of aqueous liquids is indicated by the pH scale. Numbers below the neutral pH (or 7) are increasingly acidic; and above pH 7 are increasingly basic (antiacid).

Picture Framing. A paint defect in which extra paint accumulates at the edges of a panel, darkening the edges.

Pigment. The tiny solid particles used to enhance the appearance and color, or to improve the functional properties of a paint film.

Plastic. When used as a noun, it is a broad category of materials produced from petroleum-based and agricultural plant-derived organic polymeric resins. Used as an adjective, it indicates ductility in a material.

Polarity. Every compound is made up of molecules, and all the molecules of each compound are identical and distinct in the arrangement, number, and types of atoms that comprise them. As a result, molecules have a characteristic magnet like property called polarity, which can vary from no polarity to extremely high polarity. Compounds with uniform electron distributions are described as having low polarity; those with less uniformity have slight, moderate, or high polarity, depending on the uniformity of electron distribution across the molecule.

Polymer. In organic chemistry, a chain or network formed by linkage of many repeating individual chemical structures (monomers). See also Oligomer.

Powdercoating. This type of coating is comprised of resins, pigments, and additives, but does not contain solvents. Powdercoats are thus applied dry.

Power-and-free Conveyor. A power conveyor can use a separate pusher chain unattached to paint hooks riding freely on a separate support beam (as distinguished from a continuous power conveyor). This conveyor allows for variable parts spacing and for parts to be held stationary even when the pusher chain is moving.

Power Conveyor (Continuous). An electrically driven cable or chain power conveyor is mechanically attached to hooks, onto which parts to be painted are hung. The conveyor is used to carry parts through the painting process operations. When the line is operating, all individual hooks on the line will continue to move and maintain their spacing.

Pressure Feed. Paint delivered to an application device by air pressure on a paint reservoir (pot) connected to the application equipment, as opposed to gravity flow or positive displacement pumping of paint.

Pressure Pot. Various-sized paint tanks with delivery tubes extending to the bottom inside the tank are pressurized with compressed air to force paint to the application device. The tanks, known as pressure pots, usually have bolt-on covers.

Primer. A paint formulated to be directly applied to a substrate, before any other paint is applied. Primers promote better adhesion and provide protection to the substrate, among other benefits.

Printing. When parts with soft paint films are stacked together, and the deformable paint causes the parts to stick to each other and to packing materials, or to show imprints in the paint.

Programming (Robots). Instructional guidance that tells the robot electronically what to do mechanically; programming can be point-to-point or continuous path.

Radiation-cure Coatings. Coatings that cure when exposed to radiation such as ultraviolet (UV) or electron beam (EB).

React-in-place or Nonrinsed Conversion Coating. Several pretreatments for steel and aluminum can be used when corrosion resistance requirements are not extreme. They react fully enough to avoid the need for being flushed off the surface with water.

Reactive Diluents. Substances added to a coating to reduce viscosity that during curing will cross-link with the base polymer without evaporating, thereby not producing VOCs.

Reciprocator. An automatic device to move a paint application device back and forth along a straight or slightly curved path. Both horizontal and vertical reciprocation can be used when applying coatings.

Resin. The polymer (plastic) component that when cured forms a paint film. Resin is also known by many other names, such as binder or vehicle. The term binder is used because resins must bond with or bind to the substrate.

Resin Kickout. If amines volatilize from a water-soluble paint, or if the components are separated during freezing, the pH can drift down to a neutral value, making the resin insoluble.

Resistive System. An electrostatic paint system designed to reduce the voltage if current draw rises too high. This provides worker protection and avoids repeated trip-out of the equipment.

Reverse Osmosis. A method of generating pure (or purer) water. In the natural process of osmosis, solvent is driven through a semipermeable membrane separating solutions of different concentrations. This is how water reaches the top leaves of the tallest trees, for example. Water always travels by osmosis from the less concentrated to the more concentrated solution. But in reverse osmosis, high pressures are applied to force pure water out of the concentrated solution. It is used to produce drinking water from the ocean. Industrially it finds application to concentrate contaminants in processed water for easier disposal and to generate water for reuse.

Runs, Sags, or Curtains. Scalloped or uneven paint film, especially likely on vertical surfaces caused by wet paint running down before complete curing.

Rupture Voltage. In ecoating, the point at which the voltage will cause excess current to flow between electrodes and form gases under the paint film, causing the paint film to lift off the substrate.

Sand-scratch Swelling. A defect that occurs when the solvent from a topcoat "swells" and magnifies the sanding scratches left in an undercoat.

Sealer. A paint film used to bridge large differences in the polarity between primer and topcoat.

Servo. An electrical/mechanical device that provides an output motion according to a given input signal.

Sheen. The specular gloss of a paint film at very small angles of light incidence and reflection.

Singlecoat. A single paint film that can perform all the required functions alone.

Siphon Cup (Suction Cup). When a special air spray tip is employed, the atomizing air just outside the fluid orifice creates a partial vacuum. As a result, atmospheric pressure on the paint in a container connected to the fluid line (such as a siphon cup) will force paint up out of the container into the fluid line leading to the gun tip. The term siphon is actually a misnomer; vacuum or suction cup would be a more accurate description of the action.

Slitting. Wide coils of roll-coated materials are cut into narrower widths.

Soft Paint Film. Deposited coatings cured to a hardness below the required specifications.

Solids. That portion of paint or powder made up of resins, pigments, and additives.

Solubilizer. Since water is a polar solvent and resins are usually nonpolar, often resins must be altered chemically to increase their polarity if they are to be used in waterborne paints. Water-insoluble resins thus altered will form polar (and hence water-soluble) polymer ions when they react with compounds known as solubilizers.

Solution Paint. If the resin molecules are fully dissolved by solvents, the paint is called a solution paint.

Solvent Wash. Paint voids or areas with thin paint caused by solvent condensation.

Solventborne. A paint that uses a traditional solvent to carry or disperse the resins, pigments, and additives.

Specular Gloss. The shininess or brilliance on highlighted areas of a painted part.

Spray Booth. An enclosure in which liquid or powder coatings are applied.

Static Electricity (Electrostatic). Electrons temporarily removed from various items can cause static charges. Anything with excess electrons has a negative charge; the object from which electrons have been taken will be positively charged. Electrons will tend to jump from one object to another if at all possible in order to neutralize all charges. This phenomenon differs from a continuous electrical current, or electrodynamics. Instead, electrostatics involves electron accumulation and depletion creating charged objects having static electricity.

Stripping. The removal of unwanted cured paint film(s) from a surface.

Styrenated Alkyd. Styrene is allowed to react with alkyd resin to improve the water resistance, alkali resistance, and drying speed.

Surface Tension. Liquids tend to reduce their total surface area due to unequal intermolecular attractive forces in this region. A low degree of surface tension is preferred for liquid coatings to maximize adhesion and minimize edge-pull and fisheye effects.

Surfactant. Organic surface-active agents used in organic acid cleaners to remove stains, streaks, and related blemishes.

Thermally Inverse Solubility. Nearly all solids follow the pattern of becoming more water-soluble as temperature increases. A relatively few exceptions decrease in solubility with a rise in temperature.

Thermoplastic Resin. Resins that soften and melt with heat, and can be completely dissolved with the appropriate solvents.

Thermosetting Resin. Resins that will not melt or soften to any appreciable extent when heated; they will soften in some solvents but not dissolve in any of them.

Thixotrope. By forming loosely held three-dimensional particle networks within paint fluids, thixotropes cause temporarily high paint viscosities. Agitation of the paint by stirring, pumping, spraying, etc., quickly destroys the networks and viscosity then drops sharply.

When agitation is halted, the networks reform rapidly and paint viscosity again rises. Thixotropic is an adjective describing such materials.

Toll Coater. A trade term for custom coaters or contract coaters who paint coiled stock for other companies.

Toners. Paints may need to be transparent yet shaded to a specific color. Soluble colorants, i.e., toners, are utilized to produce this effect and give tinted clearcoats.

Transfer Efficiency. The percentage of applied paint used that is deposited on the parts. The balance of the paint goes onto the booth surfaces, hooks, filters, floors, etc.

Tribocharging or Friction Charging. A method of charging powdercoatings without using a charging electrode. It removes electrons from powder particles, but is not successful with all powders.

Triglycidyl isocyanurate (TGIC). A complex chemical used to cross-link paint, especially polyester powders, to increase exterior durability.

Ultrafiltrate. The output of ultrafiltration; also called permeate and flux.

Ultrafiltration. Ultrafiltration uses low-pressure membrane filtration to separate small molecules from large molecules and fine particulate. In ecoat operations, water is extracted from the paint bath for parts rinsing using ultrafiltration.

Ultrasonic Cleaning. Vibrational frequencies slightly higher than audible are used to agitate immersion cleaning tanks. Microbubble formation and collapse in the liquid accelerates dislodgment of soils.

Unsaturated Bonds. Carbon-to-carbon double bonds in resins can react chemically with atmospheric oxygen to form a cross-linked paint film.

Value. A grayness scale that varies progressively from pure white to pure black.

Vapor Condensation Curing. A seldom-used process that employs the heat of condensing vapors to cure coatings.

Vapor Curing. A process using a room temperature vapor catalyst to speed the curing of a two-component coating.

Vapor Degreasing. A method of using condensing solvent vapors to clean parts prior to painting.

Viscosity. The ratio of the shearing stress to the rate of shear of a Newtonian liquid.

Volatile Organic Compounds (VOC). Emissions of organic compounds that are subject to governmental regulations.

Volatiles. That portion of paint that evaporates during curing, namely the fluidizers or solvents.

Voltage. A measure only of the potential difference (force or pressure) in electrical systems; it does not indicate amounts of current flow.

Waterborne. A paint that primarily uses water as the agent to carry or disperse resins, pigments, and additives.

Wavelength. When referring to light, the tiny distance between corresponding consecutive oscillation points in the electromagnetic radiation.

Weir. The (often adjustable) barrier that controls the paint depth in an ecoat tank and over which the paint flows to the circulation pump and filters.

Wicking. Air entrapment or solvent absorption along the fiberglass/plastic interfaces in reinforced plastics.

Wraparound or Wrap. In electrostatic application, the force that attracts paint droplets or powder particles around to the backside of a part being coated.

Index